In the Nature of Landscape

RGS-IBG Book Series

For further information about the series and a full list of published and forthcoming titles please visit www.rgsbookseries.com

Published

Forthcoming

In the Nature of Landscape

Cultural Geography on the Norfolk Broads

David Matless

WILEY Blackwell

This edition first published 2014
© 2014 David Matless

Registered Office
John Wiley & Sons, Ltd, The Atrium, Southern Gate, Chichester, West Sussex, PO19 8SQ, UK

Editorial Offices
350 Main Street, Malden, MA 02148–5020, USA
9600 Garsington Road, Oxford, OX4 2DQ, UK
The Atrium, Southern Gate, Chichester, West Sussex, PO19 8SQ, UK

For details of our global editorial offices, for customer services, and for information about how to apply for permission to reuse the copyright material in this book please see our website at www.wiley.com/wiley-blackwell.

Library of Congress Cataloging-in-Publication Data

Matless, David.
  In the nature of landscape : cultural geography on the Norfolk Broads / David Matless.
    pages   cm
  Summary: "Covers a great diversity of topics, from popular culture to scientific research, folk song to holiday diaries, planning survey to pioneering photography, and ornithology to children's literature"– Provided by publisher.
  Includes bibliographical references and index.
  ISBN 978-1-4051-9081-7 (hardback) – ISBN 978-1-4051-9082-4 (paper)
1. Human geography–England–Broads, The.   2. Landscapes–England–Broads, The.   3. Broads, The (England)–Social life and customs.   4. Broads, The (England)–Historical geography.
5. Broads, The (England)–Environmental conditions.   I. Title.
  GF552.E2M37 2014
  304.209426′1–dc23
                                          2014005435

A catalogue record for this book is available from the British Library.

Cover image: Museum of the Broads, Stalham, photographed by David Matless
Cover design by Workhaus

Set in 9.5/11.5pt Plantin by SPi Publisher Services, Pondicherry, India
Printed and bound in Malaysia by Vivar Printing Sdn Bhd

1   2014

Contents

Series Editors' Preface

The RGS-IBG Book Series only publishes work of the highest international standing. Its emphasis is on distinctive new developments in human and physical geography, although it is also open to contributions from cognate disciplines whose interests overlap with those of geographers. The Series places strong emphasis on theoretically-informed and empirically-strong texts. Reflecting the vibrant and diverse theoretical and empirical agendas that characterize the contemporary discipline, contributions are expected to inform, challenge and stimulate the reader. Overall, the RGS-IBG Book Series seeks to promote scholarly publications that leave an intellectual mark and change the way readers think about particular issues, methods or theories.

For details on how to submit a proposal please visit:
www.rgsbookseries.com

Neil Coe
National University of Singapore

Joanna Bullard
Loughborough University, UK

RGS-IBG Book Series Editors

List of Illustrations

Preface and Acknowledgements

The research informing this book has been undertaken over two decades or more, beginning as a sideline, moving through various divergent projects on specific themes and individuals, and coming together as a regional monograph in recent years. My thanks to the commissioning editor of the book for the RGS-IBG series, Kevin Ward, and his successor, Neil Coe, for seeing the work through to overdue publication. Their patience and comment have been appreciated. An anonymous reviewer of the manuscript also provided extensive and insightful commentary, and helped clarify empirical discussion and extend theoretical argument.

Research for the book has drawn upon many sources, and has not been without incident. In 1994 I was awarded a 'New Lecturers' grant from the University of Nottingham to begin some Broads research, with Norwich Central Library a key resource. The grant commenced on 1 August, and on the same day the library burned down. For the next few years the library and county Record Office had an itinerant existence before finding separate permanent homes, the library in a new building on the former site, the Record Office on the city outskirts at County Hall, now also housing the East Anglian Film Archive (formerly at UEA), drawn upon here for television and holiday promotional films; thanks to Katherine Mager for enabling access there. The Norwich library's Norfolk Heritage Centre has re-gathered material lost in the 1994 fire, alongside the holdings which survived, and is the key resource for rare Broadland works, common texts, newspapers and ephemera. Clive Wilkins-Jones has been an especially valuable source of advice and information concerning material held in the library, while Clare Everitt facilitated the reproduction of images from the library collection. Yarmouth public library also provided source material. The papers of key Norfolk naturalists, notably EA Ellis and Robert Gurney, have been consulted at the Castle Museum, Norwich, with current and former curators Tony Irwin, Rob Driscoll and David Waterhouse providing valuable intelligence. The Museum's displays of art and natural history remain a fine indoor introduction to the region. The archive of the Norfolk Wildlife Trust has also been a rich source of material on naturalist cultures. Richard Denyer insightfully discussed his photographic studies of

Broadland, Brian Moss answered queries on his ecological studies, Peter Marren generously shared notes on the 1965 New Naturalist Broads publication, Bridget Yates pointed me towards material on the Thurne bungalows at Potter Heigham, and Peggy Rand shared private archival material on her relative Drew Miller. Stephanie Douet facilitated participation in the 2002 Field Day excursion of artists and scientists on the Broads, and the subsequent 2004 exhibition at Waxham Barn, where I collaborated with artists Anne Rook and Chloe Steele. Simon Partridge of How Hill Trust, Lesley George of Humpty Dumpty Brewery, artist Nicholas Ward, David Waterhouse of the Castle Museum, and aerial photographer Mike Page provided access to images, as did Jenny Watts of the Norfolk Record Office, and Maria Erskine of Nottingham City Museums and Galleries. Nicola Hems of the Museum of the Broads facilitated permission for the use of photographs of the Museum's exhibits. Broadland material also occupies national collections, including the BBC Written Archives at Caversham Park, Reading, and the Post Office Archive, the Science Museum and the Linnean Society in London. I am grateful to archival staff at all of those institutions.

Research on one Broadland figure, Marietta Pallis, has been conducted with Laura Cameron, with a small grant from the British Academy generating several publications (Cameron and Matless 2003; 2011; Matless and Cameron 2006; 2007a; 2007b). Pallis sources included the Norwich Castle Museum, the British Ecological Society in London, the Northamptonshire Record Office (where Pallis's letters are included in the collection of her friend Joan Wake), the Bodleian Library in Oxford, and the King's Lynn Consortium of Internal Drainage Boards. Staff at all institutions were very helpful. Pallis's private papers are held at her former home near Hickling, Dominic Vlasto kindly allowing access to documents and images, and giving insight into Pallis's private landscape. Ivor Kemp of the Hickling Local History Group also provided valuable assistance. At a 2001 presentation on Pallis to the Group in Hickling Village Hall, questions from a primarily local audience prompted several subsequent oral history interviews, showing other dimensions of Pallis's public and private persona. I am grateful to Laura Cameron for prompting our work on Pallis, and for all our subsequent revelatory excursions and discussions.

Interviews with key individuals also inform this book. The late Clifford Smith, Phyllis Ellis and Humphrey Boardman discussed their Broadland works and lives, while Martin George provided valuable insights into the work of the Nature Conservancy, alongside informed comment on events in the region over the past 50 years. His published work, along with that of Tom Williamson, Brian Moss and John Taylor, has been an important reference point. A key interviewee, who became both a source of research material and a commentator on research as it developed, was the late Joyce Lambert, whose role is discussed in Chapter One, but to whom I am immensely grateful for her generosity and insight. One each of the interviews with Lambert and George were conducted as part of an ESRC funded research project at the University of Nottingham with Charles Watkins and Paul Merchant on post-war cultures of nature in Norfolk and Herefordshire, which also involved interviews with other key figures in nature conversation in the counties. Archival research on Hickling conducted for that project has been drawn upon in this book, alongside some of the interview findings. Project publications are listed in the References (Watkins, Matless and Merchant 2003; 2007; Matless, Watkins and Merchant 2005; 2010).

Broadland research has over the years benefitted from discussion with colleagues at Nottingham and beyond, including Charles Watkins, Stephen Daniels, Mike Heffernan, George Revill, Daniel Grimley, Mike Pearson, Hayden Lorimer, Simon Naylor, Caitlin DeSilvey, Colin Sackett and Tim Boon. Events organised by Simon Pope, Helen MacDonald, and James Mansell and Scott Anthony, highlighted new research dimensions which shaped the direction of the work. Parts of Chapter Two were presented at the 2002 Jay Appleton lecture at the University of Hull, and I am grateful to Jay Appleton for sharing his own Norfolk memories. Tim Dee enabled the broadcast of three 'Essay' talks on Broadland naturalists (Ellis, Day and Pallis) on Radio 3 in 2008. An Edward Clarence Dyason Fellowship at the University of Melbourne in 2006 allowed productive discussion on regional cultural landscape with Fraser MacDonald. Many audiences have had my Broadland research presented to them over the years, and I am grateful for all comments made in response, but the first academic audience was perhaps for a seminar in the Geography department at Lampeter around 1993, organised, if memory serves, by Chris Philo. The response to that initial presentation helped convince me to pursue such work, and 20 years later here is a book.

Research has also been shaped throughout by family support. My parents, Brian and Audrey Matless, have contributed support, advice, press cuttings and excursions, and I cannot thank them enough. I hope they will enjoy reading about an area with which they are very familiar, though one which we tended to bypass in my childhood in favour of the beach. My wife, Jo Norcup, has provided love, wit, intelligence and field accompaniment, and has put up with the book's slow finishing. The book is dedicated to her, and to our son Edwyn, whose own skills of field observation have illuminated things since 2012.

List of Abbreviations

AWA   Anglian Water Authority
BA   Broads Authority
BAAS   British Association for the Advancement of Science
CPRE   Council for the Preservation of Rural England / Council for the
        Protection of Rural England
EDP   *Eastern Daily Press*
EEN   *Eastern Evening News*
FoE   Friends of the Earth
GYPHC   Great Yarmouth Port and Haven Commission
IBG   Institute of British Geographers
IDB   Internal Drainage Board
IPE   International Phytogeographical Excursion
IWA   Inland Waterways Association
LLNB   *Life and Landscape on the Norfolk Broads*
MAFF   Ministry of Agriculture, Fisheries and Food
NC   Nature Conservancy
NCC   Nature Conservancy Council
NFU   National Farmers' Union
NNNS   Norfolk and Norwich Naturalists' Society
NNR   National Nature Reserve
NNT   Norfolk Naturalists Trust
NRC   Norfolk Research Committee
NT   National Trust
NWT   Norfolk Wildlife Trust
RGS   Royal Geographical Society
RSPB   Royal Society for the Protection of Birds

SBL	Sutton Broad Fresh-Water Laboratory
SSSI	Site of Special Scientific Interest
TNNNS	*Transactions of the Norfolk and Norwich Naturalists' Society*
UCL	University College London
UEA	University of East Anglia

Chapter One
Cultural Geography on the Norfolk Broads

A Geographical Visit

In the Nature of Landscape offers an excursion around an eastern English wetland, the Norfolk Broads. This chapter introduces the region, and gives an account of cultural geography on the Norfolk Broads, ideas from a field of enquiry put into play. For over a hundred years people have taken boat excursions on the Broads; here cultural geography goes on the Broads, investigating landscape, finding how it might shape regional understanding.

This is not the first geographical visit to the region. In 1927 Albert Demangeon's *Les Iles Brittaniques* examined the Broads:

> The peaty swamps, the still sheets of water hidden by reeds, the wide channels overhung by willows, and the lonely marshes frequented in winter by water-fowl exhibit Nature in all her wildness, loneliness, and melancholy. But in the summer these solitudes are full of holiday-makers, and the Bure, Ant, and Thurne, together with Wroxham, Salhouse, and Oulton Broads, are dotted with motor cruisers and sailing yachts. Away from the Broads and swamps, the ground is covered with grass and forms a rich pastoral district in which graze thousands of cattle. Green fields, grazing cattle, windmills, willow-lined channels, boats sailing among trees – all these remind one of the scenery in Holland. (Demangeon 1939: 282–3)

Demangeon shows an early twentieth century French regional geographic sensibility abroad, his passage signalling lines of enquiry followed throughout this book; the

In the Nature of Landscape: Cultural Geography on the Norfolk Broads, First Edition. David Matless.
© 2014 David Matless. Published 2014 by John Wiley & Sons, Ltd.

Figure 1   Map of Broadland. Source: Fowler 1970. © Jarrold-Publishing.

aesthetics of regional description, the geographies of regional discovery, and Broadland as a region whose 'curious features' are reminiscent of somewhere else (Demangeon 1939: 282; Clout 2009).[1] The landscape features itemised too warrant continued geographic scrutiny; reeds and birds, marsh pastorals, cheer and melancholy, seasonal shifts.

This chapter gives an outline of region and book, conveys the possibilities of thinking through landscape and culture, examines early accounts of regional scenic governance, and considers regional cultural landscape as a term worth revisiting for its theoretical, political and poetic potential. The chapter concludes with a survey of Broadland institutions and scholarship, and an introductory Broadland tour.

Outline

Six rivers flow, some into one another, all waters ending in the North Sea. To make up the 'Southern Rivers', the Chet joins the Yare, and Yare and Waveney meet at Breydon Water. For the 'Northern Rivers', the Ant joins the Bure, the Thurne and Bure meet, and the Bure continues, ending in the Yare below Breydon. Only the Yare keeps its name to the sea at Yarmouth, though six river waters meet the salt; which itself moves inland upstream daily for various distances according to tide. Northern and Southern systems are gathered under the regional name of the 'Norfolk Broads', though the Waveney forms the Norfolk–Suffolk county boundary, the term 'Norfolk and Suffolk Broads' sometimes used. The broads are shallow lakes distinctive to the region, between 40 and 50 of them depending on definition, filled-up medieval peat diggings whose artificial industrial origin was figured with some surprise 60 years ago.[2] Some broads sit to one side of the rivers, linked by dug channels (as with Ranworth on the Bure, or Rockland on the Yare), some occupy the river as if it had simply 'broadened' in its flow (as for Barton on the Ant).

The Broads appear in print in upper or lower case, and the conventions followed in this book can help clarify aspects of the region. There are many broads in the Broads, lower case individual lakes in a region named from them, otherwise termed Broadland. An individual broad achieves upper case when named, as in Rockland Broad, or Barton Broad. Deciding on a holiday, you might imagine cruising on the Broads (a regional experience), or on some broads (several points to visit). Such case conventions are followed in this book. One further element of regional nomenclature is worth noting, concerning fen/Fen. The Broads are sometimes regionally confused with the Fens, the former-wetland agricultural flatland in west Norfolk, Cambridgeshire and Lincolnshire, where rivers flow to The Wash. As a wetland, Broadland includes extensive fenland, but the Broads are not the Fens, and indeed there are few fens in the Fens.

Instructive questions of terminology also surround the status of Broadland as wetland and/or waterland. Broadland as waterland carries qualities of scenic beauty, land framing water, free open air, leisure for profit, territory for regulation. Broadland as wetland triggers a poetics and politics of habitat, a landscape neither-water-nor-land, refuge for flora and fauna (human included), or waste wanting reclamation, needing drainage (Purseglove 1988; Bellamy and Quayle 1990; Giblett 1996; Cameron 1997). In *The Conquest of Nature* Blackbourn (2006) traces conflicts over German water landscape, the reclamation of marsh and fen provoking both eulogy and lament concerning the transformation of landscape and German identity. Parallel regional matters of hydrology and identity shape Broadland, whether in the maintenance of grazing marsh as iconic regional landscape, contests over rights of navigation, or the defence of fen and reedbed as home for regional fauna and flora, against both human reclamation and natural succession.

Five chapters follow this one, along with two interlude studies of regional icons, the wherry and the windmill. An outline of book topography will convey the shape of argument. Chapter Two, briefer than the rest, addresses Broadland origins, but rather than begin with geological or prehistoric background, historiographic analysis emphasises contested narratives of regional landscape formation. The 1950s discovery by Joyce Lambert that the broads were flooded medieval peat diggings prompted argument over the regional standing of science, and the value of landscape features no longer deemed natural. In keeping with studies in the historical geography of science, emphasising both the geographical shaping of scientific enquiry and 'the geographies that science makes' (Naylor 2005: 3; Livingstone 2003; Lorimer and Spedding 2005; Matless and Cameron 2006; Cameron and Matless 2011), the chapter shows claims to regional authority shaping the reception of origin accounts. Definitions of and claims to the region shape scientific argument (Matless 2003a). Here as elsewhere the book examines 'geographies of authority' (Kirsch 2005), with institutions and individuals exercising claims to regional knowledge.

Chapter Three turns to conduct. From the leisure 'discovery' of the Broads in the late nineteenth century, sponsored by railway companies and boat-hire firms (and with an associated discovery of regional folk life), the region has been defined through contested pleasures, as either essentially a pleasure waterland, or a nature region threatened by such conduct, with particular sites, notably Potter Heigham, a focus for dispute (Matless 1994). The moral geographies of leisure, concerning conduct becoming or unbecoming a particular landscape, are shaped through guides, novels, films, posters, detective stories, children's literature, political campaigns and policy documents, cultural geographic excursions on the Broads demanding that connections are made between such diverse sources. Policy debate has turned on the modes of conduct deemed appropriate to the region, and the scales of authority – national, regional, local – appropriate for Broadland governance. Thus the possibility of Broadland becoming a national park brought decades of argument over conduct and the geographies of authority; what kind of region should this be, and who should exercise authority over it?

Consideration of conduct in Broadland also encompasses folk life and the comic. Broadland as waterland of leisure life is shadowed by narratives of authentic regional culture. The working lives lived by those long resident, heard as manifest in folk song and dialect, have been subject to collection and performance by those beyond and within the region. The discovery of Broadland entailed the discovery of a regional folk, in keeping with wider enthusiasms for folk culture as emblematic of national and local identity. The performance of folk life, including its self-conscious articulation by local residents such as dialect artist Sidney Grapes, could mix serious cultural labour with comic effect. An emphasis on conduct in work and leisure indeed draws attention to the comic qualities of landscape, the Broads as a space of amusement, an issue perhaps neglected in recent formulations of emotional geography (Davidson, Bondi and Smith 2007). Jokes and satire, joy and laughter, shape Broadland cultural geography.

Between Chapters Three and Four, and Five and Six, come two interlude studies of regional landscape icons, the wherry and windmill. The wherry as river vessel carrying regional cargo, or pleasure craft carrying leisure visitors, has, over 200 years, stood for the regional present and past, and since the mid twentieth century been subject to heritage salvage. Another technology of air and water, the windmill, the key mechanism for drainage until the mid twentieth century, and since subject to efforts of preservation and restoration, has likewise achieved iconic Broadland status. From their depiction in early nineteenth century 'Norwich School' painting to their restoration by present enthusiasts, the wherry and windmill work as Broadland icons, with questions of navigation, drainage, heritage, governance and beauty condensing around them. 'Wherry' offers an interlude after discussions of leisure and regional life in Chapter Three; 'Windmill' sits between the accounts of marsh and drainage in Chapters Five and Six.

Chapters Four and Five concentrate on the human encounter with the non-human world, the animal and plant landscapes of Broadland. Non-human life shapes cultures of landscape, subject to human attention, care and exploitation, acting in accordance with or across human expectations, human and non-human subjects and objects defined through relation. The title of this book, *In the Nature of Landscape*, plays on Broadland's qualities of nature, notably the prominence of the plant and animal in accounts of the region, yet also the use of nature to underwrite, indeed naturalise, human power, not least in overlapping categories of land ownership, sanctuary and reserve. The term 'nature' has been questioned for the impossibility of fixing categorical boundaries between the natural and cultural, and the ways in which things deemed distinctly 'natural' emerge from hybrid relations such that the supposedly 'pure' category quickly collapses under historical scrutiny (Latour 1993; Whatmore 2002). It is nevertheless worth retaining 'nature' and associated terms (natural history, nature study, natural science, etc) to register its complex colloquial work, whether in specialist languages of science, aesthetics and spirituality, or in vernacular and/or popular appreciation. For all its problematic conceptual status, 'nature' may retain powerful communicative coherence. Specific genres of writing and picturing may convey the non-human such that their practitioners become 'nature voices', authorities on particular species or conveyors of a general value in the natural world (Matless 2009a). The complexities of nature, for Raymond Williams 'perhaps the most complex word in the language' (1976: 184), are such that it happily problematises itself, as a working word impossible to erase.

Chapter Four considers Broadland's animal landscapes, emphasising the ways in which mammals and birds have variously appeared as objects of biodiverse value, quarry for killing, creatures for careful observation, cherished regional icons, or alien intruders. The term animal landscapes carries enquiry through fields as various as marsh, river, committee room, reedbed, museum and sky (Matless, Watkins and Merchant 2005). Naturalists and nature institutions study, document and broadcast, landowners reserve nature in private, voluntary and state bodies reserve nature for public interest, specific species such as the bittern and coypu concentrate argument over Broadland life. Chapter Five concentrates on Broadland plant life, botanical and ecological study finding scientific and cultural value in Broadland flora. Cultural geographies of the non-human have tended to concentrate on the animal, giving little attention to vegetation (Head and Atchison 2009); the investigation of plant landscapes also entails movement across marsh, committee room, undergrowth, museum, reedbed. Processes of ecological succession, advancing in part through the relaxation of human marsh management, inform dispute over what Broadland's plant landscape should be. Key sites concentrate argument, including the private Edwardian Sutton Broad Laboratory, the Wheatfen reserve established by naturalist EA Ellis, and Long Gores, home of ecologist-artist Marietta Pallis. If Chapter Three considers the collection of regional human custom, and Chapter Four the human gathering of regional fauna, here the cultural harvest of vegetation is addressed.

Chapter Six turns to the ends of landscape. Broadland's present is shadowed not only by origin disputes but forebodings of destruction, via overgrowth by wood, aquatic transformation through eutrophication, drainage for cultivation, or flood invasion by the sea. Possible futures haunt the present; as threats, or sometimes as opportunities. Stories of flood echo the narratives of plant succession explored in Chapter Five, for some an erosion of distinctive scenery and habitat, for others the restoration of natural order. The prospects for destruction are explored through historic narratives of past floods and former estuaries, and of the region thrown in and out of balance as land and water shift. Climate change concentrates minds on a possible regional end, with senses of irreversible change, interpreted as loss, increasingly governing Broadland accounts. The Broads as waterland is set in relation to the North Sea and its underwater topography, the coast a historically shifting line, the sea bed once land, the land perhaps under future sea. The possible ends of landscape – undersea, overgrown – haunt the region, Broadland's outline forever on hold.

Landscape Colloquial, Culture Resounding

It is in the nature of landscape, as a word, to move, between paint and ground, people and rock, vegetable and animal, profit and emotion, the wistful and the earthed. For Daniels, landscape's potential proceeds from its 'duplicity', 'not despite its difficulty as a comprehensive or reliable concept, but because of it' (1989: 197); the implication is that 'We should beware of attempts to define landscape, to resolve its contradictions; rather we should abide in its duplicity' (218). Duplicity emerges in part from the historical geographies of the term, with landscape's varying proprietal, communal and imaginal associations (Cosgrove 1984; Olwig 1996; Matless 2003b; Olwig 2008). For Wylie, 'landscape is tension' (2007: 1), of proximity/distance, observation/inhabitation, eye/land, culture/nature, tension making landscape a subject/object tangle. To a coinage of duplicity and tension I would add landscape as colloquial, denoting the presence of different voices, forms of attention given, and a varied cultural constitution, moving across the academic and popular, the specialist and ordinary. Thus landscape entails a colloquium of disciplines, specialisms conversing over shared interest, though in the manner of conversation sometimes talking over and past one another. Subjects from the humanities, social sciences and natural sciences chip in; this book gives a cultural geographic voice. Landscape also invites the colloquial; everyday talk of vernacular tone, skilled talk of vernacular tone, voices of other accent, quality and formality, also able to talk over and across one another, or to make deep chat. Accents of landscape, whether spoken or written, articulate multiplicity, with questions of technique and accomplishment, commonplace and slang, always pertinent, whether for broad dialect, tones of authority, Received Pronunciation, regional wisdom, formal poetics.

Sound and voice indeed register throughout this book, whether in leisure sound, regional song, scientific speech or bird call. While landscape is shaped through senses in combination, the particular qualities of sound, especially of the being-heard rather than being-seen, alert us to landscape's shaping of and through forms of address, mark landscape as colloquial. Broadland indeed shows sound's capacity at once to transgress and reinforce social division, to travel across open air regardless of the listener's readiness or willingness to hear. Voices marked by accent, timbre, intonation are given different hearing in popular culture and policy argument. Vocabularies of tradition, expertise, craft and fun shape the Norfolk Broads. For some the region is to be marked by silence (a silence of nature sound with barely a human utterance); for others a jolly cacophony belongs. In *A Voice and Nothing More*, Mladen Dolar states that: 'We are social beings by the voice and through the voice; it seems that the voice stands at the axis of our social bonds, and that voices are the very texture of the social, as well as the intimate kernel of subjectivity' (Dolar 2006: 14). The texture of voice achieves political charge, with regional accent effectively 'a norm which differs from the ruling norm': 'The ruling norm is but an accent which has been declared a non-accent in a gesture which always carries heavy social and political connotations' (20). Attending to voice regionally allows scrutiny of the processes whereby voices achieve popular and official hearing, whether authority is marked by expert language spoken without evident regional connection, or by embeddedness heightened by the self-conscious performance of accent and dialect. Differences of authority are shaped by traces of region in the voice.

Dolar also argues for attention to the materiality of voice alongside the explicit meaning of words uttered, with the voice 'the link which ties the signifier to the body', which 'holds bodies and languages together' (59–60), though paradoxically 'does not belong to either' (72). In Broadland, we find body, language and voice interplaying in complex style, in the accounts of travellers discovering the region, in scientists' and naturalists' fieldwork and findings, and in dialect performers putting region into public play. Dolar's psychoanalytic approach seeks to direct attention to the 'object voice' (4), the 'material element recalcitrant to meaning', stating of the voice: 'it is *what does not contribute to making sense*' (15). This statement should not though be taken to imply a rigid analytical division between the three senses of voice elsewhere identified by Dolar as 'vehicle of meaning', 'source of

aesthetic admiration' and 'object voice' (4). Cultural analysis would indeed emphasise the ways in which sonic qualities such as accent and noise disturb any clear
distinction of object and sense, materiality and meaning. Dolar takes his title, 'a
voice and nothing more', from Plutarch's account of a man plucking a nightingale
and finding little to eat on the bird: 'You are just a voice and nothing more' (3).
The phrase provides a departure point for a rewarding meditation on the constitu-
ency of voice, yet attention to culture might suggest the unrealisability of marking
out voice alone. The nightingale, plucked in death as flying in life, evidently was
something more; in corporeal terms the 'nothing' here is more a surprising 'not
much'. That Plutarch's incident turns on an irony, of disjuncture in body and
voice, meat and song, suggests less a lack than a narrative 'something more', culture abhorring a vacuum. The edition of Plutarch's *Moralia* cited by Dolar indeed
renders the passage as: 'A man plucked a nightingale and finding almost no meat,
said, "It's all voice ye are, and nought else"' (Plutarch 1949: 399). If 'a voice and
nothing more' emphasises wistful reflection on the tiny carcass, the alternative
foregrounds human frustration and hungry resignation. Variety in translation,
between the plaintive 'just a voice' and the hollow bombastic 'all voice', further
underlines vocal complexity.

Questions of voice are central, in different fashion, to Anne Whiston Spirn's *The
Language of Landscape*, a landscape architectural study conjoining 'the pragmatic and
the imaginative aspects of landscape language' (1998: 11). Spirn's rich and nuanced
analysis indicates however potential cultural tensions concerning voice and landscape
literacy. If Spirn offers a generous, open language of landscape, culminating in an
appeal for 'cultivating paradox' (262), openness and paradox nonetheless carry a
shadow side. When Spirn gives a general diagnosis of linguistic loss, asking whether
people (in everyday life, or the professions of architecture and planning) can any
longer 'hear or see the language of landscape' (11), the prevailing tone of generosity
turns:

> most people read landscape shallowly or narrowly and tell it stupidly or inadequately.
> Oblivious to dialogue and story line, they misread or miss meaning entirely, blind to con-
> nections among intimately related phenomena, oblivious to poetry, then fail to act or act
> wrongly. Absent, false, or partial readings lead to inarticulate expression: landscape
> silence, gibberish, incoherent rambling, dysfunctional, fragmented dialogues, broken
> story lines. The consequences are comical, dumb, dire, tragic. (22)

This passage comes within a pertinent critique of planning and development around
a buried, canalised creek, prone to return in flood, but the tone has a striking sweep.
Absent here is any reflection on the cultural constitution of landscape voice, and landscape literacy, and the shaping of formations of subject, citizen and people through
such process (Matless 1999a). Attending to the landscape colloquial might conversely
allow movement across fields of articulation, with landscape literacy always a matter
of cultural contest, and claims for unfair dismissal of voice heard. Linguistic landscape policing might be resisted. Silence, poetry, dialogue, the fragmented, gibberish,
all are heard in their fashion.

Such a hearing of voices is in keeping with attention to cultures of landscape. If the 'culture' to which geography (along with other disciplines) was deemed to have turned through the 'cultural turns' of the late twentieth century has perhaps been neglected, or taken for granted, in some recent work, its capacities and complexities still reward reflection, allowing the term to resound. Williams may have judged nature most complex, but culture was also 'one of the two or three most complicated words in the English language' (Williams 1976: 76). The complications of culture remain richly evident in recent deployments of cultural geographic thinking beyond geography's disciplinary borders, as in Sanders' work on early modern drama (Sanders 2011), and in a significant body of work on the geographies and archaeologies of material culture, as in Tolia-Kelly's studies of landscape, race and visual culture, or DeSilvey's constellations of landscape's material histories (Tolia-Kelly 2010; DeSilvey 2007; 2012). The narratives of memory and place examined in and produced through such work are echoed in Hauser's cultural histories of archaeological enquiry, where visual technology, historical imagination and impulses to collect conjoin through figures such as OGS Crawford (Hauser 2007; 2008; Johnson 2007).

The 'culture' in cultures of landscape variously indicates ways of life, habits of place, spheres of representation, material objects, forms of media, the province of a 'cultured' elite, that which is popular, that which is not nature, the modes through which nature is valued. Cultural geographic excursions into Broadland thereby necessarily move between high arts and low pursuits, ways of life and popular forms, the site of culture becoming variously the painting, the folk play, the bird photograph, the bittern, the postcard, the iconic boat, the riverside gesture, the marsh tool, the diary, the novel, the cruise, the sail, the stick of rock. Attention to cultures of landscape demands an approach both open and discriminatory, giving space to all manner of acts and artefacts, while exercising critical judgement on their enactments of power and claims to value. Culture, like landscape, entails movement across fields, sometimes obviously adjacent and conversant, sometimes ostensibly living in discrete parallel.

Cultures of landscape also denote fields of conduct. If recent geographical landscape work has brought renewed attention to direct landscape experience through phenomenological study, notably through the work of Wylie (2002; 2005; 2009), phenomenology nevertheless continues to keep culture at arm's length. In contrast, emphasis on conduct sees culture and experience necessarily conjoin, conduct registering the rituals and conventions of landscape experience, its geographical formation and 'historicity'. Seeking to elaborate 'the notion of experience' in a manner avoiding deterministic explanatory resort to economic and social context, or 'a general theory of the human being', Michel Foucault posits 'the very historicity of forms of experience', via 'a history of thought', where 'thought' is 'what establishes the relation with oneself and with others, and constitutes the human being as ethical subject':

> 'Thought,' understood in this way, is not, then, to be sought only in theoretical formulations such as those of philosophy or science; it can and must be analysed in every manner of speaking, doing, or behaving in which the individual appears and acts as subject of learning, as ethical or juridical subject, as subject conscious of himself and others. In this

sense, thought is understood as the very form of action – as action insofar as it implies the play of true and false, the acceptance or refusal of rules, the relation to oneself and others. The study of forms of experience can thus proceed from an analysis of 'practices' – discursive or not – as long as one qualifies that word to mean the different systems of action *insofar* as they are inhabited by thought as I have characterised it here. (Foucault 1986: 334–5; also Foucault 2000: 199–205)

Such a formulation of thoughtful living usefully sidesteps polarisations of practice and representation which have informed recent geographical debate (Nash 2000; Lorimer 2005; Anderson and Harrison 2010), but also speaks to earlier humanistic geographical study, where work on landscape experience, notably that of Jay Appleton, certainly proceeded from a 'general theory of the human being'. In *The Experience of Landscape* Appleton argued that landscape aesthetics could be rooted in a human behavioural preference for habitat sites combining 'prospect' and 'refuge', capacities to see and hide, spy and shelter (Appleton 1975).[3] Such facets of landscape experience are present in this book, with, for example, the bird hide giving a classic site of prospect and refuge, a particular form of human observational power thereby secured, birds coming under human view unawares. The intent here though is less to present a hide as indicative of general human aesthetic preference, than to examine the cultural and historical geography of such settings, techniques and experiences.

Scenic Governance

The first combined pictorial and written account of Broadland rivers was James Stark and JW Robberds' 1834 *Scenery of the Rivers of Norfolk comprising the Yare, the Waveney, and the Bure* (hereafter the *Scenery*) (Stark and Robberds 1834). The *Scenery* can serve here as a bridge between the discussions of landscape and culture above and regional cultural landscape below, and offers a chronological opening into the regional story. Stark's pictures and Robberds' words indicate key elements of Broadland landscape complexity, notably tensions of aesthetic and commercial value, the narration of human and natural history to inform the present, and the connections of scenic imagery and governance, with the book an intervention in aesthetic and political debate over regional identity. The *Scenery* also allows discussion of the place of Norwich in Broadland, the city's centrality or peripherality dependent on the manner in which the region is defined.

The *Scenery* included 36 engravings from paintings by Norwich artist Stark (1794–1859), pupil of noted landscape painter John Crome. Stark would later be grouped within the 'Norwich School' of landscape painters, his career shuttling between Norwich and London (between 1814 and 1819, and from 1830) (Hemingway 1979; Blayney Brown, Hemingway and Lyles 2000). Robberds, a Norwich worsted manufacturer, contributed 'Historical and Geological Descriptions' (Edwards 1965; Hemingway 1992; Beadle 2008). The 'scenic' register allowed landscape art to occupy a distinct aesthetic space, while also connecting to schemes of landscape change, notably an ethos of 'improvement' directed to navigation and commerce. As Revill

(2007) suggests in his study of William Jessop's work on the River Trent, landscape improvement, whether for agriculture or navigation, entailed a mode of landscape governance. The Broadland rivers had been subject to improving legislation from the late seventeenth century, with navigation on the Bure above Coltishall extended to Aylsham in the 1770s, and the Ant extended by canal beyond North Walsham in the 1820s (Boyes and Russell 1977).[4] The *Scenery* supported Norwich civic and commercial schemes under the 1827 Norwich and Lowestoft Navigation Bill to connect the city by river to Lowestoft, thereby bypassing and undercutting the port of Yarmouth, in the process enacting the slogan 'Norwich a Port' (Robberds 1826: iii; George 1992).

Words and pictures in the *Scenery* anticipated engineering works, notably the digging of the New Cut canal connecting the Yare and Waveney, and the construction of Mutford Lock for navigation between Oulton Broad and Lowestoft, and on to the sea. Landscape appreciation and commerce appear in navigational alignment, with pictures and text giving both a nostalgic record of scenes which might be lost through development, and projections of a fine future (Beadle 2008). Thus 'Reedham Mill' depicts the Yare riverbank with a ferry boat taking passengers from a landing stage. Trees, cottage and mill stand behind, with another mill distant through the trees. This is a sight for picturesque appreciation, but also with commercial appeal from its potential transformation: 'the proposed ship canal will shorten the connection between the Yare and the Waveney. It is intended to commence this work where the

Figure 2   'Reedham Mill', by James Stark. Source: Stark and Robberds 1834, image courtesy of Norfolk County Council Library and Information Service.

group of cows is standing, and the mill which is perceived through the trees, marks the spot where it will join the other stream' (Stark and Robberds 1834: no pagination). Reading the text, and spotting the mill across marshes cast in bright light, transforms pastoral scene into industrial prospect.

The navigation works were completed by 1833, though never achieved the expected return, traffic being further eroded by the Norwich and Yarmouth Railway, via Reedham, opening in 1844. The Norwich and Lowestoft Navigation Company was itself purchased in 1844 by railway entrepreneur Samuel Morton Peto, who sought to shape Lowestoft as port and resort through rail, extending the track from Reedham to Lowestoft by 1847, running alongside the New Cut. The navigation company was wound up in 1850, one mobility overtaken by another (Edwards 1965). Peto had acquired the Somerleyton estate in the Waveney valley, rebuilding Somerleyton Hall from 1844 as a modern mansion in Jacobean style, Peto one of several national business and political figures to remake aspects of Broadland for their own image (Pevsner 1961: 390–1; Port 2004).

Stark and Robberds' *Scenery* also projected landscape through historical and geological narrative. In 1826 Robberds had published *Geological and Historical Observations on the Eastern Vallies of Norfolk*, and his arguments on landscape formation informed his river commentary. Robberds argued from geology, archaeology, tradition, place names and historical records that the area had been an estuary in historical times. Natural history underwrote 'Norwich a Port': 'It cannot be otherwise than satisfactory to the advocates of the measure, to find, that their plans, if realised, will follow the original course of nature, by restoring what appears to have been the most frequented entrance to the ancient Gariensis' (Robberds 1826: 'Advertisement').[5] The *Scenery* presented the restoration of the ancient Lowestoft estuary entrance, avoiding Yarmouth's difficult harbour. Stark's pictured 'The Mouth of the Yare', choppy waters at a tricky harbour mouth, while 'The Lock at Mutford Bridge' was shown 'as it will appear', boats waiting for calm passage. If Yarmouth's 'feudal tyranny and chartered monopoly' had barred 'the free exercise of natural rights and advantages', frustrating Norwich's business, the light cast across the Reedham marshes for the course of the New Cut denotes the light of reason and improvement, 'the more enlightened and liberal spirit of the present age', regional rivers freed for prosperity (Stark and Robberds 1834: no pagination). Parallel questions of navigation, trade and open waters would shape discussion of Broadland governance throughout the nineteenth and twentieth centuries.

The broads and the northern rivers receive attention in the *Scenery* only in pictures of 'St Benedict's Abbey', 'The Island at Coltishall', and 'Decoy Pipe for Wild Ducks' at Ranworth Broad, Robberds briefly discussing 'many extensive and deep hollows, filled with water and forming numerous lakes, which are locally termed *broads*'. If later regional narratives are dominated by the Bure and its tributaries, this first work on Norfolk river scenery is dominated by the southern rivers, and is not framed by the term 'Broadland'. There are broads here, but this is not 'The Broads', with the lakes noted only for fish, wildfowl, reed and rush, and their name a vernacular curiosity. Stark's image of 'St Benedict's Abbey (On The Bure)' (a frequent contemporary art subject, commonly termed 'St Benet's', and discussed in the 'Windmill' section below (see Figure 34)) does however prompt Robberds to reflect on 'at once an emblem and a monument of fading glory', abbey ruins suggesting the grandest schemes might fail (Stark and Robberds 1834: no pagination). Improvement is shadowed by hubris, St Benet's effectively registering buyer beware in a scenic prospectus. The New Cut would be cut, but the navigation would fail, and a train run its embankment.

The opening of the New Cut saw celebration in Norwich, civic identity renewed by new navigation; two city pubs adopted the celebratory name 'Norwich a Port' (Thompson 1947: 37). The episode underlines however Norwich's particular Broadland role, as the major regional city, but at the navigational limit. If Stark and Robberds' *Scenery* emphasised Norwich as river city, and if its port would handle significant freight well into the twentieth century, Broadland as leisure region placed Norwich on the periphery. Norwich is on the river Wensum, upstream of its meeting with the smaller Yare near Thorpe; the Wensum loses its name in the merger, the river marked by its Yarmouth destination. Norwich is a long river journey (down to Yarmouth, and back up the Bure) from the northern river leisure centres. Norwich's yacht station remains modest, offering mooring and basic services, though the 1945 City of Norwich Plan had hoped for something more, envisaging a new station with a riverside walk to reinstate Norwich as a river city, and raise its Broads profile: 'a Yacht Station, worthy of the name, with full club facilities and properly laid-out grounds' (James, Pierce and Rowley 1945: 69). In the event the existing station continued, a riverside walk only emerging decades later. Norwich remains at the head of navigation on the lesser-used southern rivers, with leisure Broadland centred north.

Regional Cultural Landscape

Revisitation

'Regional cultural landscape' is familiar as a theme, if not always as an exact phrase, from earlier modes of geographical enquiry, and from a wider extra-scholarly topographical literature. The term might usefully be revisited given the comprehensive retheorisation, across a range of disciplines, of each of its constituent terms: region, culture, landscape. What happens if region, culture and landscape, in their various ways rethought, are brought together again? This book serves as a demonstration piece for such an exercise. This is not a matter of combining three conceptual terms into a new steady template for empirical application; rather their recombination generates a productive complexity and instability, from the cultural baggage through which they are constituted, and the various intellectual traditions shaping associated thought. As with landscape, we might abide in rather than seek to resolve such instability.

Landscape and culture have been considered above, but region requires further discussion, the distinctiveness of this study coming in part from its regional focus. This section considers regional definition and governance, cultural geographic engagements with region, and traditions of regional writing.

Running regional rule

Broadland is maintained through boundary work; barriers to keep water fresh by holding sea out, edges of jurisdiction between historic port authorities and contemporary planning bodies, symbols and markers registering Broadland, or select parts of it, as enchanted or magical; cultural efforts to proof the region, themselves forming a significant part of the regional story. All such work is provisional; in terms of provisional as tentative, always requiring rework, and in terms of provision as sustenance, food for geographic maintenance.

The term 'region' carries associations of rule, Bourdieu (1991) and Williams (1983) noting the word's root in definitional regulation, though with spaces defined as regions inevitably contested. Paasi's Finnish studies offer a substantive geographical contribution, highlighting landscape's work 'as a visual and territorial category' for national and regional identity (Paasi 2008: 513; Paasi 1991; 1996; 2003; Jones and Olwig 2008; Prytherch 2009).[6] The definition of regional scale, in authority and affiliation, itself here becomes part of the regional story. Defining a region may entail moral rhetoric, whether in the policy projection of appropriate regional conduct, or the bio-regionalist ideal 'to become dwellers in the land', where the region becomes the chosen scale for eco-critique (Sale 1985: 42; 1984; Whitehead 2003).[7] Regional cultural landscape inevitably concerns the articulation of scale and its political, economic, imaginative, emotional consequences, with the definition of a region an active and mutable component in its life. Region carries such questions in a way which 'place', for example, may not, with the term conveying a dual status of something carrying its own (contested) integrity, yet also being a region of something else: 'There is an evident tension within the word, as between a distinct area and a definite part' (Williams 1983: 264). If the Norfolk Broads are presented as a region distinct, an enticing leisure waterland or rare remaining wetland, they are shaped by flows of water and structures of governance from beyond their boundary; and while the late twentieth century brought a single planning body, the Broads Authority, other authorities continue to jostle.

If the term 'region' suited the discipline of geography in the early and mid twentieth century, this was in part for its scalar politics; in France and England it could denote political balance between nation and region, whole and constituent parts (Fawcett 1919; Vidal de la Blache 1928), while Estyn Evans could seek an Ulster 'common ground' in contested territory through a rural regionalism of peasant material culture (Graham 1994; Evans 1996). Regional cultural landscape might indeed achieve renewed twenty-first century political purchase, notably in terms of geographic identity in a future England (Matless 1998; 2000a; Colls 2002; Jones 2004). In a future of possible Scottish independence and British political fragmentation, the politics and culture of England and Englishness may surface. Within an England politically and economically dominated by London, questions of regional England might arise. Whether or not the particular issues shaping Broadland regional cultural landscape would loom large everywhere, questions of regional articulation, in forms appropriate to a given region, could well return.

Regions turned cultural

Earlier geographic engagements with the region carried a cultural and political complexity often passed over in textbook disciplinary history, with regionalism a focus for concerns of governance, landscape and citizenship (Livingstone 1992; Matless 1992; Clout 2009). Attention to culture in such work tended however to be circumscribed, whether in studies of the region as material expression of a distinct way of life, as in Vidal de la Blache's possibilist studies of France as 'a medal struck in the likeness of a people' (1928: 14; Robic 1994), or in specific attention to aesthetic outlooks manifest in leisure, as in Bryan's work on 'the cultural landscapes of recreation and the gratification of the aesthetic sense' in Michigan and north Norfolk (Bryan 1933: 341). Recent geographical works however work through the complexities of culture to fuller effect, indicating the possibilities of a revisited regional cultural landscape.

Hayden Lorimer's works on the Cairngorms and wider Highland Scotland effectively combine into a regional cultural study, encompassing field study and the geographer-citizen, landholding and animal killing, natural histories of flora and fauna, and the animal landscapes of herding Cairngorm reindeer. Lorimer indeed sets the latter within 'the treatment of locality in geography's longer heritage of landscape study', revisiting materiality and observation though different 'cultural topographics of inquiry': 'The footwork and field trudge may remain the same but the manner in which landscape is approached and expressed can be retuned' (Lorimer 2006: 515–16; 2000; 2003; Lorimer and Lund 2003). Fraser MacDonald's research similarly makes for a Hebridean regional cultural study, with folklore, photography, militarism, geography and archaeology shaping island landscape (MacDonald 2004; 2006a; 2006b; 2011). Dydia DeLyser's *Ramona Memories* likewise takes a region for scrutiny, the 'Ramona Country' of southern California, defined touristically following Helen Hunt Jackson's 1884 novel *Ramona*: 'The most important woman in the history and geography of southern California never lived, nor has she yet died, for Ramona lingers here still' (2005: 188). The sense of a 'Ramona myth' itself becomes a 'regional vernacular term' (xvii), DeLyser examining regionally iconic maps, guides, pageants and relics. If there is no equivalent Broadland literary fiction in terms of transformative effect, Broadland is certainly a region significantly made through writing, with a parallel interweaving of publishing and touristic discovery.

Reflections on the life and work of Denis Cosgrove following his death in 2008 (Cultural Geographies 2009) have tended to concentrate on two of his three monographic studies, *Social Formation and Symbolic Landscape* and *Apollo's Eye* (Cosgrove 1984; 2001), to the neglect of Cosgrove's middle 1993 *The Palladian Landscape*. The overlooking of this study may reflect not only a seemingly lower theoretical ambition, but also its geographic particularity. *The Palladian Landscape* offers however an instructive example of a late twentieth century cultural geographic working of a region, which also signals some tensions of regional enquiry. Cosgrove states: 'I have written the work as a geographer and it aims to be a geographical interpretation of a region of northern Italy in the specific historical period of the late Renaissance' (1993: xiii). The term 'Palladian landscape' made sense for Cosgrove not simply as a shorthand for the regional and international significance of Palladio as architect, but because Palladian designs worked landscape as an integral part of their effect, acting, whether in central Venice or the rural Veneto, 'quite directly to articulate the dramatic qualities inherent in the topography' (3). Palladio's work thus characterises 'a small region where, during a relatively brief historical period, particular social groups sought to refine a vision of their human place in the order of nature and to represent that vision through various forms of landscape making – in texts, pictures and maps as well as in building, agricultural practice and environmental engineering' (5; Cosgrove and Petts 1990). Cosgrove moves across the visionary and prosaic, from cosmogony to land drainage, the Veneto also shaped by visions and objects of the transatlantic 'New World', whether in utopian dreams or new crops, regional monographic study dragging attention across the globe.

The Palladian Landscape nevertheless displays tension concerning the place of theory in regional study, evident in the structure of Cosgrove's opening chapter, which moves from conceptual discussion of landscape and culture, through a biographical account of Palladio, to 'outlining in conventional terms the geography of the Palladian landscape' (Cosgrove 1993: 5). A three page account distinguishes three physiographic areas according to relief, surface geology and drainage history, and outlines historic urban morphology and contemporary post-industrial development. Questions of aesthetics are by implication curiously absent from the bulk of this 'conventional' geographic account. Occasional impressionistic seasonal evocations interrupt, and thereby confirm the conventionality of, Cosgrove's regional geographic description: 'Venice swims into a morning light' / 'uplands emerge as islands from the summer haze of the low country' / 'long-extinct volcanoes rise blue in early autumn mists' (24). The remaining description shows conventional restraint, the interruptions taking differently conventional flight, as if Ruskin were interrupting Dudley Stamp. Questions of theory and description, held awkwardly apart in *The Palladian Landscape*, are returned to in the final section of this chapter.

Regional cultural landscape also preoccupies scholarship beyond geography, with parallel questions of theory and description, and retuned tradition, posed. Matthew Johnson's *Ideas of Landscape* thus examines the 'habits of thought' of the 'English landscape tradition' in archaeology and landscape history through figures such as WG Hoskins and Maurice Beresford. Johnson crosses a presumed landscape divide between theoretical reflection and defiant empiricism: 'Landscape studies are simultaneously one of the most fashionable and avant-garde areas of scholarly enquiry, and also, paradoxically, one of the most theoretically dormant areas' (Johnson 2007: 1; Matless 1993; 2008a). Johnson, who prefaces his work with a regional exercise, 'Thinking about Swaledale', jumps any division between those pursuing pure theory, and those deploying an empiricist (anti) aesthetic of practical mud-on-boots common sense, his 'Glossary' of landscape terms offering indicative alphabetical leaps: genius loci, gentry, hachuring, hermeneutics, hundred, husbandman, ineffable (Johnson 2007: 204).

Regional landscape enquiry also navigates the historic and experimental in performance studies (Daniels, Pearson and Roms 2010). Mike Pearson's book '*In Comes I*': *Performance, Memory and Landscape*, like his subsequent 'Carrlands' project, takes a patch of north Lincolnshire to develop the possibilities of performative engagement with landscape, with performance 'a topographic phenomenon of both natural history and local history' (Pearson 2006: 3). As in the Veneto, and Broadland, drainage shapes Pearson's country; dialogue with Pearson's Lincolnshire work has shaped this study (Matless and Pearson 2012). Pearson presents documentation of and proposals for performance pieces in the field, and a performative form of writing. '*In Comes I*', its title from the introductory line of characters in folk plays, strikingly adopts longstanding geographic landscape conventions of fieldwork and mapping, perhaps most obviously in the 'neighbourhood' excursions section of the book, around Hibaldstow, Redbourne and Kirton in Lindsey (Pearson 2006: 96–141). Pearson notes of his text that: 'For periods, the aesthetic practice of performance is barely mentioned, though the text itself remains resolutely *performative*: it employs voices of different discursive register in a number of narrative styles, in juxtapositions of material from various disciplinary approaches' (16–17). Surface geology, enclosure, field labour, town lives, appear through prose of field observation, autobiographic memory and historical geographical fact, Pearson writing through different registers to show landscape as 'a matrix of related stories' (17), to be tapped in performance.

Pearson's work prompts reflection on disciplinary approaches to regional cultural landscape, made in the spirit of the 2001 invitation in Pearson and Shanks' *Theatre/ Archaeology*: 'And the folklorist, the archaeologist, the geographer are most welcome to come and stand in our field. We do not want simply to appropriate their methodologies. We want them to look, and to enable us to look through them, at performance' (Pearson and Shanks 2001: xiv). Pearson takes region (rather than author, period or genre) as his 'optic' (Pearson 2006: 3) for enquiry, noting that his journeys 'resemble acts of contemporary chorography' (xiii); it is worth pausing at the word 'resemble'. This is performance moving close to, perhaps mimicking, but not pretending to become, something conventionally set as other to itself. If performance entails reflection on the conventions, techniques and rituals via which a subject (a person, a discipline) shapes itself, geography and performance studies find themselves performing in subtly different fashion. For Pearson region carries novelty, but for the geographer the word might signal rediscovery, even reassertion; or a rut escaped. The resemblance becomes a family one, with all the complexities of affinity, anxiety, enchantment, familiarity and contempt which that term might entail (Matless 2010). Performance and geography carry different genealogies to meet over 'region', the word migrating between the novel and familiar, the avant-garde and old hat.

Regions of writing

Introducing a collection on the narration of landscape and environment, Daniels and Lorimer identify themes of textuality, temporality and locality as shaping renewed geographical engagement with narrative, with the region one variant scale (Daniels and Lorimer 2012). Wylie develops such themes in his study of Tim Robinson, whose Galway and Connemara writings and mappings appear exemplary in presenting a regional problematic of dwelling (Wylie 2012). Landscape and region have long met in regional writing, whether fictional (Snell 1998) or documentary, and consideration of such work concludes this section, indicating further possibilities for regional cultural landscape.

Across the Humber from Pearson's north Lincolnshire, on another English east coast flatland, Yorkshire's East Riding was put into fictional play in Winifred Holtby's 1936 novel *South Riding*, subtitled 'An English Landscape' (Holtby 1936).[8] One hundred and sixty-eight participant characters are listed at the outset, with Holtby's plots of love and local government set off by the arrival of young headmistress Sarah Burton. The prosaic low regional landscape holds, and sometimes struggles to contain, the dreams and ambitions of those within. Early in the novel, squire's daughter Midge Carne looks out from Maythorpe Hall onto dullness, little knowing what is to come:

> There was not a hill, not a church, not a village. From Maythorpe southward to Lincolnshire lay only fields and dykes and scattered farms and the unseen barrier of the Leame Estuary, the plain rising and dimpling in gentle undulations as though a giant potter had pressed his thumb now more lightly, now more heavily, on the yet malleable clay of the spinning globe.
>   A dull landscape, thought Midge Carne. Nothing happens in it. (29)

Holtby's regional landscape work was recognised from 1967 in the Royal Society of Literature's Winifred Holtby Memorial Prize, awarded for 'the best regional novel of the year', winners including Graham Swift's 1983 Fenland novel *Waterland*. Swift told of life by another 'River Leem', tributary of the Great Ouse, and the agrarian reclamation of a wetland, the turning of fen to productive Fens (Swift 1983). Regional work has since however appeared to run against the literary grain. In 2003 the RSL replaced the Holtby Prize with the Ondaatje Prize, an annual award of £10,000 for 'a distinguished work of fiction, non-fiction or poetry, evoking the spirit of a place'.[9] The new prize was endowed by businessman Christopher Ondaatje, also a leading geographical benefactor, with the main Royal Geographical Society lecture theatre now named for him. Literary attention shifts from region to place, and indeed to nature, as in a 2008 issue of literary magazine *Granta* on 'The New Nature Writing', also featuring Fenland stories, discussed further in Chapter Five below (Granta 2008). The prized literary region retreats.

If the Holtby prize has gone, however, the region still attracts literary study. Thus Ralph Pite's *Hardy's Geography: Wessex and the Regional Novel* argues for a transformed sense of regionalism as central to Hardy's radicalism, challenging any assumed conjunction of regionalism, conservatism and constraint (Pite 2002), while a recent collection on 'regional modernism' takes architectural 'critical regionalism' as a starting point for excavating a neglected literary conjunction of the regionalist and modernist, resisting an orthodox historical 'metronormativity' aligning modernism with the metropolis (Herring 2009: 2). The Lincolnshire stories of Jon McGregor, making up *This Isn't the Sort of Thing That Happens to Someone Like You*, demonstrate continuing life in the regional modernist meeting, vocabularies of landscape harnessed for formal nuance and experiment; geometries of drainage, trajectories of power, topographic secrets, place name litanies (McGregor 2012).

Two further examples of regional work, deploying words and pictures through a variety of forms and sensibilities, indicate how regional attention continues to play productively. Michael Bracewell and Linder's 2003 collaboration, *I Know Where I'm Going: A Guide to Morecambe & Heysham*, writes through present and past coastal Lancashire, imagery and souvenirs from lost heydays evoking and transgressing the regional guidebook form, the work introduced with the summative: 'Our region is Pop, surreal and neo-Romantic' (Bracewell and Linder 2003: 7). Named after a Powell and Pressburger film, the book exercises a fondly possessive regional authorship. Motorways, holiday camps, monastic ruins, hotels, graves, promenades and nuclear power vie for regional attention, and Bracewell posits a particular presence of history, where sites 'articulate the past within the present with particular, at times unsettling, intensity' (8). In different fashion, Colin Sackett rubs older conventions of landscape and regional survey against themselves through a series of book works around twentieth century English sources (Sackett 2004; 2008). In 'Hereabouts' (1999), Sackett takes JA Steers' post-war coastal geographies of England and Wales, moving clockwise around the coast from Axmouth and back again, re-sorting words from each of Steers' photographic captions into alphabetical order (Sackett 2004: 68–70). Acts at once systematic and arbitrary are performed on old geography. Sackett's *The True Line: The Landscape Diagrams of Geoffrey Hutchings*, takes cropped sections from the sectional diagrams and drawings of a pioneer of regional survey and field studies, showing anew Hutchings' graphically economical landscape transcription (Sackett 2006). Regional geography makes for contemporary book art, convention sparking experiment.

Recent geographic literary studies include John Tomaney's account of Basil Bunting's modernist autobiographical poem *Briggflatts*, highlighting 'the storied nature of regional identity', Bunting's work showing a 'subtle, complex, and pluralistic sense of his Northumbrian home-world' (Tomaney 2007: 355–6; cf Tomaney 2010). For Tomaney, Bunting challenges those presenting regions as relationally subordinate to the global, or classifying concerns for regional identity as atavistic: 'Bunting's poetry demonstrates the progressive potential of regional narratives while avoiding recourse to a crude metaphysics of scale' (Tomaney 2007: 356). For Robert Colls, Bunting stands in succession to the late nineteenth and early twentieth century 'New Northumbrian' movement, with an ambivalent relationship to England, simultaneous concern for the modern and historic, and attention to conventions of travel and dialect: 'In culture, the movement had to wait for over fifty years before its credo was distilled into a single work of art – Basil Bunting's *Briggflatts*' (Colls 2007: 177).[10] Bunting and *Briggflatts* can conclude this section.

Bunting (1900–85) published *Briggflatts* in 1966, its attention to biography, history and dialect, and the animal, vegetable and mineral non-human world, offering motifs for a revisited regional cultural landscape (Bunting 1968; Makin 1992; McGonigal and Price 2000). Makin stresses Bunting's blend of pantheism and Quakerism in the scrutiny of things, observation shading observance: '*Briggflatts* is an ecology of fox, slow-worm, rat, blow-fly, and weed; sheepdogs and pregnant sheep; light on water and foam on rock: things seen' (Makin 1992: 16). *Briggflatts* takes its name from Brigflatts, the Quaker meeting house in rural Westmorland, key in Bunting's autobiographic territory; Peter Bell's 1982 film on Bunting shows the poet journeying to Brigflatts by car and foot, poem extracts read over scenes of house and adjacent River Rawthey (Bunting 2009). The poem traces other historic and present Northumbrian regional edges; the coast with stars over sea, their light travelling 50 years to earth, as Bunting lives 50 years on from remembered childhood Brigflatts love: 'Then is diffused in Now' / 'The star you steer by is gone'. *Briggflatts* is a poem of region and passage, of life time and migratory journey, in Bunting's life through Europe and Asia and Northumbrian return in 1952, in historic movement shaping Northumbrian kingdoms, with 'Baltic plainsong speech' carried over the North Sea. The books of the poem proceed through seasons, from spring to new year, with fauna variously signing Bunting's masculinity; the 'ridiculous' and 'sweet tenor' bull opening the poem by the Rawthey, the slowworm as a mark of life, moving in spring, sustained by autumn gleanings, moving with love and starlight in late life (by which time the bull has turned to beef). Makin notes Bunting's tropes of webs, weaves and shuttles, commenting: 'the solidest basis is a mobility' (Makin 1992: 149). Region in *Briggflatts* marks passage from, through and back to, Bunting's middle life return posing a regional question: what would I have settled for?

The Broadland Scene

Bodies and sources

Broadland has been shaped by organisations operating at various scales, from supranational bodies such as the European Union, shaping farming in Broadland as elsewhere in Europe, to national bodies as diverse as the Ministry of Agriculture, the Great Eastern Railway and the British Association for the Advancement of Science, to specialist regional organisations such as the Norfolk Wherry Trust, concerned to preserve particular objects. All will feature at various times in this book. Other institutions however have Broadland as a whole within their remit, or shape regional landscape through their Norfolk work, and require introduction in advance. While the term 'region' may often denote a scale greater than the English county, the Broadland region is a sub-county unit (though it also cuts across the Norfolk–Suffolk boundary), and Norfolk county institutions have played an important role.[11]

The Broads Authority (BA), formed in 1978 as a joint committee of existing authorities, became the planning authority over an area granted national park status in 1989. Broadland's waters were previously under the jurisdiction of the Great Yarmouth Port and Haven Commission (GYPHC), successor to the Yarmouth Haven and Pier Commission, established in 1670. Yare jurisdiction was historically divided between Yarmouth Corporation downstream of Hardley Cross, and Norwich Corporation upstream, successive Acts consolidating responsibility for the haven and for Yare, Waveney and Bure navigation with the GYPHC (George 1992: 343–5). The work of the BA, in navigation and other fields, is considered in Chapters Three and Six, but here it is worth noting their intellectual policy deployment of landscape, whether in the early 1980s landscape classifications discussed in Chapter Six, or through their 2006 Landscape Character Assessment, identifying 31 distinct 'local character areas' within the Authority region, their qualities to be considered in planning decisions (Broads Authority 2011: 28).[12] 'Landscape' here follows the European Landscape Convention's definition: 'an area, as perceived by people, whose character is the result of the action and interaction of natural and/or human factors' (27); a definition designed to allow formal assessment and political latitude, but which stretches towards breaking point when the complexities of culture are injected.

Broadland institutions have emerged not only through governance but commerce, with holiday hire companies such as Blakes and Hoseasons shaping the region as leisure landscape, staking the Broads as core business territory. Other bodies emerge through enthusiasm; the Broads Society, established in 1956 by sailing enthusiasts but with a remit of vigilance and care over regional landscape; the Norfolk Naturalists Trust (NNT) (now Norfolk Wildlife Trust), established in 1926 to hold land for nature conservation, and with a significant portfolio of Broadland reserves; the Norfolk and Norwich Naturalists' Society (NNNS), established in 1869 to foster study and appreciation of county flora and fauna, with Broadland a key field site. The *Transactions* of the NNNS have been central to the record and dissemination of Broadland field study. Broadland science was also the province of the state Nature Conservancy (NC; precursor to the current Natural England), established in 1948 and with its East Anglia office in Norwich, and of the University of Cambridge and the University of East Anglia (UEA), established in Norwich in 1963, and whose personnel would, as discussed in Chapter Six, play significant roles in regional research and governance. Facets of all of the above institutions' work feature in the Museum of the Broads, established in Potter Heigham in 1996 and moving to Stalham Staithe in 2000. The Museum's rich displays range from a reconstructed wherry, card models of Broadland craft, boat engines, boat lavatories, gun punts, coypu traps, a coat of mole skins, EA Ellis's coypu fur hat, beer mats and ashtrays, and the borer through which Joyce Lambert established the broads' artificial origins. The Museum's span of objects crosses the remit of this book.

Existing scholarship on Broadland has shaped public regional debate, whether on culture, history or ecology. The late nineteenth century Broadland photography of PH Emerson has acquired an international art reputation, and been subject to detailed research, notably by photographic historian John Taylor (Taylor 1995; 2006). A 1986 UEA exhibition, to which Taylor contributed, reasserted Emerson's regional status, setting his imagery as a key reference point for contemporary Broadland (McWilliam and Sekules 1986). The UEA has also shaped two key regional monographs. Tom Williamson's *The Norfolk Broads* (1997) provides a regional landscape history, outlining key landscape types and emphasising issues of industrial history and land drainage. Brian Moss's *The Broads: The People's Wetland* (2001), an ecological survey within the New Naturalist series, and successor to EA Ellis's 1965 New Naturalist regional volume (Ellis 1965), gives an overview of late twentieth and early twenty-first century Broadland science, discussed further in Chapter Six. The standard reference for any Broadland research remains however Martin George's *The Land Use, Ecology and Conservation of Broadland* (1992). George played a key role in Broadland as Nature Conservancy Regional Officer for East Anglia (1966–90, following a term as Deputy 1960–6), and was awarded an OBE on retirement for 'services to conservation'. From 1991 he served on the committee of the Broads Society, acting as Chair from 1998. From the NC offices in Norwich, and a house overlooking Strumpshaw Fen, George has produced key works on the region.

Figure 3   Museum of the Broads, Stalham. Source: photographs by the author, August 2011. Reproduced by permission of the Museum of the Broads.

Archival and interview sources are noted in the Preface preceding this chapter, but a key informant, who became both a source of research material and a commentator on research as it developed, was the late Joyce Lambert, who established the broads' artificial origins in the 1950s. Lambert's contribution to this research deserves discussion here. After an initial interview at her home in Brundall, where she lived after retiring from Southampton University, I met Lambert on a number of occasions, in

Brundall and later in a nursing home on the edge of Norwich, discussing issues of shared interest within and beyond Broadland, from ecological research and landscape history to local sport and national politics. Examination of private research documents, and press cuttings concerning the origin findings, were crucial in researching origins debates. Lambert offered comment on my own Broads publications, and on articles in draft, notably a piece considering her own origins research, the article improved in both historical accuracy and level of argument through her commentary (Matless 2003a). The generous donation of her general archive of Broadland press cuttings, compiled between 1947 and 1970, proved immensely valuable, with many, though not all, references to press material from that period coming from Lambert's collection. The composition of her cuttings archive is itself worth consideration, as it demonstrates the categorisation of Broadland and Norfolk by a leading regional scientific figure. Three folders contain clippings, largely from the Norwich morning paper, the *Eastern Daily Press* (*EDP*), but also other regional and national newspapers and magazines, organised by topic and date. Folder 3, 'Newspaper Cuttings 1956–1968', paperclips articles together in six-monthly or yearly bundles. Folders 1 and 2, labelled 'NORFOLK BROADS, ETC. Newspaper Cuttings. 1947 –', have articles cut and pasted onto pages, each section ordered chronologically between 1947 and 1956, topics grouped under 21 headings. The topic list itemises Lambert's regional study:

NEWSPAPER CUTTINGS

INDEX

1. Norfolk Broads – General administration
2. Norfolk Broads – Access
3. Norfolk Broads – Riverside tenancies
4. Norfolk Broads – River crossings, upkeep of staithes, and other amenities.
5. Norfolk Broads – effect of proposed lock.
6. Norfolk Broads – Silting, reed growth, and clearance.
7. Norfolk Broads – Flora
8. Norfolk Broads – Fauna
9. Norfolk Broads – Economic utilisation
10. Norfolk Broads – Grazing marshes and drainage.
11. Norfolk Broads – Historical
12. Norfolk Broads – Miscellaneous notes.
13. Coastal erosion and protection.
14. Geography.
15. Geology.
16. Archaeology.

----

17. Norfolk Research Committee
18. Norfolk Naturalists Trust
19. Norfolk and Norwich Naturalists Society

----

20. Breckland and other Norfolk areas
21. Miscellaneous.

All of Lambert's 21 categories feature in this book. Her 'Miscellaneous' included museum news, archive stories, Fenland pieces and uncategorisable nature news; a cultural geographic miscellany might also stretch to painting, comedy, wildfowling, novels, poetry, film, television, holidays, architecture, postcards, souvenirs.

Field excursions have also shaped this book. John Barrell memorably began his 1982 *Journal of Historical Geography* essay on 'Geographies of Hardy's Wessex', examining 'the means by which … various characters and narrators explore places and come to know them', by stating: 'I have never been to Dorset – though I believe I may have passed through it on the train to somewhere else. But I make that con-fession, not to disqualify myself from writing this essay, but to indicate at the outset the sort of essay it will not be' (Barrell 1982: 347). I have been to Broadland many times, but have passed by it many more, the region thereby acquiring a peculiar kind of familiarity. For the past 25 years visits to Broadland have been accompanied by the possibility of research, first as a site for a possible element of doctoral study, never realised, then as the focus of various research papers, some on the region as a whole, some on individuals or specific aspects – animals, origins, sound, film (Matless 1994; 2000b; 2000c; 2003a; 2005; 2010; 2011; 2012). Visits in adult life have therefore always carried the academic with them, shaping the perspectives and pleasures gained, even when walking a riverbank, or chugging on a day launch. Sometimes trips have aspired to the systematic; cruising a range of rivers over a weekend, or driving select sites for a day. Before research reared, however, another geographical psychology was in play. I grew up near to the Norfolk Broads, around eight miles from the nearest broad, but had little experience of the region. This was not an autobiographical centre, but rather a region at close distance, a remote zone near home, difficult to access unless you had a boat, with little to see of the region from the road. Growing up in suburban Norwich, I recall only a couple of boat excursions on large tour boats from Wroxham. Boat hire came only after I had moved from the area, and thought of Broadland as a research site. The Broads had earlier been something adjacent, but apart, a site of special interests other than my own, less appealing than the open and free coast ten miles further on. I had no enthusiasm for fishing or birdwatching. This regional account thus emerges from a particular geographical psychology.

Broadland tour

This chapter concludes with exercises in geographic description; journeys down the river valleys in this section, and attention to six sites to end. Description in academic geography has tended to be a pejorative term. Whatever geography's etymological claim to 'earth writing', 'descriptive' has often denoted lack of analysis, a bland sur-face accounting, with any greater descriptive adequacy out of reach, as in HC Darby's 1962 diagnosis of 'The Problem of Geographical Description':

> It is a humiliating experience for a geographer to try to describe even a small tract of country in such a way as to convey to the reader a true likeness of the reality. Such

description falls so easily into inventory form in which one unrelated fact succeeds another monotonously. How difficult it is to transcend a painstaking compilation of facts by an illuminating image. (Darby 1962: 2)

Description may however carry other cultural possibility. As Svetlana Alpers suggested in her 1983 study of seventeenth century Dutch painting, there is a complex 'art of describing', Alpers seeking to recast studies of visual culture away from Italian art historic emphasis on symbolic narrative, and querying the distinction between description and narrative action: 'northern images do not disguise meaning or hide it beneath the surface but rather show that meaning by its very nature is lodged in what the eye can take in – however deceptive that might be' (Alpers 1983: xxiv; Wall 2006). Far from being a dull inventory trap, description offers complex cultural practice, at once calculating and generative, giving an account of landscape in the sense of both patient itemisation and events proceeding. Description carries a usefully dual performative sense of, on the one hand, distanced representation – the observer set back, however closely, from a scene – and on the other inscriptive enactment – the instrument describing, engraving, a line.

In 1967 the naturalist JA Baker wrote: 'Detailed descriptions of landscape are tedious. One part of England is superficially so much like another. The differences are subtle, coloured by love' (Baker 1967: 7). Baker's statement appears early in *The Peregrine*, a book which, far from avoiding landscape description, pursues a heightened poetic prose on the Essex landscape life of the peregrine falcon. Baker's successor volume, *The Hill of Summer*, extends the technique through elements of Essex landscape, April to September: 'While describing twelve separate landscapes, I have recalled the memories of many years' (Baker 1970: 8). Baker renders a consistent intensity of observation, outdoor prose concentrated in subtle attentiveness, whether at enclosed wood, waterside or open field. Mark Cocker indeed sees *The Hill of Summer* as pared down 'to one goal: how can a naturalist capture in words what he or she sees and experiences' (Cocker 2010: 11). The differing modes of description (at once, in Darby's terms, both painstaking and illuminating) informing the journals of poet RF Langley further indicate descriptive capacities. Langley's *Journals* render detailed observations of landscape, insects, plants, buildings, written at intervals since 1970, wrought prose lending quizzical observation and scrutiny to things, words shuttling between the lyrical and staggered, the reader attentively following the observing writer. Langley does not loom over words with personality, claiming a country for his own; rather he exercises a first personhood restrained, a different at-one-remove for the reader to inhabit (Matless 2009a). Langley walks a rural Suffolk 'Uncle's Lane', the name mooted from the Anglo-Saxon *Uncuo*, unknown or strange:

There is a grim sort of pleasure in this landscape, so stripped down and simplified, and the power given by this stripping down is given to what is left, such as the flints in the soil, white or black or brown, split or knobbed, and the open flashes of the butterfly wings. There are often three or four skylarks which rise and sing, sometimes two close together. At the top of their climb, they stiffen their wings, float level for a distance, then turn down and plummet. They were doing it, were they, 4,500 years ago, when people walked this track, heading for the higher ground on the northern skyline, with their illusions that matter-of-fact was not all there was to it? *Uncuo.* Strange, or unfriendly. Weird. Matter spread out almost flat.

Wings that flicker at your feet. The popping sound of swollen seed cases. All this seems to be holding back, stopped in its tracks, caught hard and dry within itself. No ecstasy, just the slightly astonishing ferocity in the lark's dive down at the earth, a somewhat desperate curving attack. Or the sense that the spread of quick colour in the butterfly wings is a blink, as you meet, momentarily, face to face. Life itself was weird, the Saxons felt. Is this a possibility? Today I like very much this all that there is. I would come back for it. (Langley 2006: 119–20)

Broadland can be a region of peculiar rhythm and visibility. The end of a Broads boating holiday is typically marked by a dramatic shift in pace, from a week at a few miles an hour to ten times the speed in a car or train, and sudden movement across the river valleys which have taken so long to windingly navigate. Waters retreat from view, become hard to spot, indeed away from the river Broadland is commonly characterised as a region of elusive waters, as in Pevsner's guide to the buildings of Norfolk: 'Sails sliding silently among fields against the low horizon are as much a recurrent sight of the Broads as expanses of reeds behind which one does not at once see the water' (Pevsner 1962: 14). Unless on a boat, broads are difficult to see. Thus Reginald Wellbye's 1921 *Road Touring in Eastern England*, in the 'Roadfaring Guide' series, could happily hymn the car's 'special outlook on the country', aiding 'a perception of geographical continuity' (Wellbye 1921: viii–xi), but found a problem in Broadland: 'It is well that the roadfarer should realize that the wide fame of the Broads is due to the facilities they offer for a water holiday, and that, unless he himself will take to the water, they will not present him with anything like the same satisfaction, since he will not be able to see very much of them' (39). Wellbye noted river viewing spots, but of the 'typical broad' stated: 'The roadfarer has to obtain his view of the Broads themselves in peeps – *delicious peeps*, it may truly be said, and by no means infrequent, though all too few' (41).

Here, the riverfarer will be followed down each valley, allowing quick introduction to the regional scene, beginning with the northern rivers, and ending at Cantley on the Yare.

Bure

Canoes might make a way from the former head of navigation at the market town of Aylsham, past Buxton to the current head at Horstead mill. Boats only ascend as far as the mill, turning again down to the riverside green at Coltishall. Woods line the long bend past Belaugh, to Wroxham's arched main road bridge and boatyards, and well appointed riverside homes downstream. Either side of Horning village a series of side broads – Wroxham, Hoveton, Ranworth, South Walsham – connect to the river through dykes, bankside carr woodland opening up below Ranworth and the Ant mouth. The ruins of St Benet's Abbey are passed on the north bank, with open marshes around the south turn of the Bure at Thurne Mouth, to Acle Bridge. Through Stokesby and down to Yarmouth the Bure winds embanked through grazing marsh, on the northern flank of flats stretching south to Breydon, the north view short into low upland, the south views long, Yarmouth the east horizon frame.

Ant

Canoes might make a way from the former head of navigation at Antingham, down the disused North Walsham and Dilham Canal, to the current head around the main road Wayford Bridge, where houseboats view the A149. Downstream branch channels head up to Sutton and Stalham, the Museum of the Broads by Stalham staithe. The Ant widens into Barton Broad, side channels leading to Barton Turf and Neatishead, the river exit passing Irstead and through open marsh and extensive reed to How Hill, an Edwardian country house, now an education centre, looking down to the river, a Broads Authority estate around. Ludham bridge is low with a right angle turn after, and a short trip to meet the Bure.

Thurne

The Thurne enters the Bure at Thurne Mouth, but has various beginnings. Journeys starting a mile from the coast on Waxham New Cut and taking in Horsey Mere (with mooring by the mill for Horsey village and the coast) and Meadow Dyke, or at Hickling and across Hickling Broad and Heigham Sound, or at West Somerton and through Martham Broad, all through wide marsh and reed, remote from road traffic, meet above Potter Heigham to pass riverbank chalets either side of modern and medieval road bridges. Past the entrance to Womack Water (for Ludham village) and Thurne Dyke fronted by a white mill, the river meets the Bure, embanked through grazing marsh.

Chet

The shortest river, the head of navigation shared between Loddon and Chedgrave, boats leaving the basin for an embanked stream, passing Hardley Flood and meeting the Yare at Hardley Cross, remote sixteenth century monumental limestone marker of the former break in Yare corporate jurisdiction between Norwich and Yarmouth.

Waveney

Canoes might make a way from the market town of Bungay or higher, down to Geldeston Lock, current head of navigation, and on through marshes past the south bank town of Beccles, following the Norfolk–Suffolk boundary all the way. Past Burgh St Peter staithe a channel goes right to Oulton Broad, centre for summer speedboat racing, framed by suburban Lowestoft, Mutford Lock connecting to the town's industrial Lake Lothing and the harbour and North Sea. Beyond the Oulton turn the Waveney goes north past Somerleyton, tracked and bridged by the Lowestoft–Norwich railway, the high ground of Lothingland close on the east, grazing marsh opening to the west and north as the Waveney and Yare valleys merge, viewing north across flat miles to the Bure. By Haddiscoe station the straight canal of the New Cut heads north-west to the Yare, taking the train alongside. The grazing marsh of Haddiscoe Island is bounded by the Waveney, Yare and New Cut, the two rivers meeting at the head of Breydon Water, the Roman fort walls of Burgh Castle on the hill to the east.

Yare

Canoes might make a way past Bawburgh and Marlingford and under the Norwich southern bypass to the University of East Anglia, with its broad dug from gravel, to the head of navigation above Trowse. Downstream heads past elements of industrial Norwich, the Wensum entering from the city, Thorpe's suburbia with a short New Cut cutting off a meander and avoiding two rail bridges, skirting recently dug recreational broads at Whitlingham. Under the Norwich southern bypass bridge, by the foamy outfall of the sewage works, by isolated riverside pubs at Bramerton and Surlingham, on to Brundall's north bank boatyards and riverside estate across from the south bank Coldham Hall pub, the valley opens from woods to the marsh and reed of the Strumpshaw reserve, with Surlingham, Wheatfen and Rockland Broads on the south side. The Lowestoft railway tracks the north side to Reedham, marsh widening and open, past the Chet mouth and Reedham's chain ferry and village, beneath the rail swing bridge and the New Cut going south to the Waveney, through wider marsh with the Yarmouth rail line on the north side and views north to the Bure and south to the Waveney, to roadless Berney Arms with rail station, mill, pub and few houses, and into Breydon. A channel between posts keeps craft from Breydon's mudflats, shown at low tide, and across the water under Breydon Bridge. The Bure joins from the north, the Yare continues south under Haven Bridge, between Yarmouth and Gorleston quays, shipping moored, with a final turn east to the sea.

Also apparent on the Yare is Broadland's most substantial riverside structure, though one often ignored in regional accounts, the Cantley sugar beet processing factory. The 1962 and 1997 editions of the Pevsner 'Buildings of England' series pass the factory by, noting only the village church, but for its architectural and arable presence Cantley deserves attention here (Pevsner 1962; Pevsner and Wilson 1997). Established in 1912 on a 40 acre site by the Anglo-Netherland (later Anglo-Dutch) Sugar Corporation, Cantley was the first beet sugar plant in England, built with Dutch expertise to reduce dependence on imported cane sugar. Sugar beet gave East Anglian farmers an autumn and winter harvest, the processing 'campaign' running from September to January. Cantley had its own rail siding, and the Yare offered water supply and barge transport. Sixteen further factories were established in the 1920s, all part of the British Sugar Corporation from 1936 (Watts 1971). Donald Maxwell's 1925 *Unknown Norfolk* described the new industry, observing winter workers living on moored holiday vessels, and illustrating a wherry in sail passing the smoking factory, and a steam lorry tipping beet into a trench, Maxwell discovering a Norfolk 'unknown' to the tourist (Maxwell 1925: 23–32). John May likewise visited Cantley in a 1952 guide, his wife having countered a suggestion to visit Langley's abbey ruins: 'People don't build ruins … Let's go on to something a bit modern – this sugar factory' (May 1952: 30). On the front cover of Richard Denyer's 1989 Broadland photographic survey *Still Waters*, Cantley's plume blends with clouds, both reflected in the river, Denyer presenting Broadland as working landscape, a flying coot helping gather the factory into the scenic (Denyer 1989).

The great majority of Broadland narratives ignore this most conspicuous regional built structure, unmissable by any Yare cruiser since 1912, the sizeable original factory (its main building 197ft long and 62ft high) later supplemented by larger silos. For accounts of Cantley the reader must instead seek out Watts' 1971 *Transactions IBG* survey of the beet industry (Watts 1971), or the 1913 *Journal of the Board of Agriculture*. Twenty years of British debate had considered 'the great possibilities of sugar beet', yet 'the first factory has been erected chiefly through the enterprise and confidence of strangers' (Chadwin 1913: 582; Robertson-Scott 1911). Cantley offers one of many Broadland-Dutch connections, returned to at this book's conclusion.

Six sites

To end this chapter, six pieces of landscape description are offered. If the preceding accounts have given quick valley excursions, these condense regional landscape and culture around specific sites. The six selected are relatively evenly distributed across the various river valleys, at pressure points within the regional cultural landscape, where tensions of landscape become acute, viewpoints can be taken, contrasting pleasures meet, anxieties surface.[13] Various aesthetic principles inform the composition of these accounts, as indeed the rest of this book. There is a democratic attention to all kinds of regional object, treating things in non-hierarchical fashion but attentive to the ways in which hierarchies (of value, of beauty, etc) operate. There is likewise a cross-regional democracy, with all kinds of location treated in non-hierarchical fashion, but with attention to the ways in which hierarchies (of service provision, of cultural distinction) operate. Landscape is worked through all senses, though with attention to the ways in which hierarchies of sense operate; the truths of sight, voices heard, sounds and textures of objects. The accounts here are monovocal, deliberately singular, unlike the remainder of the book, where landscape's colloquial multiplicity is given hearing. They respond to an assumed request to describe a landscape without reflex reference and resort to the commentary of others.[14] Singularity of voice may sometimes entail evocation of specific personal experience, but more commonly this is experience amalgamated across times and places, offering composite motion through the region.

Ranworth
Two ways to reach Ranworth Church: by car to the small parking space near the gate, or on foot from the water. From a boat moored at the staithe on Malthouse Broad (Ranworth Broad being private), the church is a quarter mile up a lane, west up a gentle rise of the land. By car you might arrive without viewing a broad.

Into the church to an open stone floor, the holy end right beyond a painted saints screen. Across the floor another door and spiral stairwell, stone inside the tower, moving by ladders through the belfry towards the top, and out of a narrow hatch.

To a rare airy prospect. Onto something unexpected (boats!), or something just seen (boats). Fields, fruit trees in line, and in the far east water towers, and turbines; some in the sea. South, in season, factory smoke. The city south-west, seven miles out of sight. Leaning over low railings to graves, tilting slightly back with perspective. West the broad with boats, north the broad without; with thatched, floating, conservation centre. Where the two join, a straight dyke cut to the Bure. East a huddle of ex-abbey; walls, with some tower. Modest landmarks. The valley north-east thick wooded, to the right open to marsh, sails evident over the flat. Yachts and mills. Rare perspective in the region.

Reedham

The train from Norwich stops, a proper station at the edge of a village barely seen. Moving on past few buildings in cuttings, off left a line goes across marsh to Yarmouth, the Lowestoft line curving right. An empty cutting heads back left to the other branch, then under a bridge and some building backs and out to bridge a surprise of wide expanse.

The Yare swing bridge, open for rare sea coasters and high masts, with its flags and operatives, viewing riverside Reedham; a quay and moorings, green and housing. The tidal Yare takes the slack of rope; cruisers wake at a different level. Boats run away with themselves approaching on the flow. A couple of bends up, the last chain ferry, cars saving miles for a fee, inn and tents by the marsh. Otherwise distance with little to lean on.

The line continuing, curving left, gaining the bank of the straight New Cut and on dead straight under high slim pylons and a concrete span, curving to Haddiscoe and the furthest east of England.

Potter Heigham

For a mile the banks lined with chalets, one deep on either side between river and grazing marsh. The Thurne a front garden to cruise, noting variations in chalet form, standard shed to quirky millstump, a verandah or a lawn, fishing spot or private mooring. The river as lounge, or chalet-ruined. After a mile a modern boatyard tower, a basin of cruisers for hire, a tight bridge arch ahead, medieval stone. Aim and duck. A high road bridge bypasses where rail ran. Chalets further on the rond.

Traffic lights over the hump, into riverside Potter Heigham, the village proper a distance on. Busy Saturdays in season; boat turnover, day hire, ices. Year-round callers for local bargains; clothing, provisions, souvenirs, maggots, rods.

The village sign towards the staithe, bench-circled, with carved scenes and history text. The old bridge and packhorse, Roman potters, reed boats, heron, bittern, drainage mill, yachts, cruiser, angler, peat cutter. The origin of the Broads explained.

How Hill

On a rise above the Ant, a house, thatched, substantial, south facing for the sun, sheltered from any north wind, taking in the valley view. A polite landscape, with lawns, topiary; a hundred years a model estate. Closed gardens nurture exotics. A boardwalk trail finds a stick in carr woodland ground, to be pushed and pulled up and down for ten feet, liquid mud beneath. By the river, restored mills of unusual type, once marsh pumps. Cut reeds stacked for transport. Electric boats carry silent tours through narrow waters, stopping at hides, noting reedbeds, their birdlife, their management. Painting classes by water. A marsh cottage restored, tight accommodation, eel tools, dampness. Education, propriety, the sustainable.

Forty years ago, visiting to play at cricket on the neat grass. Young excitement, the Broads of no interest.

Twenty years ago, an elderly gent on his young estate life, the family car coming home from the city, seeing its lights the only lights across the marsh.

Cantley

The B1152 between Potter Heigham and Acle, by Clippesby and about to dip into Bure valley, factory smoke on the south winter horizon. Sugar beet process at Cantley by the Yare, the plume seen from ridge roads and valleys beyond; Thurne, Waveney, A146.

The Yare downstream of Brundall between open marsh. The Yare upstream of Reedham between open marsh. Stumps of mills, cattle, rooks; trains along a train line. Silos and chimneys and stores stand for miles on the flat. On the reach by Langley marshes Cantley grows, a pub dwarfed underneath, the new pleasure staithe a small mooring. The eye drawn; distracted, grabbed; blot or beauty.

A station and yard. Industrial movement. Beet trucked, once by train or wherry. Arable Norfolk in Broadland.

West Somerton

The top end of the river. Through Martham Broad, the Thurne turning to West Somerton staithe. Free mooring. Permanent boats in still weed beyond. A small green, a village sign. Mill, wherry and church in black metal. Over fields the North Sea, under two miles; forceful waters. There are the dunes, just over there. The Lion, with noted double cod. Wind farmed on the hill.

For once you can stand on the waterside, look down, and see clarity. Plants visible, greenery through wind ripples. The only Broadland spot which never clouded.

Broadland Scene

Figure 4   Broadland Scene I (Top: from Ranworth Church tower, August 2011; Bottom: Carrow Bridge, Norwich, August 2011). Source: photographs by the author.

Figure 5 Broadland Scene II (Top: Potter Heigham Bridge, August 2011; Bottom left: Tide gauge, Wheatfen, August 2010; Bottom right: Reedham rail swing bridge, May 2011). Source: photographs by the author.

Figure 6   Broadland Scene III (Top: Cantley sugar beet factory, May 2011; Bottom: Berney Arms railway station, August 2011). Source: photographs by the author.

Figure 7   Broadland Scene IV (Top: Norwich Yacht Station, August 2011; Bottom: Breydon Water, August 2011). Source: photographs by the author.

Figure 8   Broadland Scene V (Top left: Heron's Carr, Barton Broad, May 2011; Top right: How Hill gardens, May 2011; Bottom: West Somerton staithe, August 2011). Source: photographs by the author.

Chapter Two
Origins

Broadland with the IBG (and RGS)

On 6 January 1959, under the headline 'Winter Broads Tour', the *Eastern Daily Press* reported a sighting of geographers:

> A BROADS TOURS craft carrying over 60 passengers caused some surprise in the Wroxham, South Walsham, Salhouse and Ranworth areas on Sunday. A familiar sight for half the year, it is rare to see one carrying sightseers in this weather.
>
> The passengers were members of the British Institute of Geography who were holding their annual conference at Cambridge and they were taken in the Marchioness through Wroxham and Salhouse Broads to South Walsham Broad and thence to Ranworth Broad.
>
> The party had travelled by diesel rail cars to Wroxham that morning. At Ranworth the Rev. E. D. Everard (vicar) met them and they ascended the 96 ft. high tower which gives an extensive view of Broadland.
>
> Among the party was Dr. Lambert, who has written extensively on the origin of the Broads, and Mr. C. T. Smith, lecturer in the Department of Geography, Cambridge University. The party returned to Wroxham for tea. (EDP, 1959a)

These early January Broads tourists travelled from the 1959 Institute of British Geographers (IBG) annual conference in Cambridge, their guides Joyce Lambert and Clifford Smith, two authors of the 1960 Royal Geographical Society (RGS) Research Series memoir No.3, *The Making of the Broads* (Lambert et al 1960). Botanist Lambert had discovered the broads' artificial origins in medieval peat

In the Nature of Landscape: Cultural Geography on the Norfolk Broads, First Edition. David Matless.
© 2014 David Matless. Published 2014 by John Wiley & Sons, Ltd.

diggings, historical geographer Smith corroborating findings. The RGS Research Series memoir No.2, geomorphologist Joseph Jennings' 1952 *The Origin of the Broads*, had held to natural origins (Jennings 1952); the 1960 memoir (to which Jennings also contributed) identified broads as human creations. Origins work made the Broads worth a 70 mile winter geographical trip.

Lambert and Jennings' geographical engagements also encompassed the Geographical Association (GA), Jennings explaining origins to the Norfolk branch in December 1951 (EDP 1951c), Lambert speaking as Norfolk GA President in 1960, arguing that the proposed UEA should have a geography department 'high among the priorities': 'geography was no longer dull and deadly and had become the indispensable link between the arts and sciences' (EDP 1960c). No such department was established; the UEA is discussed further in Chapter Six. Before the UEA's foundation in 1963 however Broadland fell within the intellectual field territory of the University of Cambridge, with regional ecological and geographical studies conducted by figures such as Jennings, Lambert, Marietta Pallis and Harold Godwin, Cambridge scientists working through Norfolk scientific institutions including the NNNS and the Norfolk Research Committee (NRC). The NRC, based at the Castle Museum, had been established by Cambridge archaeology student Roy Rainbird Clarke, son of Breckland topographer WG Clarke, in 1934, inspired by the Fenland Research Committee (Matless 2008b). As Castle Museum Curator Clarke would direct NRC co-ordination of archaeology, history, geology, natural history and ecology, acting as Secretary until his death in 1963 (Edwards 1984). NRC presidents included Castle Museum naturalist EA Ellis (1950–1), Lambert (1952–3), Godwin (1957–8) and Cambridge geographer JA Steers (1960–1), Cambridge and Norfolk conversing across disciplines.

This chapter addresses the geographies of science shaping landscape narrative, with Jennings and Lambert's work considered against preceding theories, and scientific and public debate around artifice examined, including the ways in which science informed policy proposals for landscape intervention. Original theories worked present and future Broadland.

Origin in Transgression

Early origin theories jostled water and ice. In 1871 geologist JE Taylor addressed the NNNS positing glacial origin, the broads as ice-scooped hollows filled by water (Taylor 1872; Nature 1871). Taylor's glacial theory was quickly dismissed, becoming the counterpoint to an origin orthodoxy of estuarine relics, informed by Robberds' 1826 argument, noted in Chapter One above (Jennings 1952: 9–14; Robberds 1826). Henry Stevenson's 1872 NNNS Presidential Address, published in the same *TNNNS* issue as Taylor, criticised the glacial theory, noting Brady and Robertson's 1870 *Annals and Magazine of Natural History* study of tidal river fauna as having identified the broads' 'very recent origin', from a shallow estuary's 'gradual silting-up': 'they will at no distant date cease to exist' (Stevenson 1872: 3–4; Brady and Robertson 1870; Woodward 1883).[1]

JW Gregory's 1892 'The Physical Features of the Norfolk Broads', published in *Natural Science*, summarised the estuarine orthodoxy. Submergence had turned three rivers into an estuary, with subsequent silting inland as sand bars blocked river mouths. Alluvium had blocked side channels to form side-valley broads, while deltaic outgrowth formed ronds to cut off bypassed broads (Gregory 1892; Reid 1913). Gregory, in a national scientific journal, became the standard citation for the next 50 years, though some sensed there might be more to say. Marietta Pallis, giving the first academic scientific analysis of broads ecology in 1911, saw Gregory's 'extremely suggestive' theory giving only 'partial explanation', with 'evidence' indicating 'other causes have probably also had a considerable influence in giving origin to the broads of East Norfolk. The topic is not, however, ripe for further discussion' (Pallis 1911a: 222).[2]

Pallis's 'other causes' likely included marine transgression, sea incursion at the broads' origin. Jennings' *The Origin of the Broads* noted Pallis nearing this explanation ('though Pallis did not publish such a deduction' (Jennings 1952: 53)), concluding: 'the transgression is responsible for the formation of the Broads and largely explains the present physiography' (49). Jennings began his Broads research at Cambridge in 1938 at the suggestion of geographer Steers (NNNS President in 1940–1), with ecologist Godwin guiding pollen analysis and peat stratigraphy, Jennings sending his 1940 report on 1938–9 Ant valley fieldwork to Godwin 'from an Army camp' (1). Jennings resumed research in 1947, appointed as Lecturer in Geography at University College Leicester, also undertaking comparative fenland research (Jennings 1950). *The Origin of the Broads* was single authored, though Jennings collaborated with Lambert from 1947, the NRC having put them in touch, its 1949–54 Bulletins reporting their researches. *The Origin* also included an appendix 'Table of Mollusca' by Ellis, classifying remains in middle Bure alluvium (Jennings 1952: 62–3).[3] Ellis conveyed Jennings' transgression argument to a regional public in his daily *EDP* 'In the Countryside' columns, describing 'Old Shore Lines' (Ellis 1952a), 'Submerged Forest' (Ellis 1952c) and 'Land Drowning' (Ellis 1952e), while Sainty registered the theme in *TNNNS*: 'It is thus clear that the prime cause of the formation of the Broads is the sea' (Sainty 1948a: 372).

Jennings retained Gregory's classification of 'side-valley' and 'bypassed' broads, but saw the estuary origin analysis as having 'little foundation in physical geography' (Jennings 1952: 12). Extensive boring instead showed two marine transgressions leaving clay layers, with peat laid down between, and bypassed broads formed behind clay flanges. Boring prompted landscape visualisation:

> We must envisage the appearance of these valleys in the Early Iron Age or in Romano-British times somewhat as follows. There were reed-swamps on either side of a main tidal channel into which the tidal current, not very saline here, swept fine mineral sediment to be deposited at the slack of high water. In this way, clay banks were gradually built up which were probably uncovered at low water. Areas of open water, fed from the upland and also by water filtering through the reedy banks at high water, developed between these clay banks and the valley sides. (49)

Jennings laboured with 'a hand peat borer of Hiller design, with a chamber 50 cm. in length and extension rods which give it a normal maximum length of 13.5m' (16–18), the work 'shared by numerous friends, students and senior schoolboys' (1). Diagrammatic sections with exaggerated vertical scale accompanied location maps, vertical layering of peats and sediments showing regional underlay. Pollen diagrams confirmed post-glacial origins, a woodland history of pine, birch, elm, oak, lime, alder, beech and hazel, Jennings concluding that 'the Broads had attained their essential form before the end of the Middle Ages' (48; Jennings 1955). The broads take shape from tidal waters, formed in nature, boring accounting for origin.

Artifice Discovered

Peat digging had long registered in origin accounts as partial explanation. Samuel Woodward in 1834 and Horace Woodward in 1882 considered Barton partly or wholly artificial: 'Barton, and perhaps also Hickling, may have been formed by the cutting of turf, and they are, therefore, more or less artificial' (Woodward 1883: 459). In 1948 Sainty noted peat cutting may have 'enlarged the broads' (Sainty 1948a: 373); in October 1951 Ellis commented: 'it is now an accepted fact that some of the Broads existing today owed their origin to peat-cutting' (Ellis 1951). Jennings' *Origin* mentioned peat cutting on several occasions, extensive cutting in undrained fens making it 'more difficult to find areas which have not been used in this way than areas which have, and even the possible creation of "broads" in this manner cannot be excluded' (Jennings 1952: 3). Barnby Broad in the Waveney valley was 'probably artificial' (6), while at Woodbastwick Fen: 'According to local information small broads had here been incorporated in peat cuttings. There was an obvious value in investigating the stratigraphy of such an area, as the origin of some broads has from time to time been ascribed to peat cutting' (24). Woodward's 1834 observations of peat ridges along the Barton–Irstead parish boundary prompted Jennings to comment: 'It is not difficult here to walk round in the broad on the firm top of the peat' (44). Walking around the water did not however suggest general artificial origin, peat cutting remaining complication rather than explanation: 'a general theory based on peat cutting appears to be unacceptable' (52–3; Jennings and Lambert 1951).

Artifice became general theory in 1952. In the NRC's summer 1951 Bulletin 3 Lambert reported sections at Surlingham showing 'extensive deep hidden peat excavation' (Lambert 1951: 5); the undated Bulletin 4, likely published in 1952, noted similar findings:

> in the other Surlingham sections, at Strumpshaw, at Buckenham and Hassingham, and possibly also at Carleton. It is now a moot point whether in fact any part of any of the Yare valley broads can be considered to have a natural physiographical origin, and the whole question of the origin of the broads in general is now under reconsideration in the light of this new evidence. (Lambert 1952a: 7)

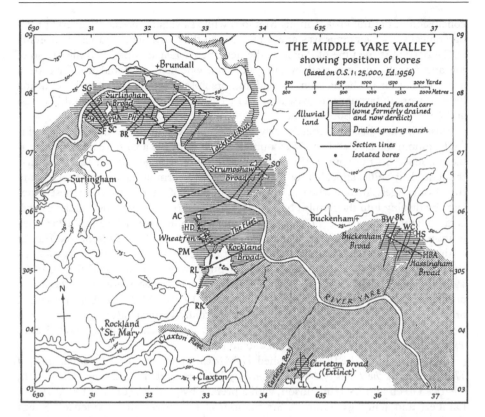

Figure 9   Bores in the Yare Valley. Source: Lambert et al 1960. © Royal Geographical Society (with the Institute of British Geographers).

Lambert was based at Newnham College, Cambridge from 1950, moving from lec-turing at Westfield College, London on obtaining a Leverhulme Research Fellowship to investigate Broadland vegetation (Nature 1950); she took up a lectureship in botany at Southampton in summer 1953. Lambert recalled how 'the penny suddenly dropped', dashing downstairs at Newnham to explain to historical geographer Jean Mitchell. Lambert and Jennings had Godwin to dinner, who initially said 'nonsense' but was convinced by the evidence; Steers initially said 'nonsense' but was persuaded by Godwin, and would communicate the new findings in the US *Geographical Review* (Steers 1954).[4] In March 1953 Jennings and Lambert published a note in the RGS's *Geographical Journal* on the revised theory, dated 13 November 1952, Lambert's closely lateral bores for Yare vegetation study showing 'undoubted excavations', and reconsideration of Ant and Bure evidence supporting a general artificial explanation; the 'greater part of the story of the natural evolution of the valleys' was however unchanged (Jennings and Lambert 1953). Lambert recalled the note as 'an emer-gency measure to enable Joe [Jennings] to refute, as quickly as possible, his main conclusion of natural broads in his Memoir, while salvaging the more general physi-ographical work'. Jennings was about to move to the Australian National University at Canberra: 'At the time of the note, the new theory rested entirely on the say-so of Joe and myself, unsupported by any published evidence. It would not have been accepted but for Joe being seen, by lead authorship, to take the initiative in rebutting his own work in favour of the new idea.'[5] The two agreed that Lambert would be lead author in subsequent publication.

Ellis, who had been Lambert's early mentor, gave a hint of the new theory in his *EDP* column for 29 October 1952: 'human activities in the area have had considera-ble influence on the course of events, possibly over a longer period than was first thought likely'. Ellis appealed for 'local evidence concerning the demand for peat in the Middle Ages. Anyone who can unearth something of interest affecting these prob-lems by delving into the family chest or burrowing into a lawyer's dusty archives, will achieve fame' (Ellis 1952d). The first substantive account came in the July 1953 pub-lication of Lambert's NNNS Presidential Address on 'The Past, Present and Future of the Norfolk Broads' (Lambert 1953a; also Lambert 1953b).[6] The Address, given on 19 April 1952, was 'partly amended' for publication, a note dated 5 December 1952 signalling 'significant new data, calling for a major reconsideration of the whole problem of the origin of the broads' (Lambert 1953a: 223). Lambert presented broads in individual peat-lined basins, with 'virtually vertical sides' (233) and peat peninsulas, islands and ridges, indicating historical deep excavations, flooded after abandonment, with later navigational cuts to rivers.

In February 1960 the RGS published *The Making of the Broads: A Reconsideration of Their Origin in the Light of New Evidence* by JM Lambert, JN Jennings, CT Smith, Charles Green and JN Hutchinson, with a preface by Godwin (Lambert et al 1960). Lambert and Jennings outlined 'Stratigraphical and Associated Evidence', Smith 'Historical Evidence', Green and Hutchinson 'Archaeological Evidence'. Historical geographer Clifford Smith, Assistant Lecturer in Geography at Cambridge from 1951, was put in touch with Lambert through Mitchell. Charles Green, Archaeological Consultant to the Ministry of Works, moved to Norfolk in 1951 for the Caister Roman town excavations, having curated Gloucester Museum since 1932, and succeeded Lambert as NRC President for 1953–4; he would undertake excavations at Burgh Castle between 1958 and 1962. Hutchinson was Assistant Resident Engineer of the Consulting Engineers to the Central Electricity Generating Board, overseeing power station construction on the Yarmouth South Denes, with Green as 'official observer', gathering borehole data (Lambert et al 1960: 113). A 1965 *Geographical Journal* paper by Green and Hutchinson would list Hutchinson as a Cambridge research student (Green and Hutchinson 1965).

Smith and Lambert recalled varying connections and affinities within the research team, with the RGS playing no role beyond publication: 'they had nothing to do with it really'.[7] Smith was close to Lambert, knowing her in Cambridge, jointly presenting evidence to the NRC in June 1954 (EDP 1954c), and publishing a joint summary in *New Scientist* (Lambert and Smith 1960). Smith knew Jennings from having been Assistant Lecturer in Geography at Leicester from 1948–51, but rarely met Green and Hutchinson; Lambert recalled she 'never in fact actually met Hutchinson, though

I knew Green well'.[8] Lambert, Smith and Green presented their findings to the 1961 British Association for the Advancement of Science (BAAS) annual conference in Norwich, *Nature* reporting: 'As a piece of scientific detection, the story unfolded to the Norwich symposium was a masterpiece' (Kidson 1961). Lambert also summarised the work for the BAAS in its published survey of *Norwich and its Region* (BAAS 1961a: 54). The three also appeared, with Clarke (who also summarised the work for *Norfolk Archaeology* (Clarke 1960)), on the BBC Midland Home Service, recorded in Norwich on 9 January 1957 and broadcast at 10.15 pm on Friday 15 February, discussing 'The Mystery of the Broads'. The programme was chaired by Norwich journalist and essayist Eric Fowler; *Radio Times* promised 'the qualities of suspense and discovery of a good detective story' (Radio Times 1957).

Detection becomes a key research trope, Godwin introducing the 1960 Memoir as 'this latest and most liberal "whodunit."':

> The ruins of St. Benet's Abbey, seen as one sails quietly down the Bure to Thurne mouth looking for an evening mooring, seem at once so unaccountable and isolated that they force upon even the least sensitive an uneasy awareness that they belong to a lost medieval world.
>
> In a quite analogous manner, Mr J. N. Jennings and Dr. J. M. Lambert ... became incredulously aware of the possibility that these serene stretches of water were not natural, but lay in hollows excavated by long-forgotten generations of turf-cutters. (Lambert et al 1960: iii)

Broads and abbey become medieval landscape mysteries solved. Lambert and Jennings' stratigraphic account described extensive and arduous fieldwork, 340 bores for the 1952 Memoir but for 1960 'a combined total of approximately 2150 bores, distributed between the valleys roughly as follows: Yare-1120; Bure-440; Ant-260; Thurne-240; Waveney-90' (Lambert et al 1960: 4). Sections crossed and figured Broadland valleys, over 30 long lines of enquiry. The 'unexpected and puzzling features' (6) discovered made sense as: 'former great peat pits, usually between 3 and 4 metres deep and quite distinct from shallow surface cuttings. On this hypothesis, the apparent anomalies fall into place' (22). Smith sought socio-economic reasons for digging, working through uncatalogued cathedral and priory records in Norwich Cathedral tower, the St Benet's records found in a tin box.[9] Peat was a key fuel in populous medieval Norfolk, powerful institutions recording an industrial 'specialized form of land use' (77), with extensive thirteenth and fourteenth century workings. Smith found however 'a long and losing struggle with standing water' in 'a period of great storminess and flooding in the North Sea basin' (99): 'Once flooded, the turf-pits were impossible to drain' (102). Smith also looked for Dutch comparison, finding similarly regular lakes of recent origin: 'The districts north of Gouda and west of Amsterdam have a multitude of such lakes, the former group comparable to the broads not only in their origin, but also in their present function as resorts for pleasure craft' (106; Smith 1966).

Smith asked how pits, even if kept free from 'disastrous flooding', could remain dry for excavation, suggesting seepage held up by clay, with 'the permeability of peat ...

less than has been supposed' (Lambert et al 1960: 107). Green and Hutchinson gave another answer, of changing relative land and sea levels explaining dryness during excavation and subsequent flooded abandonment. In 1956 Green gave the *EDP* and *Times* his solution to 'The Birth of Broadland': 'we have here, at last, just what was needed to explain how the Broadland peat-digging could be done' (Green 1956a; 1956b). Yarmouth excavations suggested a Romano-British 'Great Estuary' contracting through land emergence, but slow submergence from the late thirteenth century flooding peat excavations. South Denes bores showed mussel and barnacle colonies alongside thirteenth century pottery, and a mussel colony above a late thirteenth century silt band. Green and Hutchinson extrapolated dramatically lower sea level: 'Though the evidential value of these records is clearly variable, the sum of the inferences leads to one conclusion only. The coastal region ... stood some 13.0 ft. (4.1 m.) higher in relation to the sea towards the end of the thirteenth century than does the land today' (Lambert et al 1960: 140). The authors noted current estimates of postglacial isostatic submergence implying a change of only 3.5 feet over 700 years, but posited variation in rates, with rapid late medieval sea rise and slower recent change.

General public debate over artifice is discussed below; Green and Hutchinson's arguments however drew doubt from within the research team, detectable even within the Memoir. Lambert and Jennings suggested minor water level changes could explain reduced flood risk, with more dramatic shifts not needed, and inland changes seemingly not 'comparable in magnitude' to those indicated at Yarmouth (Lambert et al 1960: 57). Smith recalled being puzzled by Green and Hutchinson's work, seeing recorded severe fourteenth century flooding as sufficient to explain pits becoming unusable.[10] In 1965 Green and Hutchinson acknowledged 'controversy' around their work, though held to their estimates (Green and Hutchinson 1965). By the early 1980s scientific consensus found Green's estimates wrong, the likely sea level difference over 700 years being no more than one metre (Funnell 1979; Cornford 1982; George 1992: 94–8). If however for Smith and Lambert this was a point of research vulnerability for the new theory, other commentators missed such internal doubt in their eagerness to dismiss artifice outright.

Artifice Disputed

On the 1957 *Mystery of the Broads* broadcast, after Lambert's contribution, Fowler commented: 'When Dr. Lambert's astonishing theory became known, there were highly critical letters in the local press; and several stalwart amateurs rushed in to take issue with the scientists who'd upset all they'd ever believed about the origin of the Broads.'[11] The 'local press' was the *EDP*, for whom Fowler wrote under the dialect pen-name 'Jonathan Mardle', 'mardle' being rambling conversation. The *EDP* fits Ian Jackson's characterisation of the twentieth century 'provincial press', carrying expectations of accurate record, critically supportive of key institutions, scrutinised by locally knowledgeable readers, with the correspondence column a space for those 'who feel deeply involved in the subject that prompts them to write' (Jackson 1971: 152). The *EDP*, founded in 1870 as a morning paper covering Norfolk, was in the 1950s a broadsheet of 8–12 pages, less populist than its companion *Eastern Evening News*. Letters appeared on the editorial page, alongside Ellis's 'In the Countryside' column.

On 7 July 1953 the *EDP* front page stated: 'New Theory on Origin of the Norfolk Broads. Dr JM Lambert Suggests a Series of Huge Peat Cuttings' (EDP 1953j). Lambert was billed as of Newnham College, Cambridge, her theory 'put forward … with proper scientific caution': 'ecological research has opened a new line of investigation into documented history' (EDP 1953k). Sainty supported the new theory in a 13 August article on 'Norfolk's Waters': 'So seemly are they in our landscape that it is with something like a shock that one learns how much the great majority owe to man's handiwork; the beauty is artificial rather than natural' (Sainty 1953b). A week later however artificiality was challenged: 'Peat Cutting Theory on Broads Challenged by Norfolk Man' (EDP 1953m). The headline catches science challenged by local authority, 'Norfolk Man' against Cambridge scientist. Debates over science proceed through a newspaper of local record, with claims to the local a touchstone of truth (for a full account see Matless 2003a). Mr R Baden Gladden of Sutton Hall, Broadland poultry and dairy farmer (and former Cambridge student), whose 1960 *EDP* obituary would note 'a life-long study of sea erosion along the Norfolk coast and of the origins of the Broads' (EDP 1960b), set his 'lifelong interest' against Lambert, citing 'early writers' against new theory (EDP 1953m). An editorial dismissed Gladden's 'interesting criticism' (EDP 1953n), and Lambert wrote from her family Brundall address, inviting Gladden to see 'the full evidence' (Lambert 1953c). Gladden replied welcoming any 'opportunity to meet Dr. Lambert' (Gladden 1953); Lambert recalled a polite yet unproductive meeting.[12]

Gladden returned to dispute in April 1955, giving the *EDP* 'an interesting account of his researches', his 'life-long study of the Broads' suggesting excavation impossible under watery conditions. A subheading stated 'Ghost Fleet': '"There is also a ghost story, handed down from generation to generation, that the Danish fleet appears on certain nights on South Walsham Broad," continued Mr. Gladden. "The early writers saw nothing improbable in these records, even if they were untrue"' (EDP 1955c). Clarke and Ellis wrote in response for the NRC against:

> somewhat uncritical and romantic interpretations of local history and geography. Mr. Gladden is, of course, entitled to believe anything he likes on the subject, but he should not proclaim it to the world as a solution of equal validity to that advanced by Dr. J. M. Lambert, F.L.S., Lecturer in Botany at Southampton University, who, with other trained scientific workers, has given most careful attention to all the evidence cited by him.

Lambert had 'earned the right to give the professional answer to this scientific problem':

> If Mr. Gladden is to advance a theory worthy of consideration, he must first challenge the basic facts on which Dr. Lambert builds, and that he can do only by a large-scale programme of deep boring. So far as we are concerned, Mr. Gladden voices the popular misconceptions of yesterday: we had dared to hope that these were decently interred by now. (Clarke and Ellis 1955)

Ellis and Clarke, committed to education through the Castle and the media, were hardly against the popular, but sought popular local knowledge grounded in science, superseding folkloric tradition. Two constitutions of local knowledge clash, one setting the local against the scientific, the other incorporating the non-scientific as one form of local knowledge to be registered through modern scientific understanding, as object of interest rather than source of wisdom.

Gladden was defended by regular *EDP* correspondent SG Wheeler of Coltishall, speaking for 'Mr. Gladden and the considerable number of those who support his view' (Wheeler 1955a). Lambert countered from Brundall against 'idle speculation', setting out regional credentials as 'one who is Norfolk bred, whose home has been in Norfolk for over 30 years, and who has tramped, squelched and rowed over more of Broadland than a great many broadsmen' (Lambert 1955). Wheeler replied, locally trumping Lambert: 'Finally, if it has any bearing on the subject, and including progress by perambulator, the present writer can claim to have tramped, squelched, rowed and sailed over Broadland (and other places) and to have counted Norfolk as his home for more than double the 30 years that Dr. Lambert quotes' (Wheeler 1955b). Gladden again cited past 'learned men of integrity', alongside marsh knowledge: 'The marshmen will tell you that in the days when peat was cut, the holes would fill with water as soon as the hovers to a depth of 2 ft. were removed' (Gladden 1955). Clarke, Ellis and Lambert replied to Gladden and Wheeler together as the Norfolk Research Committee, seeking to close debate – 'Though we do not intend to prolong this correspondence ...' – asking people to wait for the scheduled New Naturalist volume on *The Broads*, then due but only published in 1965: 'Despite the local interest, we feel that the correspondence columns of your paper are hardly the place in which to carry out an adequate discussion' (Clarke, Ellis and Lambert 1955).

The following day Fowler offered a Mardle essay on 'Theorists and the Broads', defusing controversy from his position as scientifically informed amateur writing under a dialect name, between Lambert's 'extraordinarily interesting theory' and 'two old residents of the district', 'valiantly upholding tradition against science'. Mardle concluded, perhaps with a view to the evidently close intellectual community of the NRC: 'I hardly blame the scientists for getting annoyed with us, and I await my wigging with due humility' (Mardle 1955b), but what appeared as provocation was in effect defusion via the discursive position crafted by Fowler-as-Mardle. In 1961–2 Fowler would serve as NRC President, and also introduced the British Association's 1961 *Norwich and its Region*, alongside Lambert, Ellis, Clarke, Steers and others (Fowler 1961). In February 1960 Mardle reviewed *The Making of the Broads*, the 'astonishing theory' becoming fact through a 'Norfolk botanist' and her colleagues: 'It may now be considered beyond doubt that the Broads were man-made – and their existence has surely become more rather than less interesting on that account' (Mardle 1960). Doubt still emerged, MG Grapes querying 'that the learned societies have decided that the Broads were man-made' (Grapes 1960), and 'Anonymously' stating: 'I am very glad to see another defender of the Natural Broads ... there can be no doubt that to try to trace their ultimate origin in man-made pits can only belittle them in the public eye' ('Anonymously' 1960). In 1960 doubts were however countered by local correspondents writing as ordinary readers, suggesting sceptics should read the research: 'speculative reconstruction will no longer do' (Castell 1960).

The May 1965 launch of the New Naturalist volume on *The Broads*, discussed further in Chapter Five, edited by Ellis with contributions from Lambert, Jennings, Smith, Green, Sainty and Clarke, prompted further dispute (Ellis 1965). River engineer KE Cotton counter-claimed 'Broads Formed Naturally' (EDP 1965f), while detective novelist Alan Hunter devoted a column on 'The Broads On Early Printed Maps' to 'A Jolt for the Turf-cutting Theory', a 1749 map showing the broads treble their current area: 'Thus, if the turf-cutting theory of the origin of the Broads was embarrassed by the magnitude of the operation it had to suppose, there are grounds here for treble embarrassment, and perhaps for rather more comprehensive field work' (Hunter 1965).[13] Ellis's *Broads* book-signing in Jarrolds of Norwich was reported under the heading 'Not a Fairy-Tale – Mr. Ellis': 'Never mind what anybody says ... Dr. Joyce Lambert's theory of the origin of the Broads is not a fairy-tale' (EDP 1965d). Hunter, profiled by the *EDP* in 1955 as 'a man of direct and firmly held beliefs' (EDP 1955d), returned to dispute in 1968 after receiving Petch and Swann's *Flora of Norfolk*, which stated artifice as 'beyond all doubt' (Petch and Swann 1968: 19): 'I feel convinced that a more thorough and comprehensive investigation is necessary ... and that it may not be amiss to begin by rechecking Dr. Lambert's results' (Hunter 1968). Scepticism takes on the air of defiant, obstinate entrenchment, grudging against Lambert's authority.

Lambert held a 'turning-point in local acceptance of the theory' to have been the Nature Conservancy's 1965 *Report on Broadland* presenting 'accepted fact', with subsequent local official reports and educational texts doing likewise (Nature Conservancy 1965: 12).[14] In 1958 the widely used textbook *The Norfolk We Live In*, written by City of Norwich School teachers, had given detailed discussion, with 'a strong case for believing the Broads to be of artificial and not natural origin' (Blake et al 1958: 14); the 1974 edition stated: 'Although it is now widely accepted that the Broads are of artificial origin this in no way detracts from the charms of this unique series of lakes' (Blake et al 1974: 24). Lambert, responding from a nursing home in 2002 to my own draft paper on the origins debates, would reflect: 'I felt the idea of man-made broads had really become locally "respectable" a few days ago. The subject of the broads came up in casual conversation here, and an 83-year-old co-resident sitting next to me turned to me and said: "Of course, all the broads are really old peat pits. Didn't you know?"!'[15]

An Equivalent Pool

In 1911, researching at Cambridge, Marietta Pallis had produced the first academic scientific account of Broadland ecology (Pallis 1911a). By 1953 Pallis was living on a remote marshland property at Long Gores, near Hickling. Pallis's life and Broadland research will be discussed in Chapter Five, but in late life she re-engaged with regional origins. Since 1911 Pallis had shifted from scientific research to painting and travel, but maintained ecological interests, publishing on

The General Aspects of Vegetation in Europe (Pallis 1939). Pallis's ecology was increasingly vitalist and spiritual, matching her religious Greek Orthodox devotion, and in 1953 she marked Long Gores as a devotional landscape through excavating a swimming pool around islands shaped as a double-headed Byzantine eagle with imperial crown, the two-barred cross of the patriarch of Constantinople, the three-barred papal cross, and the Greek initials MP (Matless and Cameron 2006; 2007b):

> In the summer of 1953 I dug a bathing pool at Long Gores, Hickling, Norfolk, which with its large island, covers a good three-quarters of an acre. I fell naturally into the medi-aeval method – the obvious one – for extracting peat below the water-level, by digging pits, and isolating them from each other, leaving bars between to be cut later when re-uniting the waters. What I wanted was a clear six or seven feet of free water for swimming, measured from the summer water-level. (Pallis 1956: 3)

Pallis died in 1963, and is buried on the island alongside her companion Phillis Clark. The pool echoed Pallis's extraordinary book *Tableaux in Greek History*, published for the 500th anniversary of the fall of Constantinople, the pool a symbolic supplement in peat, water and reed (Pallis 1952). Excavation also however prompted Pallis to reconsider broads origins, and to meet Lambert.

Pallis read Lambert's Presidential Address in *TNNNS*, writing to naturalist Catherine Gurney in August 1953: 'I read Dr Lambert's address with much interest ... The work had been very well done, she does not seem to have missed very much. About the central idea, that the broads are artificial I cannot agree – because if peat cuttings are too deep – peat cutters do not go deeper than 3 to 4′ (at the very outside), because of the floodings of the cuttings.'[16] Pool digging however changed Pallis's mind, suggesting peat was impermeable, and that the broads could have been excavated without flooding: 'When I began digging the pool I had taken for granted that peat was permeable, now I do not think so. The impermeability of the peat is quite sufficient for the extraction of pressure-peat to a depth of eight to ten feet, and more, provided there are adequate bars' (Pallis 1956: 4–5). Between August 1953 and August 1961 Pallis wrote to Lambert, beginning after reading her Presidential Address: 'Thank you so much for giving me your paper. I would much like to see you – Mr Ellis gave me your message – and now, while I am doing a dig, for I think it might interest you – quite big trees we find, and in quantity. If you care to come one day, and the sooner the better, as I do not think I will be at it much longer. Do bring sandwiches.'[17] Lambert visited in early September 1953, after which Pallis wrote regularly, with almost daily reports in October 1953:

> The Double-Headed Eagle is finished, and I can tell you that with 4 men I could reach a depth of 10′, have gone a full 8′, and I have little doubt 12′ – if 6–8 men, I think 14′, about the depth to the sand at Long Gores, or even 20′, if that is with stepped sides. With solid bars, not less than 2′ wide ... the water is easily held and easily drawn without the need of a pump ... So this is added confirmation for your findings.[18]

Figure 10   Double-Headed Eagle Pool, Long Gores, Hickling. Source: Top: aerial view, photograph by Mike Page, 2006, © Mike Page; Bottom: ground view, 1953, © The Pallis/Vlasto Archive.

On completion Pallis declared: 'The Double-Headed Eagle of Byzantium in Norfolk has turned out beyond my wildest expectations … towards the end, we used all the Norfolk marsh tools – of which I have a complete collection … There is no reason to suppose that any others were used in the cuttings of the 13th century – all adequate, no <u>detestable</u> modern contraptions necessary.'[19]

Pallis wrote supportively as Lambert's scholarly predecessor, fellow female scientist and indeed former Newnham researcher, re-energised by Lambert's work. On 1 April 1957 Pallis commented: 'I have been seeing the correspondence about the Broads – what nonsense they talk and some, for instance Wheeler, whoever it is, is hardly coherent.'[20] Lambert would later recall Pallis's 1911 research as having been her 'starting point' and 'bible', though she found her later writings 'odd': 'I think she was a serious scientist but with a rather shall we say airy fairy side to it', non-scientific elements to be taken 'with a pinch of salt'.[21] Pallis criticised not only Wheeler but Green, writing to Lambert in 1956: 'I saw the article in the *Times* of March 14th on the Norfolk coast line, in which you are mentioned.' Pallis saw her pool countering Green: 'As regards the artificial nature of the Norfolk Broads – abundantly proved by the vertical sides – there is no need to invoke heroic measures of oscillations of level, it can be proved quite simply by the impermeability of peat, nothing more is necessary, though changes of level are in themselves interesting.'[22] Pallis's letter was effectively a draft for, or adapted an early draft of, her 1956 pamphlet *The Impermeability of Peat and the Origin of the Norfolk Broads*, supporting Lambert and questioning Green: 'I think I can offer a simpler, more immediate explanation, which makes fluctuations of level redundant, even if they exist as Mr Green thinks' (Pallis 1956: 3). Lambert thanked Pallis for her 'long and interesting letter' with its 'comments and observations', noting she had argued with Green 'for some time' over sea level change.[23] Pallis's 1961 pamphlet *The Status of Fen and the Origin of the Norfolk Broads* supported Lambert and Smith while critiquing Green and Hutchinson, 'who think that mean sea level in relation to the land surface was then lower by some 13 feet' (Pallis 1961: 1). Evidence from mussel beds was inadequate due to 'constant changes in the North Sea bed whereby deposits are piled and hollows made often very rapidly': 'To sum up: I do not believe in Messrs. Green and Hutchinson's evidence as proof of subsidence and elevation' (18).

In April 1956 Lambert informed Pallis that Mr KB Clarke, engineer to the Lowestoft Water Company, had also deduced impermeability from reservoirs above Fritton Lake: 'It is most useful for me to have this practical demonstration from both you and Mr. Clarke of the possibility of digging deep peat even today, and I hope to incorporate a paragraph about it in writing up the work for publication.'[24] *The Making of the Broads* stated:

> For instance, an ornamental swimming pool has recently been excavated in the Hickling marshes to a depth of nearly 3 metres entirely by intermittent hand labour without the use of elaborate pumps, with the men working well below the level of the water in the nearby dykes; and it is estimated that even greater depths could have been reached without difficulty. (Lambert et al 1960: 57–8; also Ellis 1965: 52)

Lambert referenced Pallis's 1956 pamphlet, and sent Pallis the 1960 Memoir. Marginalia and underlinings in Pallis's copy indicate careful reading; when, on page 104, Smith estimated medieval peat digging at Flixton of one acre to a depth of 10 feet in 6.5 years, Pallis added: 'and I in 3 months'.[25]

New Broads

In July 1956 the Danish newspaper *Politiken* picked up on new origins, noting the likely role of ninth century Danish settlers, a photograph of Oulton Broad accompanying a cartoon of Vikings digging, sweating and quenching thirst: 'The old peat-pits, where … Vikings and their descendants worked hard to procure fuel for the English winter, are now preferred by the Britons who choose to take off the bowler hat and umbrella and celebrate their holidays in a primitive way' (Politiken 1956).[26] *Politiken*'s comic labour-holiday contrast resonated with contemporary debate on the possible excavation of new broads *for* leisure, artificial origins held to imply potential new creation.

The dominant post-war Broadland policy concern was a reduction in open water, from closure or encroaching vegetation. Chapter Three examines such debates further, but the new origins theory here gained policy purchase. Scientific research expanded upon earlier topographic commentary, such as that by Anthony Collett in 1926 that: 'the Broads are among the most transitory attractions of English scenery; they were formed late, and they promise to vanish early' (Collett 1926: 83). In 1949 Jennings and Lambert documented 'The Shrinkage of the Broads' over 100 years for the *New Naturalist* journal (Jennings and Lambert 1949), while Jennings' 1952 Memoir stated: 'this problem in physical geography is of more than theoretical interest for it is widely recognized that the nature, and even the very existence, of the Broads is critically dependent on physical processes at work to-day as in the past' (Jennings 1952: 1). Historical research moved into political discourse during a 1957 House of Lords debate on national parks, former Labour Minister of Town and Country Planning Lord Silkin commenting: 'The position to-day is just this. Either something is done immediately about the Norfolk Broads, or the Broads will disappear.'[27] For Lambert the dating of artificial origin revealed the rate of accretion, with accelerated loss of water area in the previous 80 years as broads shallowed to a 'critical level' (Lambert 1957a). Artifice however might open conceptual space for intervention. Writing in the *Times* in July 1957 on 'Diminishing Broads', Lambert drew policy implication, the 'true broads' simply requiring 'a well-planned programme of selective dredging' to restore deep water, accretions removed and shrinkage halted (Lambert 1957b). *EDP* correspondents also took new history to imply present action, FH Denton commenting:

> It is strange how loath many people are to accept the idea that the Broads were largely man-made ... Can there be a psychological reason for the reluctance to accept the theory shown by many who derive pleasure or income from the water? ... Is it that to admit that the Broads were largely man-made is to admit the corollary – that they could and should be man-maintained? Not everyone interested in using the Broads has the honesty of a member of the navigational authority I know of. After hearing Dr. Lambert lecture, and being asked his opinion – 'Well,' he said, 'I'm convinced, and the moral for us, I suppose, is that if our forefathers dug 'em out, surely we can keep 'em clean.' (Denton 1956)

The Nature Conservancy's 1965 *Report on Broadland* followed its statement of artificial origin with a proposed 'Expansion Programme' of 'New Broads and waterways': 'A series of new Broads suitably spaced throughout the region would provide this interest and variety and help to achieve a better distribution of activities throughout Broadland' (Nature Conservancy 1965: 59). New broads would also 'allow the cycle of natural succession by wild life to begin again' (44). The NC envisaged a new cut linking Bure and Yare, creating new routes for round trips, costed at £552,000, five miles long, 66 feet wide, six feet deep, from Tunstall Bridge on the Bure to the Reedham end of the New Cut on the Yare, a meandering line along the marsh-upland boundary. The 1971 Broads Consortium *Broadland Study and Plan* rejected the cut as likely to draw traffic north rather than south, but similarly projected 'Possible New Broads' in marshland along all the rivers, with 'large new areas of water' possible at Whitlingham, Postwick, Oulton Broad, Somerleyton and Beccles (Broads Consortium 1971: 44/maps 21–22). None were made, though a University Broad would be dug from gravel at the UEA in the 1970s, and two broads dug at Whitlingham Country Park near Norwich for water recreation, Whitlingham Little Broad opening in 1997, Whitlingham Great Broad following a few years after.[28]

Origins debates also prompted reflection on potential sea flood and marine transgression, discussed in Chapter Six below. The primary policy emphasis from artifice was however of potential modern equivalence. In his 1967 *Portrait of the Broads*, discussing 'How the Broads Began', James Wentworth Day reflected: 'It is a sardonic twist of the wheel of history that a Government Enquiry should now solemnly, and rightly, recommend that many of the grown-up broads and low-lying marshes which were shallow turf-diggings, should once again be excavated and flooded to make new or larger broads. That should not be too difficult with modern tools and dredgers' (Day 1967: 23). Artificial origin becomes a blueprint for regional restoration.

Chapter Three
Conduct

Terry and Lydia

In summer 1966 a small motor cruiser, the 'MV Woppy', carried the Player's No. 6 banner around the Broads. A five week tour, with skipper Michael Fox and 'Personality Girls' Diane Terry and Sue Longhurst, took in regattas and waterside pubs, promoting John Player cigarettes: 'sample packets of 3 were passed with a fisherman's landing net to other boats'. The tour was launched by television comic double act Hugh Lloyd and Terry Scott, then appearing in Great Yarmouth summer season. Player's magazine *Navy Cuttings* pictured Lloyd giving Longhurst a 'farewell kiss', Scott squinting, from sun or cigar smoke, glass in hand, the end of a chapel giving cross-river background (Navy Cuttings 1966).

Scott, a fixture in later *Carry On* films (*Loving, Camping, Up the Khyber*, etc), would become a national suburban icon in the long duration BBC sitcom *Terry and June*. With Terry and Hugh, Diane and Sue, and skipper Michael, Player's stamped the Broads with suburban glamour, tapping the region's modern leisure reputation. Scott on the riverbank marks a particular mix of modernity, technology and conservatism preoccupying 1960s Broadland, and a mode of conduct around which cultural battle would engage; back to the landscape, squinting at camera, dizzy with events and alcohol. Landscape is the backdrop to conduct which relishes necessary surroundings (air, breeze, wash), but enjoys no obligation to landscape engagement. The chapel is a pleasant adjunct, but Scott has no Pevsner in his pocket.

In the Nature of Landscape: Cultural Geography on the Norfolk Broads, First Edition. David Matless.
© 2014 David Matless. Published 2014 by John Wiley & Sons, Ltd.

Figure 11   Terry Scott, Hugh Lloyd and Sue Longhurst at the launch of the MV Woppy. Source: *Navy Cuttings* 1966, image provided courtesy of Nottingham City Museums and Galleries.

Another glamour came to Broadland 100 years earlier:

> The sun was sinking in the cloudless westward heaven. The waters of the Mere lay beneath, tinged red by the dying light. The open country stretched away, darkening drearily already on the right hand and the left. And on the near margin of the pool, where all had been solitude before, there now stood, fronting the sunset, the figure of a Woman. (Collins 1995: 265)

The Woman is Lydia Gwilt, in Wilkie Collins' 1866 *Armadale*, a 'sensation novel' plotted around two Allan Armadales, one inheriting a Norfolk estate at 'Thorpe Ambrose', unaware of his true (non)identity, the other his friend Midwinter, the actual Armadale, and aware of horror in their past. Gwilt also knows the secret; both men fall for her flame hair. Plots of tragic affection move between London, Norfolk, the Isle of Man, Devon, Germany and the Caribbean, with the Broads seeing a formative encounter. Collins' chapter 'The Norfolk Broads' introduced a region of 'strange and startling anomalies', a 'hidden labyrinth' of waters (1995: 244–5), a picnic boat trip leading both Armadales to Gwilt, standing by 'Hurle Mere', Horsey thinly disguised. The Broads lent the sensation novel a mysterious nature, indeterminacy of land and water matching scandalous indeterminacies of plot: 'The shore in these wild regions was not like the shore elsewhere. Firm as it looked, the garden ground in front of the reed-cutter's cottage was floating ground, that rose and fell and oozed into puddles under the pressure of the foot' (255). An appendix noted: 'the "Norfolk Broads" are here described after personal investigation in them' (678), the region a curiosity, as distinct from ordinary lives as *Armadale*'s sensational events.

This chapter moves from Lydia Gwilt to Terry Scott, from strange glamour to familiar fame, from a region unknown to one discovered and remade, an obvious place to promote cigarettes via suburban celebrity. The story is of conduct becoming and unbecoming landscape, shaped through class and gender, claims to local knowledge and assertions of happy ignorance, schemes of orderly planning and improvised holidaying. Silhouetted Lydia and squinting Terry show two of many forms of landscaped self-fashioning producing the region, the Norfolk Broads made through moral geographies of leisure conduct.

Regionally Self-Conscious: The Broads Discovered

Discovery beguided

The 1870s *Geography of the County of Norfolk*, by Rev D Morris, states: 'The low-lying country bordering on the east coast abounds with considerable pools of water locally called "broads."' Morris's local terminology carries no leisure reference, signalling only 'sedge and bulrush, ... fish, and ... fowl' (Morris 1995: 7–8). By 1925, in Gordon Home's *Through East Anglia*, the Broads are classic yachting space: 'If a committee of geographical and topographical experts had at some remote epoch sat down at a round table, and with modelling clay had set to work to lay out an ideal Britain, ... can one imagine that the minds of any set of men could have created anything so perfect as the Norfolk Broads?' (Home 1925: 132). From local fish pools to perfect national pleasure ground in 50 years.

Norwich City Librarian GA Stephen's 1921 *Books on the Broads: A Chronological Bibliography*, registered the region's textual emergence (Stephen 1921). After single entries from 1834, 1845, 1865 and 1866, a steady topographic run begins in 1871, books working alongside posters, photographs, paintings, railway and boat hire promotion in regional discovery. RH Mottram would classify the period as 'The Broads Become Self-Conscious', with the sense of a region discovered (or awaiting exploration) itself shaping regional identity (Mottram 1952: 93; Clarke 2010; McWilliam and Sekules 1986; Taylor 1995). The first commercially successful guide was Norwich solicitor George Christopher Davies's 1882 *Handbook to the Rivers and Broads of Norfolk and Suffolk*, covering all rivers and advising on transport, fishing, yachting, sport and natural history. Handbook illustrations denoted changing Broadland style; from six first edition picturesque drawings showing the secret places of a mysterious region, a dead-end 'Dyke near Coltishall' on the cover (Davies 1882), to later editions showing tonally lighter, clean lined airy openness, yachts prominent in a Broadland well-visited. Davies produced further Broadland volumes, some illustrated with his own photographs (Davies 1883a; 1883b).

Davies was known for his 1876 Broads boy's adventure story *The Swan and Her Crew*, discussed in Chapter Four below, where the only public leisure is a local Wroxham regatta (Davies 1876). The *Handbook* however put the Broads, and Davies, on a national leisure writing map, Davies becoming 'the man who found the Broads' (Clarke 2010: 17–20, Mottram 1952).[1] Davies included notes on conduct, with 'hitherto unwritten rules' on speed, litter, plant picking and noise: 'Do not, in the neighbourhood of other yachts or houses, indulge in songs and revelry after eleven p.m., even at regatta times' (Davies 1891: xiv–xvi). Others however gave visitors different encouragement, London-based Ernest Suffling's 1885 *The Land of the Broads* evoking a 'Water Party' around the piano: 'After dark, with the lamps lighted, and the merry party gathered around this instrument, many a happy hour is passed away; and, as the Licensing Act does not hold sway here, I am afraid that the small hours often arrive ere the saloon is closed' (Suffling 1893: 11–13; Clarke 2010: 29–30).

In Davies's 1882 *Handbook* one boatyard advertised hire; by 1891 there were 'Yachts and Boats For Hire' at Oulton Broad (14 boatyards), Wroxham (7), Thorpe (4), Norwich (2), Potter Heigham (2), Brundall (2), North Walsham, Coltishall, Horning Ferry, Filby, Coldham Hall and Cantley (Davies 1891). From 1908 Harry Blake established the London-based Blakes agency to channel hire demand. Pleasure wherries with attendants, and smaller yachts without, offered indulgent or self-reliant leisure for the upper and middle class, with weekly 1908 peak season prices from 30/- (a yacht for two) to 16 guineas (a wherry sleeping ten with three attendants) (Blake 1908; 1916). The structure of individual yards letting through agents continued through the twentieth century, with post-war business dominated by Blakes, based in Wroxham from the 1960s, and Oulton Broad-based Hoseason (Snoxell 1956).[2]

Broadland was opened to rail visitors via lines from Norwich to Yarmouth via Reedham in 1844, extended from Reedham to Lowestoft in 1847, and with the Norwich–Acle–Yarmouth line opened in 1883. The Norwich to North Walsham line through Wroxham opened in 1874, while the Midland and Great Northern line from North Walsham to Yarmouth, built 1876–81, and from 1883–1959 part of a direct link from the Midlands to Yarmouth, served Stalham, Catfield, Potter Heigham and Martham. Passengers could alight for the boatyards at a wooden platform halt by the Thurne (Gordon 1968). Rail companies put the visual arts, notably photography, into leisure service (Middleton 1978). John Payne Jennings' Broads photographs appeared from 1884 on panels within Great Eastern trains, actual and potential destinations seen in transit (Mottram 1952: 131–5; Norden 1997). Jennings' *Sun Pictures of the Norfolk Broads*, an upper Bure-heavy 100 photograph set, showed regional life within a scenic picturesque (Jennings 1892).

Painting also conveyed pleasure ground. Yarmouth watercolourists Stephen Batchelder and Charles Harmony Harrison, discussed in Chapter Four below, traded on regional popularity (Collins 1990: 25–7), while Davies's *Handbook* noted London exhibitions by 'Mr E H Fahey and Miss Osborn' (Davies 1891: xii). Emily Osborn, known for 'proto-feminist' work exploring the situation of women in art (Nochlin 1988: 88), painted on the Broads between 1886 and 1906, reviews of shows at London's Goupil Gallery in 1886 (*The Norfolk Broads*, from an 11 week wherry trip) and 1887 (*The Bure Valley*) suggesting impressionistic landscape (Morning Post 1886; Cherry 1993: 98; Yeldham 2004). Painting also illustrated guides, as in Norwich publisher Jarrold's 1912 *Broadland* volumes, with colour watercolour plates by George Parsons-Norman, praise flowing throughout: 'Belaugh might very aptly be named "Inspiration Land"' / 'For one who can read Nature's open book Coltishall is enchanted ground' (Parsons-Norman 1912: 9–10).

Regional discovery could quickly turn to over-familiarity, novelty worn. In 1887 *Nature* reviewed 'A Batch of Guide-Books to the Norfolk Broads', including Suffling and the ninth edition of Davies's *Handbook*: 'Surely no spot in the British Isles has been so "beguided" as the Norfolk Broads.' Works borrowed from one another with 'the most barefaced plagiarisms', maps bore 'a more than family resemblance' (Nature 1887: 457). In 1901 William Dutt, within his 'Highways and Byways' guide to East Anglia, asked:

> Can anything new be said about the Norfolk Broadland? If so, I shall be only too glad to hear of it, so that I may be able to set down something here which will claim attention. It is an unthankful task to follow in the steps of one writer who has set his mind on doing justice to a district; but when one has a hundred predecessors, each of whom has, by the conditions of travel, been compelled to follow practically the same river-routes, there cannot be much for a late-comer on the scene to deal with. (Dutt 1901: 133)

Two years later Dutt published his 373 page *The Norfolk Broads*.

Class at play

Guides shaped and commented on class at play, with Broadland defined through the leisure performance of class identity, working class day tripper or wealthy cruiser (Taylor 1995).

Working class Broadland leisure could enact cultures of self-improvement, as when *The Graphic* recounted 'A Week In A Wherry', the University College Hospital Working Lads' Institute taking 40 East End boys on the Bure in 1893, doctors supervising 'a branch of prophylactic medicine' in air and sunlight (The Graphic 1893: 545; Matless 2000b). Working class landscape encounter more typically came via steam launch day cruises, Mottram recalling a factory workers' Sunday cruise from Norwich: 'Some actually looked at the view, or told each other the names of the places they passed, some lent an ear to the band which ... played the Blue Danube, Two Lovely Black Eyes, and Sousa marches' (Mottram 1952: 30–1). Contemporary guides listed 'Steamer Cruises in Broadland' from Yarmouth, Lowestoft, Norwich and Wroxham (Ward Lock 1924: 53–5). Yarmouth passengers could take an outward trip to Wroxham or Norwich, a train between the two, to return on the other river, the Yarmouth holidaymaker encountering much of Broadland in one day, appreciating at leisure.

Yarmouth is presented in dual fashion, as pleasure resort from which seaside culture might take a day break to Broadland, as in Paul Martin's 1892 photographic collection *A Yarmouth Holiday* showing Broads trips alongside crowded sands (Haworth-Booth 1988), or as port town whose true character and interest predated the leisure industry. The latter version takes Yarmouth as maritime extension of, rather than seaside counter to, Broadland, with the typical view a prospect across Breydon, rather than along the prom. For Dutt old Yarmouth offered different 'entertainment' than the 'popular pleasure-resort' (Dutt 1901: 86–7), while Donald Maxwell's *Unknown Norfolk* (1925) passed over 'the usual sea front of any large seaside resort' (Maxwell 1925: 49) to find the true Yarmouth, the view from Breydon prompting citation of Ruskin: 'for seven hundred years Venice had more likeness in her to old Yarmouth than to new Pall Mall' (34). Yarmouth becomes 'The Venice of East Anglia' (42); a sandbank settlement, a tight-spaced town, an international trading port, its churches catching sunrise from the sea.

The bulk of Broadland guidebook and travel writers were middle or upper middle class men, though few promoted luxury in leisure; the region was instead a place to rough it, exercising masculinity. A contrast came in Anna Bowman Dodd's 1896 *On the Broads*, where luxury and femininity take the region at a telling tangent. Dodd was a New York journalist and travel writer, in Broadland on a hired yacht with her husband and two hired hands. Walter Rye would dismiss Dodd as 'an American lady ... giving us the benefit of her three or four days' experience' (Clarke 2010: 49), but *On the Broads* renders a significant regional culture of passing luxury acquaintance, worthy of attention. Dodd finds in Broadland theatrically civilised wildness, framing the English 'game of pleasure' (Dodd 1896: 14). The working class steam past on the 'Belle of the Yare', violin and harp music playing (53), though in Yarmouth Dodd avoids the seafront 'Northern Margate' (258–9). Dodd's global travel meets that of Broadland residents, her wherry skipper Davy interspersing tales of seven worldwide years at sea with 'unintelligible Norfolk jargon' (56). Davy also prompts landscape thought, as Dodd arrives late for dinner after watching the sunset: 'Why were towns-people so given to waste time in looking at sights as commonplace as a setting or a rising sun? was the note of Davy's unspoken contempt. It was that scorn of his class for all such aspects of nature as confirm the recent theory that adoration of natural beauty is the sign of a decadent rather than of a robust or a primitive taste' (208–9).

Dodd's Broadland of sunsets and frivolity is a happy extension of cosmopolitan London. At Horning:

> As we sailed past, London 'At Homes' were suddenly brought up before our eyes. The air was filled with the tones of the clear English voices, and with certain questions and answers which seemed as much a part of English interiors as the wall-paper in their houses. 'Do you take cream or lemon?' and 'The cake, please.' 'Thanks awfully; I don't mind if I do.' (84)

Broadland as home-from-At-Home is registered in Dodd's Prologue, where a 'hot London night' at Mrs Gower-Granger's is stirred by Violet Belmore, back from the Broads with 'breeze', 'freshness' and 'country colour' (2). On the Broads, Dodd meets up with Belmore and Renard, an American artist from Paris in knickerbockers and beret, introduced at Horning Ferry Inn. 'A Little Dinner at the Ferry' offers a late Victorian scene of Broadland life and landscape to set alongside images of wherry-men and eel catchers, Dodd's diners a smartly dressed spectacle: 'the yachts' crews' rough voices in the outer tap-room made the perfect stillness of the elements more marked; and the slow, heavy footsteps of the rustic waitresses, as they came and went over the bare unpainted floors, were a reminder of the distance of that world from which we had fled' (107).[3] Sound crosses yet underlines social boundaries. Service is different here, underpinning holiday civilisation. Renard monologues on beauty, and dinner runs to coffee under the stars: 'The inn-waitress blinked even more sleepily as she followed us out into the night' (116).

Broads authorities

Broads narratives enact a geography of the telling voice, accounts of the region gaining a hearing through the status ascribed to the author. Regional discovery sees figures such as Walter Rye, William Dutt and PH Emerson emerging as Broads authorities, their standing derived from field experience, and from amateur rather than academic learning.

Such authors staked claims beyond passing acquaintance, setting themselves in the scene, and avoiding narrative conventions such as the discursive skipper's log, a trope of touristic joviality typified by one of the earliest published Broads tours, CA Campling's hearty *The Log of the Stranger* (Campling 1871; Mottram 1952). Regional authority did not however accrue only from serious writing. *Nature*'s beguiled reviewer lauded the work of Dodd's critic Walter Rye (1843–1929), noted Norfolk antiquary whose distinction included the capability to parody others, comedy here a sign of geographic authority rather than flimsy connection. Thus Rye's 1887 *A Month on the Norfolk Broads* contained 'exceedingly clever satires on the writings of a well-known author of "Broad" books' (Nature 1887: 459). Both Davies (1883b) and Suffling (1893) had described local whirlwinds or 'Rogers', and Rye spoofs a Roger lifting his boat clean over a bank. Rye also constructed a comic historical superiority over his American genealogist friends the Grices, who at the Roman Burgh Castle want 'to break off pieces of the walls to take back to America' (Rye 1887: 22). Rye was, like Davies, a notable figure in the Norwich professional class. Born in Chelsea, first visiting the Broads in 1867 and owning the yacht Lotus from 1885, a solicitor dividing time between London, Sussex and Norwich, moving to Norwich in 1906 (and to nearby Buxton in 1910), Rye served as Conservative Lord Mayor in 1908–9, restored historic Norwich buildings, and gave his antiquarian collection as the basis of Norwich City Library. Autobiography and obituary underpinned Rye's cultural authority by highlighting eccentricities becoming an antiquarian, Rye spending 'months on the Broads at all seasons of the year', taking extraordinary walks ('His longest walk in one day was eighty-six miles'), and indulging 'epicurean tastes ... Many a bird of the heath, the marsh, and the mudflats, usually considered uneatable, was made a delicacy' (TNNNS 1929: 723–4; Rye 1916).

Rye included a Broads journey in his 1885 *History of Norfolk*, discussing natural history, customs and architecture, writing for a metropolitan audience arriving on the London line: 'those for whom this book is written will probably enter by Thorpe Station' (Rye 1885: 205). Rye began: 'It is painful for one who has known and loved the Broads as long as I have, in common honesty, to say that their charms have been grossly exaggerated of late' (256). Unflattering comment serves to confirm Rye's discerning eye. Criticism also marked Rye as a guide against gullibility, with visitors warned against a dubious 'brass of the last abbot' in a St Benet's cottage (265–6). Authentic local knowledge was however tapped as Rye, 'loafing about at Reedham', noted from a wherryman an 'absolutely unique and hitherto entirely unpublished list of the forty-eight reaches ('raches' he called them) between Breydon and Norwich, which I subjoin for the benefit of posterity' (281). Forty-eight names followed, upstream from '1. Borrow [Burgh] Flats' to '48. Cut' (280–1), truth picked up in the field.

A parallel pursuit of Broadland topography and antiquity appeared in the work of William Dutt, who sought 'a real knowledge':

> trudge the river walls; associate with the eel-catchers, marshmen, reed-cutters, and Breydon gunners; explore the dykes unnavigable by yachts, and the swampy rush marshes where the lapwings and redshanks nest; spend days with the Broadsman in his punt, and nights with the eel-catcher in his house-boat; crouch among the reeds to watch the acrobatic antics of the bearded titmice, and fraternise with the natives at the staithes and ferry inns. (Dutt 1903: 27)[4]

Dutt suggested: 'to *know* Broadland and understand fully what a unique and interesting district it is, a man must see more of it than is visible during a summer cruise' (Dutt 1901: 134; Dutt 1906). Real knowledge came in and out of season. Internal regional geography offered Dutt aesthetic stepping stones, the 'simply pretty' Bure preparing cruisers 'to appreciate the more unique and primitive charms' of Ant and Thurne (Dutt 1903: 133/154). Dutt quoted an unnamed Thomas Hardy from *The Return of the Native*, suggesting taste might turn to bleaker scenes: 'When the time this writer considers imminent arrives, the attractiveness of such a line of sandhills as lies between Waxham and Palling should be fully appreciated' (186). Dutt's books for publisher Methuen, like those of Arthur Patterson, discussed in Chapter Four, were illustrated by Frank Southgate (1872–1916), based at Wells in north Norfolk. Southgate's water-colours showed bird life, scenery and land work, occasionally derivative of Emerson's photographs of Broadland working lives. Visitors do not appear, the reader looking past leisure presence to see Broadland in authenticity (Collins 1990; Nature 1916).

Dutt and Rye remain occasional reference points in regional histories, but the work of another late Victorian Broads authority has become visual shorthand for a Broadland before commercial leisure, and deserves extended attention here. Writer and photographer Peter Henry Emerson (1856–1936) made his reputation through the region, his photographs re-emerging in the late twentieth century as international art market objects, sites for academic commentary, and regional icons, including in Broads Authority postcards and publications.[5] In 1975 Nancy Newhall's major 'Aperture' monograph registered a pioneer of photographic modernism: 'And let us face the greatest importance of P. H. Emerson: *he was probably the first true photographer-poet – the first to whom the beauty of the image on the ground glass, the moods and emotions it aroused, were more important than the subject, however important in itself*' (Newhall 1975: 4). In 1974 Peter Turner and Richard Wood had established Emerson's importance in photographic history and as a 'photographer of Norfolk' (Turner and Wood 1974); Cliff Middleton would further emphasise Emerson's regional subject matter as (with Davies and Jennings) one of *The Broadland Photographers* (Middleton 1978). A major 1986 UEA exhibition asserted Emerson's regional and photographic significance, with sociological, art historical and newly emerging cultural geographi-cal understandings of landscape, labour and leisure brought to bear on his work (McWilliam and Sekules 1986), forms of scrutiny extended by John Taylor in a 2006–7 exhibition at the National Museum of Photography, Film and Television in Bradford, and the J. Paul Getty Museum in Los Angeles (Taylor 2006).

Emerson circulates in Broadland and across the Atlantic, now as in life. Born in Cuba, Emerson was schooled in England from 1869, and qualified as a physician at Cambridge, his peripatetic life taking in Southwold, London, Beaumaris, Kent, Lowestoft, Oulton Broad, Bournemouth (1903–19) and Falmouth (from 1924). Emerson's genealogical *The English Emersons* included a third person autohagiography, recalling taking up the camera in 1882, his 1883 Southwold photographs 'destined to revolutionize photography' (Emerson 1898a: 124). On his first Broadland cruise in 1885 Emerson met painter Thomas Goodall, with whom he would co-author *Life and Landscape on the Norfolk Broads* (Emerson and Goodall 1886; hereafter *LLNB*). Solo works combined photography and text, several with a Broadland element or focus; *Pictures from Life in Field and Fen* (Emerson 1887a), *Idyls of the Norfolk Broads* (Emerson 1887b), *Pictures of East Anglian Life* (Emerson 1888), *Wild Life on a Tidal Water* (Emerson 1890a), *On English Lagoons* (Emerson 1893) and *Marsh Leaves from the Norfolk Broad-land* (Emerson 1895a), the latter reprinted without photographs as a short story collection (Emerson 1898b). Emerson issued deluxe and ordinary editions, with even the latter expensive, all over £1, and *LLNB* £6 6s for ordinary (limited to 175 copies) and £10 10s for deluxe (25 copies) (Taylor 2006: 156–7). Emerson sought distinction in an elite market, inhabiting and appealing to a higher class fraction than Davies (Taylor 1995). A hundred years on, Emerson's limited editions would become common regional currency.

LLNB derived from a yachting cruise from South Walsham in the second half of 1885. Emerson promoted his work within the region, a March 1886 illustrated lecture at Yarmouth Emerson articulating historical and aesthetic arguments later published in *Naturalistic Photography* (Emerson 1889a; Yarmouth Mercury 1886; Yarmouth Independent 1886). Thirteen of *LLNB*'s 40 photographs are credited to Emerson 'alone', the others to an 'ideal partnership' with Goodall (Emerson and Goodall 1886: preface).[6] The majority carried generic titles, the book an aesthetic elaboration on regional subject matter rather than a formal tour. Texts accompanying single or grouped photographs were credited to Emerson or Goodall, while Goodall provided an essay on 'Landscape' around five plates showing 'suggestive and paintable material', 'perfect little bits of pure landscape': 'A Rushy Shore', ''Twixt Land and Water', 'Evening', 'An Autumn Morning', 'The Fringe of the Marsh' (Emerson and Goodall 1886: 71–81; Emerson 1890b: 279–82). These are inhabited spaces, indeterminate land and water with background dwellings, though with figures less prominent than elsewhere in *LLNB*.

Emerson and Goodall rejected as contrived the photographic 'pictorialism' of Henry Peach Robinson; here instead were 'literal transcripts', 'designs presented by nature' (Emerson and Goodall 1886: 73; Scharf 1986). Emerson argued for single shot photography, rejecting Payne Jennings' combination of negatives into 'sun pictures' (Taylor 2006: 22), and presenting differential focusing as mimicry of the eye working outdoors, the observer thereby viewing the photograph as they would view the scene.[7] Goodall and Emerson set themselves in a naturalistic art tradition from Dutch landscape through Constable and Crome to Corot and Bastien Lepage, 'greatest master of all', whose British followers included George Clausen and Henry La Thangue, who Emerson met in Horsey in 1885 (Emerson and Goodall 1886: 76;

Emerson 1890b: 76–7; McWilliam and Sekules 1986; McConkey 1986). Naturalist aesthetics shook off 'traditional conventionality', to seek 'beauties in facts as they are' (Emerson and Goodall 1886: 74). The Broads, as a 'modest lowland district' appealing to 'the lover of simple landscape' (71), becomes a region to cut through conventional seeing, the kind of landscape in which to realise that 'Truth is Beauty' (77), to be found 'in the life and landscape of to-day' (81).

Naturalistic authority rested on proximity and duration. Photographs were 'taken directly from nature' (Emerson and Goodall 1886: preface), while 'all my descriptions are *written on the spot*' (Emerson 1890a: xii). Landscape photography also demanded living in 'the country' for 'long periods', diligent practice set above organised photographic trips, which Emerson disparaged as akin to game shooting parties: '32 "Ilfords," 42 "Wrattens," 52 "Pagets," etc' (Emerson 1890b: 245–6). *On English Lagoons* presented a year on Emerson's converted wherry 'The Maid of the Mist', from '15 September 1890 to 31 August 1891', observing across the seasons, through icy winter, a 'Log of the "Maid of the Mist"' giving precise notes on weather and natural history (Emerson 1893: 273–98). Journeys went beyond usual cruising grounds, reaching heads of navigation at Aylsham, Antingham, Bungay, Loddon, New Mills and Waxham, with the Muck Fleet navigated in a small boat. Emerson sold the wherry in 1893, in 1895 acquiring a Daimler motor launch, 'in which he cruised about the Broads' (Emerson 1898a: 135).

Emerson and Goodall claimed intimacy with work, Broadland landscape showing seasonal labour, and descriptions of plates outlining sequences of actions not necessarily apparent in the moment of the image. Technological change also registers, 'The Old Order and the New' showing a sail boat, a derelict windmill and a steam pump house, drainage transformed: 'Thus nature, aided by science, has changed an arm of the sea into fertile fields' (Emerson and Goodall 1886: 35). Broadland also appears as Emerson and Goodall's own working landscape, though their labours sit in blithe contrast to land work. 'The First Frost' shows boats iced, snow on frozen waters, Emerson describing their yacht frozen in at South Walsham, and watching a family gathering ice to sell: 'so the first frost brought gold to the Broadsman and his family' (39). The artists then commence their own labours: 'We also were busy, for after breaking a channel to the shore, we started to look for pictures, and came across this little nook' (39). The labour is incomparable (in wage, in necessity), yet Broadland is Emerson's workplace too: 'Shortly after we had secured this plate a picturesque old man appeared, and with trembling hands drew some of the faggots from under the snow, cut them up, and carried them home' (40).

The look and sound of regional life

Occupants

Emerson and Goodall's 'Life and Landscape' title indicates Broadland people as objects for scrutiny. Broadland discovery was preoccupied with authentic regional conduct, on the part of inhabitant as well as visitor, working long-established conventions of seeing the rural worker (for Emerson the 'peasant') as both the bedrock of England and something internally exotic in a modernising country, at once English essence and internal

Figure 12   PH Emerson, 'The First Frost'. Source: Emerson and Goodall 1886, image courtesy of Norfolk County Council Library and Information Service.

other to urban modernity (Howkins 1986; Boyes 1993). Watching Norwich day trippers on the Yare at Bramerton, Emerson reflected: 'When the land shall be all built upon and enclosed, and the peasant is no more, then may old England go grovel before the world' (Emerson 1893: 27). Working lives, customs and dialect signalled cultural value, though their pursuit could be for varied political ends, and different cultural effect.

Accounts of the 'Broadsman', and accompanying broads-women and children, commonly reference Richard Lubbock's 1845 *Fauna of Norfolk*, discussed fully in Chapter Four below. Lubbock's fauna stretched to the Broadland human:

> When I first visited the broads, I found here and there an occupant, squatted down, as the Americans would call it, on the verge of a pool, who relied almost entirely on shooting and fishing for the support of himself and family, and lived in a truly primitive manner. I particularly remember one hero of this description. 'Our broad,' as he always called the extensive pool by which his cottage stood, was his microcosm – his world; the islands in it were his gardens of the Hesperides, – its opposite extremity his *ultima Thule*. Wherever his thoughts wandered, they could not get beyond the circle of his beloved lake; indeed, I never knew them aberrant but once, when he informed me, with a doubting air, that he had sent his wife and his two eldest children to a fair at a country village two miles off, that their ideas might expand by travel: as he sagely observed, they had never been away from 'our broad.' I went into his house at the dinner hour, and found the whole party going to fall to most thankfully upon a roasted Herring Gull, killed of course on 'our broad.' His life presented no vicissitudes but an alteration of marsh employment. (Lubbock 1879: 129)

Lubbock's hero is a lower class nature curiosity, living a local frontier life, his limited geography a matter for amused admiration across a cultural distance of classical allusion. The passage above would become shorthand for traditional Broadland,

Figure 13   PH Emerson, 'A Broadman's Cottage'. Source: Emerson and Goodall 1886, image courtesy of Norfolk County Council Library and Information Service.

quoted by commentators including Davies (1883b: 9–10), Maxwell (1925: 60), Mark Cocker (2007: 92–3) and Brian Moss, whose 2010 textbook *Ecology of Freshwaters* (2010: 322) sets Lubbock's broadsman in counterpoint to ecological decline.

In *LLNB* Emerson showed 'A Broadman's Cottage', home, boats and broad, old and young men holding ends of a bow net. Emerson's description echoed Lubbock, to whom he refers elsewhere, with the cottage home to an 'idyllic Bohemian', simultaneously 'labourer, waterman, sportsman, naturalist, and philosopher', with his wife and mother, 'a family of practical philosophers'. The Broadman achieves local sufficiency, shooting, catching eels, fishing, cutting reed, hay and gladdon, selling ice, taking eggs, keeping bees, growing garden herbs to 'season the coot stew': 'Thus, from the beginning to the end of the year, he lives a life bound by no hours, and subject to no master save nature' (Emerson and Goodall 1896: 13–15). Other plates documented the Broadman's tasks, Emerson shooting 'picturesque and typical' figures in everyday clothes (Emerson 1890b: 251). 'Gathering Water-Lilies' has become iconic of Emerson and Victorian Broadland, taking the front cover of Newhall's 1975 monograph, a man keeping a rowing boat steady while a woman reaches to pick a lily. 'Setting the Bow-Net' shows the same pair, Goodall's text explaining lilies gathered as bait to lure tench; Taylor comments that far from, as sometimes assumed, a leisure idyll, 'Gathering Water-Lilies' shows a 'workaday event' (Taylor 2006: 42).

Emerson pictured Broadland women and men to convey strength, in physical labour and character. Taylor notes Emerson's varying photographic and textual treatment of regional women, with Yarmouth the site for duplicitous erotic encounters with visitors (Taylor 1995: 112–18; Emerson 1890a), while land-working Broadswomen are idealised from a close superior distance, as in the short story 'Ellen', where a girl fishes with alluring grace, Emerson watching from the long grass (Emerson 1889b: 3–8). Conservative commentators attacked Emerson for seeming criticism of landlords, notably in *Pictures of East Anglian Life*, with the poor akin to 'slaves' and 'the whole countryside ... in league against the peasant' (Emerson 1888: 81). Emerson's 'peasant' sympathy was however within strict hierarchy, an ideal social order of peasant, farmer and benevolent landlord (Knights 1986). 'Haymaker with Rake', from *Pictures of East Anglian Life*'s chapter on 'Peasant Types', showed one of two Norfolk varieties, 'a specimen of the smaller darker type of peasant' (Emerson 1888: 141). Physical rural anthropology worked within a wider racial imaginary, Emerson's story 'Visions' recounting a dream of a country racial rising: 'it is said that one day the men of the marshes intended to arise and burn all the red houses, and take the beautiful women therein and breed up a new race, beautiful and strong, brave and hard, growing like the oaks' (Emerson 1898b: 63–4). Emerson's autohagiography in *The English Emersons* concluded: 'All weak and diseased should be forbidden marriage by law, and that for the sane and healthy polygamy is advisable' (Emerson 1898a: 138).

Emerson also found Vikings in Broadland, in keeping with a common ascription of Danish origin for people in the Flegg district, the place name suffix -by showing Danish settlement; Filby, Hemsby, Scratby, Mautby. Dutt suggested: 'you may to-day, if you study carefully the faces and inquire the names ... discover descendants of those viking settlers who fished the waters of the Norfolk estuary a thousand years ago' (Dutt 1903: 4; Suffling 1893: 257–8). Emerson spotted Viking in his wherry servant Jim: 'a light-haired, blue-eyed, unconventional waterman, whose ancestors, I suspect, came over with some Danish rovers. He never spoke of Friday, but of Frida; never said lucky, but always lucka, and so on' (Emerson 1893: 4). Cosmopolitan Emerson discerns Broadland as an intensely local mixture of historical and geographical connections. International inheritance was taken to atavistic extreme in Emerson's 1925 'crime story' collection *The Blood Eagle*. The title story, set in 'Seamouth', concerns the ritual murder of Owen Owen, a Welsh Breydon fowler: 'The clothes had been hurriedly cut away from his torso, and the chest was exposed and a great hole, the shape of a bird with outspread wings, had been cut through the ribs, and the lungs had been pulled through the opening' (Emerson 1925: 22). Owen is the killer of Frank Johnson's father. Inspector Olaf Henson investigates, recognising Frank as a 'tall fine Viking': 'Good God, the Blood Eagle, is it hereditary? Strange. Well, he's a brother Nordic. And Owen killed his father. Strange, strange!' (28). The crime is unsolved, Frank remaining free, Henson confessing 'dereliction of duty', overlooking evidence to protect Frank: 'An extraordinary fine specimen of the old, *the* breed ... he had kept true to type all through the ages. I at once knew he was the murderer, and I determined that it should not be brought home to him' (31).

Singing folk

If Emerson took labour, physique and murder as possible regional life signs, he also noted verses of folk song as indicator of authentic culture (Emerson 1888: 21–3; 1898b: 139–43). As Boyes and others have shown, late nineteenth century folk song collection was presented as an act of regional and national cultural salvage, and such practices shaped Broadland discovery (Boyes 1993; Matless 2005). Song serves as standard cultural aside in guidebooks such as Suffling's *Land of the Broads*, noting 'old ditties of a bacchanalian character' sung by 'East Norfolk Natives' (Suffling 1893: 268). Song also however attracted self-consciously serious cultural labour. Thus Kate Lee, collector, concert singer and co-founder and Secretary of the Folk-Song Society, collected during Broads holidays, fellow FSS founder AP Graves noting: 'To get hold of some folk songs which she knew were reserved for the ears of the frequenters of a country inn in The Broads, she obtained admittance as a waitress at the ordinary table, … got those precious folk songs into her head, and kept them there for the benefit of society' (Graves 1913: 179; Lee 1899; Boyes 1993: 48).[8] Lee harvests culture across class divides, collecting in disguise, an inn scene to set alongside Dodd's dinner.

Lee's story alerts us to gendered folk collecting strategies; male companionable drinking, female covert service in masculine space. Both entail parallel class journeys, with the Broads pub a cultural jewel-box, with some items gathered up for metropolitan performance, Norfolk life gone national. Ralph Vaughan Williams published Broads gatherings in 'Songs of Norfolk' in the *Journal of the Folk-Song Society* (*JFSS*) in 1910 (Broadwood, Vaughan Williams and Gilchrist 1910), while four Broadland songs would appear in the 1959 *Penguin Book of English Folk Songs*, compiled by Vaughan Williams and AL Lloyd (Vaughan Williams and Lloyd 1959). The songs seldom have regional lyrical content, and the site of singing is not taken for a song's origin; rather Broadland is a region where song, if buried elsewhere, still surfaces. Vaughan Williams made three Broads collecting expeditions with composer George Butterworth in October 1910, December 1910 and December 1911, gatherings made out of summer season (Dawney 1976). Collection also informed translation, with folk tunes turned to classical performance, crossing genre and gathering other cultural capital. Composer Ernest Moeran, whose father had been vicar at Bacton on the Norfolk coast, collected in Broadland between 1913–15 and 1921–7, publishing 17 'Songs Collected in Norfolk' in the *JFSS* (Moeran et al 1922), and arranging *Six Folk Songs from Norfolk* for voice and piano (Moeran 1924; Hill 1985; Self 1986; Palmer 2003). Norfolk here is the Thurne valley, with singers named from Potter Heigham, Hickling, Sutton, Catfield and Winterton.[9] The pub was Moeran's key collecting site, *Six Folk Songs from Norfolk* acknowledging 'Mr George Lincoln, landlord of the "Windmill," Sutton' (Moeran 1924: preface; Moeran 1948). Moeran listened, noted and reworked, tunes informing classical compositions carrying evocative regional titles; the piano pieces 'Stalham River' (1921) and 'Windmills' (1922), the orchestral 'Lonely Waters' (1931). The latter ends with voice or cor anglais rendering a song collected by Moeran from Walter Gales and Robert Miller in Sutton, published in *Six Folk Songs from Norfolk*: 'Then I'll go down to some lonely waters / Go down where no one they shall me find / Where the pretty little small birds do change their voices / And every moment blows blustering wild' (Moeran 1924: 14–15).[10]

Moeran's 1947 BBC radio programme *East Anglia Sings* included a 'home-made song' by landworker and basket maker Harry Cox (1885–1971), given the title 'Barton Broad Ditty', recorded in the Windmill Inn.[11] Cox was born at Barton Turf, lived at Potter Heigham from 1899–1927, and later between Catfield and Hickling;

Moeran arranged his 'Down by the Riverside' and 'The Shooting of His Dear' in *Six Folk Songs From Norfolk*, and arranged for him to record for Decca in 1934. Cox became important in the post-war folk revival, recording two LPs, singing in London, visited by collectors and the BBC. Folk magazine *Ethnic* featured 'Harry Cox the Catfield Wonder' in its first 1959 issue. Heppa notes that Cox himself sought to preserve songs, learning from his father and other singers, Cox as singer becoming cultural curator (Heppa 2005).

Transmission of the evidently regional voice could hit cultural barriers. Stanley Bayliss commented in the *Penguin Music Magazine* on *East Anglia Sings*: 'E J Moeran introduced some recordings of folk-singers recently made in Norfolk. This was a most interesting broadcast, but not altogether an enjoyable one. It proved that collectors like Mr Moeran have been faithful and accurate in noting down these traditional songs; but let me confess that I found the timbre of the voices of all the singers extremely raucous and almost unbearably ugly.'[12] The geography of folk culture entails negotiation of the regional and national, the valuing of difference in tension with its comprehension. As regional life goes national, the temptation is to untangle tensions either through translation into polite rendition, or by applying a comic label, the regional voice to be laughed with, or at. Comic objectification would dog the career of Allan Smethurst, briefly famous as 'The Singing Postman' in the 1960s, delivering self-penned songs in broad Norfolk voice, commonly set on the north coast, but singing in 'Wroxham Broads': 'Fare thee well my friend, I'll tell yew by an' by, Cors now I go to where the bittern fly, ... Carry my bones ter Wroxham Broads an' put them down right there' (Skipper 2001: 90–1; Barker 2004).

Guides to a discovered Broadland commonly registered one specific public song. 'John Barleycorn', sung by Horning and Stokesby children, became part of a public leisure landscape:

> never a yacht or a pleasure-boat goes by Horning Street without being favoured from the bank, by boys and girls alike, with a somewhat garbled and inconsequential version of 'Hey, John Barleycorn,' sung prettily enough to a running accompaniment for coppers. ... future generations of antiquarians will wonder extremely at the enormous quantity of copper coins of the reign of Victoria I. which will be found there. (Rye 1885: 263)

Folk performance becomes canny business, Davies's *Handbook* passing Horning: '"Ho! John Barleycorn: Ho, John Barleycorn, / All day long I raise my song / To old John Barleycorn." That is all. It is simple and effective, and extracts coins from too easily pleased holiday-makers' (Davies 1891: 61; also Davies 1883b: 64; Jennings 1892: 28; Parsons-Norman 1912: 16; Patterson 1923: 40–1). Rev GN Godwin's early twentieth century description of Stokesby choir's annual outing on the 'Queen of the Broads' to Wroxham indicates the song as wider village convention: 'Disembarking at Stokesby, we were received with much singing of "John Barleycorn" by those who had stayed at home' (Godwin, nd).[13] Private leisure diaries also record song, and mourn decline: 'For the first time on record there was no "John Barleycorn" at Stokesby this year, either time we passed it.'[14] A 'Child's Diary' for July 30 1925 laments: 'We passed Stokesby but the children did not sing John Barleycorn', though on July 29 1926: 'Passed Stokesby, one enterprising babe actually attempted to sing John Barleycorn it was rather discordant.'[15] Sonic custom appears petering out.

Broadland talk
The sound of regional life also encompassed dialect. Patrick Joyce highlights the connections of language, class and social identity in the late nineteenth and early twentieth centuries, with dialect a focus of class judgement and valuation, whether in upholding distinctive working class identity, putting down the linguistically provincial, or cherishing diversity within a common English, especially in assertions of county identity (Joyce 1991).

Dialect, like song, could be a picturesque guidebook adjunct, as in Suffling's 'Glossary of Norfolk Words' used by 'the natives' (Suffling 1893: 249–54), and could underpin the authority of commentators such as Emerson and Rye. Emerson asserted: 'All the provincial words used *I have heard spoken*; they are in use therefore to-day. They are spelt phonetically' (Emerson 1890a: preface). Dialect conversation with servant Joey presents Emerson and Goodall as able to speak across class, though Joey cannot move the other way (Emerson 1890a). Emerson's 1892 dialect novel, *A Son of the Fens*, narrated by 'Dick Windmill', signals the intensely local, yet also movement beyond the region, in Dick's working life at sea and on land, and his sister Miriam in America (Emerson 1892).[16] A commercial market for dialect novels was tapped by James Blyth, born Henry Clabburn, son of a Norwich industrialist, writing 60 novels from his Fritton cottage 'On the Edge of the Marsh' from 1903 (Blyth 1903: vi; Earwaker and Becker 1998). *Juicy Joe: A Romance of the Norfolk Marshlands* presents a Waveney valley beyond 'modernity' and 'civilisation' (Blyth 1903: vi), Blyth claiming 'personal knowledge' of marshland superstition, vice and hypocrisy. Dialect marks a sensational literary realism: 'Those of my readers who have faith in that rustic courtesy which is so dear to the novelists of country life will, no doubt, be shocked and disgusted at having a true account of the dwellers on the fens laid before them' (viii). Blyth hoped Norfolk words might also shape the national literary sales map: 'I hope that they will find as much beauty in the language of the Norfolk fens as there is in the Kailyard tongue' (xii).

Living at Hickling after World War I, Marietta Pallis gathered local dialect, asking servants to transcribe folk songs, and drafting a bizarre unpublished novel concerning a village school phonetics teacher ('So the gems of native speech are everywhere being reconstituted into synthetic pebble') coming to a sticky end as children revolt for local linguistic authenticity (Matless and Cameron 2007a).[17] Pallis's friend Hope Allen, co-owner of their Chelsea house, gathered dialect for more considered ends (Hirsh 1988). Allen, an American scholar of medieval English mysticism, collected from Hickling residents, recalling in 1949: 'I have note books full of the speech of old Mrs Gibbs, who died April 1, 1940 at 90 ... This background may explain why I was so overjoyed at the chance to do Margery – which involves such a synthesis of real Norfolk life.'[18] 'Margery' is *The Book of Margery Kempe*, authored in fifteenth century King's Lynn, identified by Allen in 1934 as a classic of female medieval literature, Allen finding speech affinity across classes and centuries (Allen and Meech 1940; Kempe 1985; Hirsh 1989).

Allen also recalled: 'In an interval of holiday at Hickling in 1922 I sent to the *Atlantic Monthly* a story based on the continual talk of the wonderful old woman to whose home (then in a mill on the marshes) Miss Pallis had taken me in June 1911. The story was instantly accepted and more asked for.'[19] Alongside the medieval mystical, Allen translated Broadland into American literary modernism, publishing 'Ancient Grief' (Allen 1923a) and 'A Glut of Fruit' (Allen 1923b) in the *Atlantic Monthly*, and 'The Fanciful Countryman' (Allen 1927) in *The Dial*, stories echoing her own transatlantic life, and that of Hickling informants, as noted by Allen's friend, historian Joan Wake: 'I like old Charlie Newman best. Told me his father remembered two wherry loads a year of Hickling people going off to Yarmouth to emigrate to America by a sailing ship that went twice a year.'[20] Regional identity is marked by mobility all ways. 'Ancient Grief' and 'A Glut of Fruit' describe American visitors to a thinly disguised Hickling and Sea Palling, 'Ancient Grief' beginning:

> The solitary alien hesitated when he alighted – looked ahead at the sand-dunes from which the sea roared out, and behind, down the little street of brick old and new, which soon ran into the dim marshes. He attracted some attention, for he was obviously a double alien: he was not only a 'wisitor,' but a transatlantic one, such as was rarely seen in this corner of the coast. (Allen 1923a: 177)

The graveyard shows blood ties, familial surnames frequent: 'He felt a strange kinship to the whole village – even to the marsh' (183). The pub shows ethnic distinction: 'He saw many squat brunette types ... – remnants, he was told, of a primitive race cut off by the fens; and he saw also many tall fair ones, descendants of the Norse invaders who had landed on the coast. The rise and fall of the speech of all of them went to the same haunting tune' (180). With the parson (who turns out to be his grandfather) the educated visitor is on different social ground, discussing the fine situation of the church, appreciating local scenery.

'The Fanciful Countryman' explores geographies of love and language through rural–urban rather than transatlantic migration, with Billie Appleyard moving to London, returning to farm work after unrequited love, yet fulfilled when his London girl visits Norfolk as a servant. Courting proceeds through Jennie gaining awareness of Billie and his landscape. In an Emersonian scene, Jennie cannot hook white water lilies, but Billie can: 'he knelt, reached out his supple brown arm and caught some; ran and cut an alder sapling into a hook, and hooked others' (Allen 1927: 497). Love grows as they quant among lilies and bolder (common club-rush): '"The lila grow mighta comfortable-like along o' the old boulder," said Billie, as if he liked to see the mingling of green and white, of fibrous and of suc-culent' (499). The final paragraph is simply: 'So their courtship began' (500). Allen gives extensive dialect speech, without glossary, American readers to take what comes. On the facing page of *The Dial* is an unrelated 'Wash Drawing' by ee cum-mings; other 1927 *Dial* authors included DH Lawrence, Gertrude Stein and WB Yeats. Broadland finds a modernist literary home, through stories interrogating the constitutive language of the region.

Twentieth Century Pleasures

The leisure industry dominated the economy of twentieth century Broadland; sailing and motor boat hire increased, companies prospered, and the region was reshaped around holidays. In 1951 Doreen Wallace commented: 'The industry of the Broads region now is pleasure' (Wallace and Bagnall-Oakeley 1951: 80). This section attends to twentieth century pleasures in Broadland through regional guides, controversies of conduct, criminal stories, cultures of sailing and motor cruising, tropes of regional theatricality, and the twentieth century leisure site of Potter Heigham.

Shapes of adventure

The 50th and final edition of Davies's *Handbook* appeared in 1930, text largely unchanged since 1882, and overtaken by publications such as Blakes' annual *Norfolk Broads Holidays Afloat*, with 100,000 issued by the late 1930s. The guidebook scene shifts, national and regional companies and individuals issuing rival guides for an expanding hire market.

Two regional initiatives offer an initial entry into the mid twentieth century Broadland leisure landscape. From the 1930s, Norfolk-based Hamilton's maps and navigation charts of the Broads detailed river topography, naming all reaches and surveying channel depths by survey pole (Hamilton's 1938) and later echo sounding (Hamilton's 1955; EDP 1952d). Blank 'Official Log' pages encouraged reader commitment and loyalty, with two guineas for 'Best Log submitted' (Hamilton's 1938: 4). Guide photographs of the Lion Inn at Thurne showed facilities and landscape enfolded, a bar interior painted with mills and water (84). Drew Miller's 1937 *What to Do on the Norfolk Broads* also noted the interior, 'modernised and rebuilt 1935': 'The Broadland Lounge in "The Lion Inn". The frescoed walls depict Broadland scenes' (Miller 1937: 69). Landscape decorates, attracting outside visitors in. Miller, an American former banker, issued annual shilling guides from 1935. After World War I Miller had converted a derelict Bureside mill near Horning into a lighthouse-like dwelling ('Miller's Mill'), and made himself a Broads authority (East Anglian Magazine 1936). Miller's story collection *Seen from a Windmill*, discussed below, appeared in 1935. *What to Do*'s extensive advertising made for a celebration of local commerce, from Arnold Roy's Wroxham 'Largest Village Store in the World' (Miller 1937: 15), to the Bure's floating shops. Systematic maps showed river reaches as straight lines for efficient movement, while 'How to reach the Broads by Motor Car' had national road lines converging on Norwich. Miller died in 1936; *What to Do* continued to be published by his companion Sylvia Miller.[21]

Guides and organisations continued to enact moral geographies of leisure. Conduct could be shaped through morality defined in religious terms, as when in the 1950s Maurice Rowlandson organised 'Venturers' boys cruises, with skippers and stewardesses, prayers and services (May 1952: 132–4). The 1952 adventure story *The Cruise of the Clipper* likewise found faith in landscaped narrative, three sixth formers converted through a crime adventure, with evening shanties and prayers, and God realised in nature (Derham 1952). If such spiritual adventure emphasised quietude as a path to fulfilment, there have been other Christian sonic geographies. In 1993 ITV's *Gloria!* (the title alluding to God, and presenter Gloria Hunniford) toured the broads, mixing mission and Bure travelogue, including the large oak cross at the former high altar of St Benet's Abbey, consecrated in 1987 to restamp Christianity. *Gloria!* took a mock steamboat, guests including Bobby Ball and Alvin Stardust belting gospel into open air, passing cruisers bemused at the moral noise.[22]

Loudness in Broadland seldom signals morality. Twentieth century moral geographies consolidated earlier strictures, finding national readership in Arthur Ransome's *Coot Club* (Ransome 1934) and *The Big Six* (Ransome 1940; Brogan 1984). Ransome, well known for Lakeland children's adventures, offered two Broadland environmental morality tales. Maps situated the action, with geographical orientation a prized quality in characters and readers. Ransome allied the adventuring children of the professional class with good locals against bad locals and bad visitors. The 'Coot Club' includes visitors Dick and Dorothea Callum, with Tom Dudgeon, son of a Horning doctor, defending bird life against 'Real Hullabaloos' (Ransome 1934: 42) on board the large cruiser 'Margoletta', a mixed upper middle class company in 'white-topped yachting caps and gaudy shirts, and berets and beach pyjamas' (347), with deck gramophone and cabin radio: 'A very loud loudspeaker was asking all the world never to leave him, always to love him, tinkle, tinkle, tinkle, bang, bang, bang' (65). Landscape is set against light music, the cruisers' commercial amusement against the children's nature care and alertness. Tom sets the Margoletta adrift when it moors by a coot's nest, and battle begins, the Margoletta ending sunk on Breydon Water, the Hullabaloos rescued by the 'Death and Glories', three boatbuilder's sons from Horning. Ransome stages a Broadland Titanic, the loudspeaker 'pouring out its horrible song' as the boat sinks (332), the radio randomly short-circuiting: '"We now switch over. ... The Hoodlum Band. ... Relayed from. ..." There was a pause, and then a sudden torrent of noise that broke off short almost as soon as it had begun' (342).

Ransome's moral sonic geography echoes into the Broads Authority's 1989 longing for 'quiet enjoyment', evoking a time 'when the bittern's boom was more often heard than that of the ghetto blaster' (Clark 1989). Formal surveys of Broadland sound are rare, though Susan Walker's 1978 UEA study assessed boat users' perceptions of 'noise' and 'anti-social behaviour'; 46.1 per cent of hirers reported medium noise, 31.3 per cent encountered anti-social behaviour 2–3 times per week (Walker 1978: 160). Quantitative categorisation carries qualitative judgement, aggressively and unacademically made in James Wentworth Day's 1967 *Portrait of the Broads*, evoking an earlier Broadland not 'exploited, capitalized, vulgarised, transistorised' (Day 1967: 36):

Birds and animals are the angler's familiars. Since he is a quiet fellow, they come to him. He is part of the landscape. He promises no menace. Therefore he sees a lot that is denied to the chattering hiker, the petrol-propelled motor-boat fiend, the moronic Beatle-disciple with his long-player and longhair. These people make a noise. Therefore they repel nature. (150)

Broadland publications tend towards the anti-Hullabaloo, the region defined against noisy intrusion. Comic author Keble Howard's 1928 *The Fast Gentleman* however offers Broadland as if written by Ransome's Hullabaloos, sitting in relation to Ransome as Anna Dodd to Dutt. *The Fast Gentleman* followed Howard's 1924 motoring honeymoon tale, *The Fast Lady*, the preface telling of Howard's Harmony II, a 'motor-cabin-cruiser' built to explore the region: 'There is nothing like it in all England' (Howard: 1928: 7). City gent Leonard Rabbidge takes wife Muriel, son Stanley, comical maid Dulcie, and friends Lindsay Mountford and Olga (a boyish City girl with cigarette and monocle) for a spring cruise on the 'Naughty Nymph'. If Leonard is fond of butterflies and maps, his party are, to a responsible 'Potter Ferry' bungalow couple, nothing but 'common trippers ... I came here for a little quiet fishing – not for all this damned hullabaloo!' (103–4). Broadland becomes a region of comic japes and landscape satire, including:

Tidal Toupee
Leonard's father-in-law loses his toupee in the Thurne, local advice suggesting: 'It'll come back about ten o'clock tonight. If it doesn't, write to the River Commissioners.' Tides return the hairpiece: 'It must have been soon after nine-thirty that I noticed a dark-looking object in the middle of the stream.' (121–2)

Somnolent Swan
Fearing a Ranworth swan, Leonard wraps six aspirins in bread, the swan feeding, to fall asleep: 'the first scientific victory of Man over Swan! I should certainly have to write an account of the whole proceeding for *Feathered Life* (with which is incorporated, I think, the *Ornithological Gazette*).' (109)

Guidebook Gesture
'The Bullock of St. Benet's Abbey' finds Leonard on a solo excursion to abbey ruins: 'Just imagine all that happening in 1066, and here was I, Leonard Rabbidge, in 1927' (136). Chased by a lively bullock, Leonard hurls his guidebook in defence, finding refuge from cow terror on the walls: 'the guide-book had become impaled on one of his horns ... The pages ... turned this way and that. I remember noting quite a pretty picture of "Sunset over Hickling Broad," to face page 26.' (139)

Adventures in Broadland were also plotted through crime fiction. The frequency of Broadland as crime story setting, if not indicative of high regional murder rates, indicates elements of regional styling; somewhere well enough known to serve as a setting, where adventurous conduct, careful observation and landscaped mystery could be set in harness. Post-war children's stories effectively added crime to *Coot Club*, with Cold War secrecy shaping Kathleen Fidler's *Brydons on the Broads*, the Brydon children exploring a 'Secret Broad', finding lost secret military equipment (Fidler 1955), while Gordon Catling's *The Gang on the Broads* saw public school children break scientific spying, infra-red bird photographic equipment catching the thief of gravity-defying 'Metal X' (Catling 1958). Gladys Mitchell's *Holiday River* (1948) has youthful landscape knowledge outwitting diamond smuggling, with three girls and two boys, aged 15–17, holidaying from London on a cruiser and yacht.[23]

The crooks, ill-mannered on the London train, stand out at Wroxham: 'they're dressed all wrong for the river! They look awful in natty lounge suits with the padded shoulders' (Mitchell 1948: 22). Having bought 'all the likely maps they could lay their hands on' (64), the children can track the gang, climbing Ranworth Church tower: '"I say, that's a wizard notion!" exclaimed Roger. "Let's do that. We could take the maps up with us, and check them with what we can spot from the tower. It will give us a wonderful idea of the lay-out."' (97). Gillian spots the villains: 'Slowly she circumscribed the four sides of the tower, stopping on each side to train the field-glasses on the wide land-scape' (100). Roger unfolds the map, 'and they all held it against the breeze', oriented to pinpoint crime (102).

The Broads as region of observational solution also drew the fictional detective. CP Snow's first novel, *Death Under Sail* (1932), presented a modernist landscape of detection, reflexive in convention and self-consciously advancing on Edwardian boating adventure. Murdered holiday skipper Roger Mills is mocked for his hearty log: 'A record of a friendly expedition like this is usually merely an excuse for the keeper of the log to demonstrate an execrable prose style' (Snow 1963: 69). The party are confined in a Potter Heigham bungalow, while amateur sleuth Finbow, devoted to science, exposes the killer's ruse to make an initially obvious murder appear suicide. Alan Hunter, critic of Lambert's origins detective work, also exercised detection in his Inspector Gently stories, some set on the Broads.[24] In *Gently Down the Stream* (1957), the murderer, his broadside bungalow furnished with 'a few original pictures, a pair of Seagos, an Arnesby Brown, a Peter Scott, and a group of six water-colours of Broadland birds by Roland Green' (Hunter 1957: 66), fakes suicide in a burnt-out cruiser. Gently's dialect skills detect the genteel killer disguised in a disreputable houseboat colony, where police dydle dykes for discarded dentures. OS maps help Gently deduce in mysterious Broadland: 'the river was shut-in all the way to the dyke and the shack where the jacket had been found. Snaked roots of alder reached out from either bank, screens of reed, bramble and wild currant formed a bar-rier to the eye. The carrs were a secret place' (156). Broadland carries mystery enough to exercise reason, landscape legible to the judicious observer, whether solving murder, defending coots, or plotting holiday adventure.

The sailing life

Robert Gillmor's dust jacket illustration for Brian Moss's 2001 New Naturalist volume on *The Broads* contrasted the yacht and motor cruiser, the left side dense with colourful wildlife and bright sail, the right side grey with a cruiser's wash, showing 'leisure activities, careless of environmental considerations' (Marren and Gillmor 2009: 224–5). Gillmor activated longstanding cultural judgements of sailing over motor cruising, as regional spectacle and means to know landscape.

Ransome's *Coot Club* showed yacht technology fostering alert, dextrous minds and bodies, sailing demanding a tight alliance of boat and human, children improved by experience, whereas motor cruisers carried the lazy and Hullabaloo. Interwar diaries, published or private, also conveyed such sailing culture, with additional amused acknowledgement of sailing ritual, adults and children living out conventions of conduct and narration (Clarke 1923; Dyer 1924; Lorimer 2003). A 'Child's Diary' of 1924–8 summer holidays anticipates Ransome with unwelcome noises off, as the 'Golden Hind' plays 'a most awful wireless set ... The only real use is to hear Big Ben and to know the time.'[25] Hullabaloos are anticipated: '<u>Note</u>. "Enchantress" came and moored beside us. <u>Horrid creature</u> <u>sssssssss</u>.'[26] From 1935–40 EHT (Trevor) Jukes logged middle class Broads cruises on hired yachts with family or male friends, his private diaries, now held in Norwich public library, giving an instructive glimpse into a 1930s sailing life.[27] Jukes, who lived in Barnet ('About ten o'clock a black Morris Ten nosed its way on to the Barnet Bye-pass' (September 1936)), allotted holiday nicknames from Lewis Carroll (Hare, Hatter, Dormouse), with holiday convention framing commentary on selves and others. Two men in shorts spout 'hearty commonplaces', expressing 'whole-hearted concurrence with Mr. Blake's view that it is The Holiday that is Different'. A cruiser carries a 'bizarre family' with 'an aura of Hackney and Camberwell about them', who 'would have been much more at home on the beach at Blackpool than in the little village street at Neatishead' (June 1935). Suburban Barnet belongs, inner London is ill-placed. Tour boats too inhabit another world:

> Overheard while sailing, on being overtaken by a crowded Tour of the Broads.
>   'We have to give way to them you see, because they can't stop'
>   'Wot, can't they stop at <u>all</u>, do they always have to keep going on?'
> What we have heard today
>   The homely hooting of the coots. (See The Holiday that is Different). (September 1936)

Self-mockery also makes the holiday: 'Unwarranted statement – The Hatter: "I am now going to melt into the landscape...."' (September 1936).

In 1937 Jukes added a two page philosophical reflection on the nature of holiday time, dated 29 May, worth reprinting here:

> Time plays us strange tricks. For nine months this Log has stood upon my shelf, & I have fingered it & read from it from time to time. It became the symbol as it were of a care-free happiness like that of childhood. Ah, the Broads, I have thought, and the very thought has brought a sense of security & well-being. I have dreamed often of sailing. And now for a short week it has been my companion on another voyage, and we are home and it is time to put it back on the shelf again. But as I do so, I feel the loss of a part of my life. This week was lived, I feel, in another world, and I take up my usual life not after a week, but from the exact spot where I left it when I drove up the road that sunny Saturday morning and sang on the road to Royston. That is the strange thing, the trick of time. The timeless time upon the boat can never become a part of my ordinary self. It is not the time-ridden me of Barnet & Whitechapel, the self which is always haunted by the passing of time & the change & end which must come to all things mortal. That self is dead or asleep for a week, it is laid away in the garage when I get out the car for the Broads, at the moment when the other timeless me comes into existence, & it is instantly waked to life again when the time comes to put back the Log upon its shelf. The two can never meet, can never be reconciled & fused into one. All that I can do is to escape for a space from this world that races far too quickly towards the grave, escape & live for a spell in another world, just as I do in dreams. Yet dreams are truly timeless, & the Broads world has a time of its own, though a different sort of time from the real time.

For Jukes the Broads represented another time, not in terms of being a region set in a primitive past, but in their being a world removed, generating a holiday temporality of slow sailing movement, a rhythm apart.

Alongside holiday hire, local associations regulated private sailing, via the Norfolk Broads Yacht Club, formed in 1876 from the Yare Sailing Club, and the Norfolk and Suffolk Regatta Association, founded in 1894 to ensure 'uniformity of action and promote clean sailing' (Dutt 1903: 324). The Broads Authority continue to uphold sailing as 'part of Broads history and landscape, as well as being in line with the philosophy of quiet recreation' (Broads Authority 1993: 73–4; Broads Authority 2011: 59), with regattas as events where, if aspirations of quiet may overturn for a day, excitement and spectacle proceed in good order. Sailing history is commonly evoked via early nineteenth century Norwich School paintings such as Joseph Stannard's 1825 'Thorpe Water Frolic', depicting an annual regatta organised from 1821 by Thorpe estate owner John Harvey, with 20,000 attending in 1824. Spectators, barge races and rowing matches fill Stannard's scene of orderly crowding, pleasure crossing yet respecting social divides, with Harvey pictured at the 'hub of the occasion', the gentry on the north bank, and working people on the south, a mutual class spectacle (Hemingway 1992: 287; Fawcett 1977; Blayney Brown, Hemingway and Lyles 2000). In 1903 Dutt found similar cross-class harmony at Wroxham:

> Given fine weather, every one, from the chairman of the regatta committee to the juvenile occupants of the dilapidated marsh-boat which invariably gets in the way of the racers, is in good spirits. Yachtsmen don their whitest flannels and ladies wear their prettiest dresses; music is heard above the whispering of reeds and the fluttering of flags.

Here was 'the great day of the Broadland season', locally overseen, a space to be seen, a jolly county show (Dutt 1903: 147–8). Jonathan Mardle similarly described the 1951 Barton 'water frolic', its haphazard groups ('I saw the very image of one of Stannard's rollicking boat parties in a motor boat by the staithe') and boating modernity: 'it is one of the charms of Barton regatta that there is nothing deliberately antiquarian about it. It accepts the times as they are, motor-boats, loud-speakers and all, and makes the best of them. It would be less true to the spirit of the old water frolics if it tried to concoct an artificial one in the present' (Mardle 1955a: 90).

If regattas fit Broadland heritage, the BA find powered boating spectacle not 'compatible with the concept of quiet enjoyment' (Broads Authority 1993: 91). Mechanical speed breaks limits of conduct, with zoning a tactic for containment; occasional water skiing permitted on the Yare and Waveney, and Thursday evening summer speedboat racing on Oulton Broad. Speedboat tension is longstanding, the Norfolk Naturalists Trust forbidding speeding on Hickling and Barton in 1965, with the *EDP* suggesting lower river alternative 'reserves for the human nature that revels in noise and speed' (EDP 1965g). Oulton Broad has seen power boat racing since 1903, with the Oulton Broad Motor Boat Club founded in 1933, summer events part of Lowestoft's resort attraction. In 1952 May described 'a snarling pandemonium of speed and spray', meeting boatyard mechanic Alan Darby, British champion at 17, 'one of the water jockeys who race hydroplanes at breakneck speeds on Oulton Broad' (May 1952: 48–9).

Surveying Oulton, the BA would in 1993 grudgingly acknowledge that power boats had 'become an accepted part of the boating scene in that part of the Broads' (Broads Authority 1993: 72).

Motor cruiser culture

Motor cruisers receive different evaluation in two early 1950s books issued by publisher Robert Hale, Mottram's 'Regional Book' *The Broads*, and Wallace and Bagnall-Oakeley's 'County Book', *Norfolk*. For Mottram the motor cruiser was: 'simply an automobile that goes on water instead of tarmac. It may be an enjoyable experience … But it cannot in itself enable anyone to see the Broads as I want to see them' (Mottram 1952: 2). Doreen Wallace however questioned the yacht-cruiser cultural hierarchy:

> It is of course vulgar to use a motor-boat on the Broads. All the best people go sailing, poor dears, and on the narrower waterways waste much time tacking from bank to bank … But more and more people are wisely using motor-cruisers, which are safer, take less skill to manage, and get there with less fuss. (Wallace and Bagnall-Oakeley 1951: 84)

Given that in recent dominant regional narratives the motor cruiser is, at best, tolerated, it is worth excavating a counter-history of cruiser celebration, whether by hire agencies, guide authors or enthusiasts. Oulton Broad boatyard owner Jack Robinson's 1934 *Motor Cruising on the Broads*, a follow-up to *Broadland Yachting* (Robinson 1920), included a cruise diary by Norman Game: 'Now, what other holiday gives such varied scenery, such amusing experiences, and such happy times with so little effort' (Robinson 1934: 55; Griffin 1953). The heyday of motor cruising as effortless pleasure comes post-1945, with Blakes of Wroxham and Hoseasons of Oulton Broad chief rivals in market expansion. The Nature Conservancy's 1965 *Report on Broadland* estimated a threefold 1947–64 increase in all craft, with holiday visitors increasing from 100,000 in 1938 to 200,000 in 1955, and 240,000 in 1961, and 1717 motor cruisers for hire in 1964. Graphs and charts illustrated craft types on ascending lines and bars, motor cruisers surging ahead of rowing boats and yachts (Nature Conservancy 1965: 30). The Broadland hire holiday moved down the social scale, to the lower middle and upper working class, while 19 large passenger launches, including those operated by Charles Hannaford's Broads Tours from Wroxham (whose passengers included the 1959 IBG party), and three river steamers also conveyed 260,000 1960s day visitors annually (Nature Conservancy 1965: 33; Hannaford c.1950). Broads Tours continue, promoting craft including the Mississippi-steamer-style 'Vintage Broadsman', cruising 'this enchanting landscape': 'With live commentary along the way, light refreshments and a licensed bar – what more could you ask for?'[28]

Guides conveyed the motor cruising landscape. John May's 1952 *The Norfolk Broads Holiday Book and Pocket Pilot*, written 'by and for the once-or-twice-a-year amateur Broadsman' (May 1952: vi), countered 'snobbery' (17) by describing both sailing and motor cruising, photographs showing a carefree colour idyll of minor holiday moments; cruisers at anchor, reading on deck, coming into moor. For some, this new social landscape signalled regional destruction, Day finding tripper scum: 'a boatyard yacht-basin littered with ice-cream cartons, candy floss sticks, empty potato crisp bags, cigarette ends and the thin grey scum of contraceptives on the water' (Day 1967: 36). For others the cruiser became a fascinating Broadland object. Designs made for differing presence in and command over the scene, from pre and post-war wooden hulls with raised forward cabins, to lower 1970s bright fibreglass designs with low front helms, to recent central raised canopies behind peaked bows. In 1995 Jill and John Hawkes produced *The Spotters Book of Broads Hire Cruisers* ('Lists Over 1390 Names!'), with cover showing parents and children spotting: 'The holiday-maker, the local resident and the boat enthusiast will all find an interesting and enjoyable pastime ... The producers of the book have been doing this for years!' (Hawkes and Hawkes 1995: 1). Fifty-four pages list from 'Ace of Hearts 1' to 'Yare Twilight 11'. Pictured craft give a nominal sample:

> Balmoral Castle, Caprice, Commodore I, Connoisseur C40, Classic Safari, Ferry Fusileer, Silver Phantom, Rio Grande, Ladymore, River Derwent, Sandstorm, Fair Senator, Solan Goose, Brinks Melody, Brinks Romany, Royal Crusader, Midnight Willow, Moonlight, Diamond Light.

Names evoke (within standard designs) adventure, command, exoticism, enigma, royalty, mild nature mystery; glamorous or chivalric feminine and masculine style on the waters. Only one here carries any regional landscape echo, 'River Derwent' (one of 21 in a river series from Stalham Yacht Services), a 1970s model, a type now rarer, low in the water, in orange, blue or yellow. Contemporary cruisers (Silver Phantom, Rio Grande, Diamond Light) tend to sleek white.

Such mainstream midstream pleasures have been subject to satire. The 'Watership Alan' episode of television comedy *I'm Alan Partridge* (1997) has Steve Coogan's former television presenter, now reduced to dawntime local radio (*Up With the Partridge*), fronting a promotional Broads video for 'Hamilton's Water Breaks': *Water Way to Have a Good Time!* Partridge praises on-board toilet facilities and delightful landscape: 'with the melting of the polar ice caps, most of East Anglia will be under water in the next 30 years, so make the most of her stunning fens before the floods come'. Partridge adds: 'The Norfolk Broads offer the true peace and tranquility of the English countryside, a million miles from the urban decay of the Manchester Ship Canal, and the pot-smoking whore-ridden waterways of Amsterdam.'[29] The spirit of Wentworth Day twists.

Partridge modernity parodically echoes both contemporary corporate promotion and post-war hire publicity. In Blakes' films *Broadland Adventure* (1954), *Norfolk Broads Holidays Afloat* (1957), and *Carefree Cruising* (1965), and Hoseasons' *Broadland Panorama* (1961) and *Let's Get Away From it All* (1971), working family lives take a carefree break, the boat a novel yet familiar domestic space, where men steer and women cook. *Broadland Panorama* sees two families hire 'The Caribbean', finding modern facilities, historic sites, bird life, 'first class bathing', dinner dances ('plenty of gaiety on the Broads'), Yarmouth shows (Charlie Drake, Cyril Fletcher), and 'so many splendid pubs'. Ladbrokes Holidays enlisted variety star Roy Hudd, performing in Yarmouth for the 1971 summer season, to present a 1972 promotion of Norfolk coastal holiday camps and Potter Heigham cruisers, their 'free and easy open air life' set to easy listening sounds. Here was a British national leisure modernity to trump the Spanish Costas, where everyone 'talks the same language', and there's 'none of your funny money'.[30]

Company maps also conveyed cruising territory. On the early 1970s *Hoseasons Holiday Map of the Norfolk Broads and Seaside*, foaming jug glasses marked public houses, the Broads a beer heaven. The map embraced all Broadland architecture, vignettes showing both the 'remains of St. Benet's Abbey', and the 'Cantley Sugar Beet Factory', and registered company coastal chalet exoticism; 'Hawaii Beach Coral' (at Scratby), 'Palm and Miami Sun Beach' (at California). Hoseasons' Broadland could however raise concern, the *EDP* attacking a London showing of *Broadland Panorama*, finding 'too many superlatives', with everything 'just that bit too marvellous for belief' (EDP 1960d). Similar 1962 commentary likely refers to the same film's Norwich showing:

> A few months ago there was shown at the Noverre cinema a short film which was intended to be an appreciation of the Broads, but which most Norfolk people disliked. It showed a succession of motor boats (some of them with sails hoisted for ornament) moving against a suburban background of bungalows, faked-up windmills and jazzed-up inns, to the accompaniment of canned music. The rhapsodies of the commentator awoke no echoes in Norfolk hearts, but we have a sad suspicion that we disliked the film because it showed us the truth about what is happening to the Broads. (EDP 1962a)

Cruiser culture cannot speak to (yet helps erode) another version of the region.

Colour guides allow tracking of the cruiser's regional status. Jarrold's 1970 *The Norfolk Broads* by AN Court, and *Broadland in Colour* by Eric Fowler, are dominated by ground/water level cruiser-full pictures, water and sky vivid blue, the cruiser guiding the eye. Court includes a photograph of Womack Water, 'a small but charmingly secluded Broad lying just to the south of the village of Ludham. This is a favourite subject for artists and photographers and one of the least commercialized stretches of water in the county' (Court 1970: 11; Fowler 1970). The image shows not deserted waters, but cruisers come to relax, one arriving, others moored. One woman sits on a foldable chair, another on a blanket lets sun onto pale shoulders, a man lies back, a white poodle takes the chair's shade. This version of people-in-nature fades in later publications, where wildlife scenes prevail, indeed such images become styled as anti-nature. By Jane Bulmer's 1987 Jarrold *The Norfolk Broads in Colour*, cruisers are in retreat, Broadland now overhung by a narrative of environmental damage, examined in Chapter Six below (Bulmer 1987). By 2002, Clive Tully's *The Broads: The Official National Park Guide*, with photographs by Richard Denyer, absents the cruiser. Only three images in an image-packed book of over 100 pages show moving hire cruisers, including 'Congested traffic at Wroxham, River Bure' (Tully 2002: 106), and a large craft captioned: 'A foreigner (non-Broads), River Yare' (70). From being an everyday part of the scene, the visitor's way into the region, the cruiser becomes an awkward intruder.

Broadland in revue

A counter-history of motor cruiser culture can be supplemented by excavating a Broadland of unquiet enjoyment, the region a space not of calm and contemplation but of the brash and loud, Keble Howard's fast fun taken to theatrical excess.

Souvenir tea towels today tend to show windmills, reeds, yachts and birds; have become, in Löfgren's postcard distinction, 'scenic' rather than 'comic' (Löfgren 1985: 90; Orwell 1961). In 1993 a Potter Heigham shop, since demolished, sold a different souvenir, 'We Survived the Norfolk Broads', garish red and yellow fun, topographically as well as socially low humour. Eyes down, look in. A windmill shines over spilt ice cream, a dozing camper, an errant angler, a randy lepidopterist, a thirsty hiker. Those here for nature are cross or absurd, foamy beer toasts other fun, a boat careers in 'Gay Abandon'. Quiet enjoyment could not survive here. The Broads become licensed premises, though this humour may itself be running the rule over holiday lives less proper. If, as Andy Medhurst dissects in *The National Joke* (Medhurst 2007), comedy can be intimately bound with national identity, the tea towel points to regional humour, with the laughs on leisure. In 1935, Broadland as theatrical carry-on was conveyed by *What to Do* author Drew Miller in *Seen From a Windmill: A Norfolk Broads Revue*, 'Revue' releasing Yarmouth variety sensibility, and Miller conjuring a comic and voyeuristic Broadland. In 'The Adventure of the River Censor', a prurient 'self-appointed censor of the Bure' assessed the 'modern bathing suit' through 'high-powered binoculars', penning 'luscious articles on the subject of "Nudism on the Broads". These were eagerly read by many, who got great enjoyment from the shocks they received' (Miller 1935: 112).

Companies have traded on the comic, perhaps alluded to in the BA's 1989 judgement that: 'In the past, some holiday companies have marketed Broads holidays with scant regard for the unique character and fragile nature of the area, encouraging visitors to come to the Broads for the wrong reasons' (Broads Authority 1989). Thus Blakes' 1922 *Broad Smiles. Or How Not to Do a Norfolk Broads Holiday*, complementing their *How to Do the Norfolk Broads and What to Take*, gave music hall caricature of motor hogs, lecherous binocular men, jazzing youths, hopeless sailors (Blakes 1922). In 1948 Blakes published Dennis Rooke's *Let's be Broad-Minded!*, a 'Bunkside Book of Brighter Yachting' (Rooke 1948); *Let's be Broad-Based: The Bunkside Book of Brighter Motor Cruising* followed in 1964 (Rooke 1964). The consumption of alcohol is assumed throughout. In *Broad-Minded!*, couples and families share cabin-fevered boat space, with stereotypes paraded; the 'running commentator', the 'nautical-vocabulary critic', the 'fashion-parader', the antiquarian enthusiast for 'q. t. f.-t.' churches (quaint thatched, flint-towered) (Rooke 1948: 93). In his church connoisseur, Rooke offers knockabout comic equivalent to Roland Barthes' mid 1950s analytical critique of the touristic *Guide Bleu*, which 'hardly knows the existence of scenery except under the guise of the picturesque', views the human world in types, and where 'one travels only to visit churches' (Barthes 1972: 74–5). Barthes' 'Blue Guide' essay appeared within *Mythologies*, his critique of the 'general semiology of our bourgeois world': 'What I claim is to live to the full the contradiction of my time, which may well make sarcasm the condition of truth' (11–12). Rooke's publications, with their deployment

Figure 14 'We Survived the Norfolk Broads', tea towel, purchased 1993. Source: photograph by the author.

of satirical stereotypical excess, issued in part as lure for the tourist who might act in self-conscious distinction or parallel, indicate sarcasm's presence within the economy of middle class pleasures, in leisure's reflexive pursuit.

Another take on regional pleasure appeared in 1932, John Arrow's novel *Young Man's Testament*. Arrow, who would later lobby for a Broads national park and anthologise *The Pleasures of Sailing* (Arrow 1951), had in 1930 written a defence of DH Lawrence (Arrow 1930), and *Young Man's Testament* offered a Lawrentian Broadland, attempting to sidestep both prurience and sauciness via a serious sensuality. Three couples cruise from Essex to the Broads and back again. Boat owner Jocelyn, controlling and organisational, loses sensualist girlfriend Cecily to Mark, who has been held back by repressed girlfriend Gertrude, who falls for Jocelyn. Thelma (Cecily's sister) and Jim share a cabin as lovers, but fall out, Thelma later pushing Jim overboard to drown at sea: 'by imposing its restriction of space, the boat had accelerated the emotional changes that had lain dormant' (Arrow 1932: 177). Gertrude, against pre-marital sex, is 'an absolutely indefatigable sightseer': 'To-day she had bought some post cards. She rather liked them. They were photographs of the broads, produced with a refined and devastating sterility' (141). The tourist gaze represses, Gertrude and Jocelyn 'examples of over-civilization set down in an environment almost utterly pagan' (188). Mark and Cecily are contrastingly heroic, beyond any river censor, realising Broadland and themselves through open-air sex at Potter Heigham. Cecily dances naked in the field for Mark, though crouches as walkers pass, freedom still hitting social limits: '"Do you think they saw?" "No, no, they were deep in conversation about bacon, or something"' (170).

Potter Heigham

A September 1933 postcard, 'On the River Thurne, Potter Heigham', sent home to Yarmouth ('I could not write this last night as it simply poured with rain after tea'), shows a riverbank quietly bungalowed, with yacht and mill sails. Served by a riverside rail halt since 1877, the interwar development of Herbert Woods' Broadshaven boat-yard, and of riverside chalets on the rond, the bank separating river and marsh, made Potter Heigham a touchstone not only for novelistic excess but for planning debate, akin to the plotlands discussed by Hardy and Ward in their *Arcadia for All* (Hardy and Ward 2004; Matless 1998). Versions of pastoral clash on the riverbank, shack idyll versus riparian order; Arrow would condemn Potter Heigham's 'unsightly bungaloid growth which, in the best interests of Broadland, must be removed' (Arrow 1949: 98), while the 1947 Broads Conference concluded: 'Potter Heigham is a lesson in what not to do on the banks of the Broads' (Broads Conference 1947: 5). Broadland planning is considered below; here we examine the landscape culture of those making the Potter Heigham riverbank, which offers a study in the production of vernacular landscape.

Herbert Woods (1891–1954), born in Brundall, lived in Potter Heigham for most of his life (Woods 2002). Woods' father Walter bought the former Norfolk Broads Yachting Company premises by the bridge in 1922, Herbert taking over in 1929, building up the hire fleet to 154 craft by his death. In 1930–1 two acres of marsh were

Figure 15   Potter Heigham, River Thurne. Source: Top: postcard, dated 1933; Bottom: photograph by the author, August 2011.

dug by hand to 8–10 feet for Broadshaven marina, on the north bank downstream of the medieval bridge, Woods promoting modern leisure under the motto 'Progress Broads-Haven'. Arrow, a yachting friend, saw a 'stimulating example' of possible 'nodal point' development (Arrow 1949: 98). A three storey Tower Building, with 'HERBERT WOODS' and 'BROADS-HAVEN' around the top, became a landmark advertisement. The interwar leisure landscape also featured a branch of Roy's of Wroxham, and Gerrards stores, cafe restaurant and assembly rooms, with 'Jolly Dances every Tuesday and Thursday during the Season' (Miller 1937: 58). Woods bought and demolished the village manor house, half a mile from the river, in 1934, designing its replacement, 'Cringles', sprawling with external beaming, four car garage and front garden sign stating: 'This is Potter Heigham and the River is straight ahead' (Woods 2002: 53–4). Potter Heigham is marked with self-made success.

Leisure modernity also came from Broadshaven guests. Variety and film star George Formby, in Yarmouth for summer seasons, cruised with wife Beryl on the 'Lady Beryl', designed and built by Woods, with a smaller runabout 'Baby Beryl' (Randall and Seaton 1974). Formby prepared for his 1951 West End hit *Zip Goes a Million* at Broadshaven, inviting staff to the London theatre, adlibbing yard references into the show (Woods 2002). Formby also owned a Wroxham riverside home, 'Heronby', which, scheduled for demolition in 1991, saw unusual heritage protest, when George Formby Society members 'took along their ukeleles and sang Formby classics on the riverside behind the boarded-up luxury home' (Eastern Evening News 1991).

The Thurne bungalow vernacular, a mile either side of Potter Heigham bridge, received varying valuation, from planner critique to Arthur Patterson's reflection: 'A huddle of sorry sheds and huts is springing up here like thistles and nettles, but why be critical? The poor man needs his cottage, while the rich man has his halls' (Patterson 1930a: 31). Di Cornell's *A History of the River Thurne Bungalows*, celebrating 60 years of the River Thurne Tenants Association, reflects on debate, including Day's 1950 *Marshland Adventure* commentary on a 'bungaloid slum': 'It appears that Mr. Wentworth Day did not find the bungalows endearing. No doubt Mr. Wentworth Day had a splendid brick built house in some quiet leafy suburb of "Middle England"' (Cornell 2008: 60). Day termed the bungalows 'a disgrace to the Broads', though also found Horning's well-appointed riverside dwellings 'out of place' in their 'smugly suburban trimness' (Day 1950: 77). Cornell traces the first bungalow to 1892, with most appearing before the 1932 and 1947 Town and Country Planning Acts, some finding continuities of ownership and use for holidaying or longer term dwelling. Names personalised plots, including: Bathurst, Owl, Carlo, Mindanao, Lazidays, Happidays, Jollidays, Mi Style, Crystal Haven, St Elmo, Linga Longa. If most were relatively standard constructions, a common postcard subject was 'Dutch Tutch', part of the helter-skelter from Yarmouth's Britannia Pier, brought to Potter Heigham in 1910. Agencies, including Blakes, Hoseason and Jack Robinson, let bungalows from the 1930s. Electricity and sewage installation was completed after World War II. In 1968 a Tenants Association survey recorded 204 bungalows and chalets, 2 caravans, 2 inhabited windmills, and 21 houseboats.

The Association's formation followed River Board discussion on tenancy termination (EDP 1948a; 1948b; 1948e); successor landowning and planning bodies the Anglian Water Authority and the BA also considered erasing the bungalow landscape. When the BA's 1983 *Potter Heigham Bridge Local Plan* proposed bungalow removal and re-landscaping, the Association co-ordinated public protest, presenting embattled small owners combating powerful planning and conservation interests (Cornell 2008: 70–82). Riverbank owner Anglian Water sought to sell as one freehold property, and argument continued into the 1990s, with demolition plans overturned, and a Tenants Company buying a 99 year lease. Property values increased overnight, many bungalows subsequently being improved. Riverside property settles into propriety, and a different re-landscaping proceeds.

Potter Heigham's vernacular landscape also saw the performance of vernacular regional identity. Bridge garage owner Sidney Grapes (1888–1958), who let 32 lock-up garages to bungalow tenants, performed on stage and radio, in smock or dress, as dialect comedian, and from Christmas 1946 Grapes sent dialect letters to the *EDP* as the 'Boy John'. Two posthumous collections were published, Mardle introducing 'Sidney Grapes (Humorist and Philosopher)' (Grapes 1958a: 3–6), and Grapes chosen to frame an advertisement highlighting the *EDP*'s cultural and political range: 'Dialect or Dialectics' (Grapes 1958a: 65; 1958b: 65; 1974). Grapes' adult male 'Boy John' ('boy' being dialect for 'man') wrote of himself, 'Aunt Agatha', 'Granfar' and 'Oul Mrs W', affectionate parody recognised by those whose voices carried family resemblance, though potentially impenetrable to those at cultural remove. Granfar attends the annual St Benet's service, and sails with young ladies at Potter Heigham regatta, while Boy John meets Mardle at Hickling, who 'wornt a bit like what I thort he wuss like. He hearnt got no trousers on (he'd got sum shorts on), an he looked an torked juss like a yotten gentleman' (Grapes 1974: 76). When 400 *EDP* dialect letters published in early 1949 were issued by Mardle as *Broad Norfolk*, the Boy John penned an accompanying advertisement for Norwich Union insurance: 'Du Yow Know Wot, Tergether? – iv'ry time I see a pitcher of yar Norridge Cathedral, I fare as 'ow I think abowt insurance' (Mardle 1949a: 12).

Dialect performance shifts between regional cultural truth, ironic identity play, and stereotype to trade on. The sense of the almost lost being retrieved recurs; from the salvage work of Moeran and Lee, to *EDP* letters on lost words, to Mardle in 1967 finding dialect 'undergoing one of its periodical discussions', with local comic performers following Grapes, Norfolk's 'Singing Postman' Allan Smethurst in the pop charts, and Arnold Wesker's 1959 Norfolk dialect play *Roots* revived in London. Mardle noted that dialect revival had 'even provoked a group of young humorists at Beccles to invent, in mockery of the folk-lore societies, a rural sport called flonking the dwile', teams dancing and hitting one another with beer-soaked towels, parody in the vein of Kenneth Williams' 1960s national broadcasting comedy rustic 'Rambling Syd Rumpo', with his ludicrous rusticity of cordwanglers, moulies and grussetts.[31] Mardle wondered 'whether, when a dialect becomes as self-conscious as this, the very desire to preserve it is not a confession that it is dying'. Dialect was 'best spoken by those remaining country people who are unconscious of speaking Broad Norfolk, because it has never occurred to them to speak any other kind of English – and they are getting fewer every year' (Mardle 1967).

Grapes could also combine dialect and standard English for political effect. In the 1950s Norfolk Highway Committee planned to demolish Potter Heigham's medieval bridge, though railway closure in 1959 would allow the A149 to be re-routed along the old track, with a new Thurne road crossing, and the old bridge saved. Proponents of demolition cited congestion and accidents, opponents cited picturesque amenity, flood control, and conservation, the low bridge barring large craft from passing upstream (EDP 1953e; 1953f; 1953g; 1955b; 1955g). Grapes, with his bridge-side garage interest, wrote to the *EDP* under his own name, with 'Our Bridge' a 'historic asset': 'A local man who has spent all his life by the river said, "This here oul bridge she act as a sorter damper tew tha' tides. Teark har away, an shuv a new 'un up, an yow'll fine as how yow'll hev Hicklen flooded all winter, an in tha summer, there ount be enuff water ter float a little but, let alone a big 'un."' (Grapes 1953). Potter Heigham's proprietorial performer deploys dialect for hydrological wisdom, defending historic landscape, trading on bridge traffic, performing regional life.

Policy Conduct

Broadland as national park

From the 1930s, planning helped define Broadland, public authority seeking to guide and control private interests and leisure activity. Status equivalent to a national park was achieved in 1989, discussed in Chapter Six below, but the national parking of Broadland was debated from the 1930s, tensions of scale and questions of conduct shaping regional policy (George 1992; Ewans 1992). Governmental deliberations, including the Addison Committee (1931) (Mair and Delafons 2001) and the Scott Committee on Land Utilisation in Rural Areas (1942) supported a Broadland national park, though John Dower's key 1945 Ministry of Town and Country Planning report placed the Broads on a reserve list, citing complications of authority: 'both of drainage, navigation, etc., and of existing misuses and disfigurements ... It may prove better to deal with the area on some *ad hoc* scheme of combined national and local action, which should include the protection of substantial areas of mere and marsh as strict Nature Reserves' (Dower 1945: 9–10; Matless 1996). The sense of a distinctive region demanding distinct regulation would shape Broadland authority for decades.

Geographer Vaughan Cornish proposed a Broads national park in 1930, citing potential educational use in 'the New Geography' (Cornish 1930: 60). Regulation of conduct was however required, Cornish reporting for the Council for the Preservation of Rural England (CPRE) to the Norfolk (East Central) Joint Planning Committee in 1933 on a 'vulgarity of noise': 'the circumstance that smooth, inland waters lend themselves to holiday pursuits has brought the blare of gramophones to the once silent Broads' (Cornish 1937a: 78; 1937b: 97–102). Hullabaloo commentary achieves policy translation. Despite Dower's reservations, the July 1947 *Report of the National Parks Committee (England and Wales)* recommended 'The Broads' as 'a potential National Park which seems to belong to another world', offering 'ecology and natural history' and 'quiet adventure'. The 'less pretentious type of sailing holiday' might be encouraged, with motor cruisers limited (National Parks Committee 1947: 114–18). Minister of Town and Country Planning Lewis Silkin toured in October 1949, inspecting between Wroxham and Potter Heigham one day, and Reedham and Beccles the next, assessing park prospects (EDP 1949f; 1949g). Modern national parked landscape was outlined by Arrow in the *Architectural Review*, with Hickling Staithe already exemplary, 'functional, beautiful, ship-shape, simple and in every way a fit object for its job and position' (Arrow 1949: 96). Leisure might nurture 'self-reliance': 'There are no mechanical amusements in Broadland whatever. No cinemas, no pin-table saloons, no 'popular' restaurants, and only one dance hall, unsophisticated and not much patronized by the visitors' (94). Arrow proposed tree clearance to 'free the wind' (98) for sailing across an open 'real Broadland' (89); drawings by the

Architectural Review's townscape artist Gordon Cullen evoked a landscape restored, the Bure opened to pleasure breeze.

National park debate sparked tensions of conduct and authority. Norfolk County Council's 1945 and 1947 Broads Conferences suggested leisure zoning of 'Cinemas, Dance Halls and such like': 'Some of the visitors naturally want this sort of entertainment, and some would do anything to get away from it and be quiet … such facilities should be grouped in certain places …, and so enable the two tastes to be met' (Broads Conference 1947: 5). Noting national parks abroad, the Conference suggested: 'The term "National Park" as we understand it does not fit an area like the Broads' (6), and proposed a co-ordinating Broads Committee instead. Horsey landowner-naturalist Anthony Buxton drafted the summary report, also published in Jarrold's 1948 *What to Do on the Broads*, and wondered if the National Park as 'Pleasure Park' might paradoxically lessen holiday freedom, having 'a (prophetic) vision of Government notices, Government forms, Government posters and Government "ologists"' (Buxton 1950: 156–7). Day went further, detecting 'the insidious beginning of land-nationalisation', and its 'only addition … to the fauna of Broadland … a new race of rats – bureaucrats' (Day 1950: 5–6).

Day's was an extreme view. The prevailing tone in intense county debate was welcome for the general national park idea, but wariness of national authority; Mottram saw local bodies seeking 'to "National Park" themselves before authority in London could move' (Mottram 1952: 241). Exploring 'The Future of the Broads: National Park, Private Reserve or Commercial Exploitation', Mottram preferred the park, though stressed: 'No one wants to "nationalise" the Broads: "The Broads for the People" is nonsense' (246; Mottram 1951). A geographical tangent to discussion came from May, proposing 'a Norfolk Broads Association, with its roots among Broads holiday-makers all over Britain' (May 1952: 186). Instead of 'a local movement trying to grapple with problems of national size', here would be 'a *national* body vitally interested in the good of a locality', campaigning for a co-ordinating 'Broadlands Conservancy' (188–9). May's idea came to nothing, perhaps unable to conjoin committed care and carefree holidaymaking.

The national park itself stalled from concern over cost and multiple authority, with the 1958 Bowes Committee of Inquiry into Inland Waterways, accepted by government in February 1959, favouring 'reconstruction of the Port and Haven Commissioners on a more representative basis', with a national park not 'appropriate' (Committee of Inquiry into Inland Waterways 1958: 73). Only the Youth Hostels Association and Inland Waterways Association (IWA) continued a national park campaign, the YHA presenting a 1200 signature petition in January 1960 (EDP 1960a). The National Parks Commission stated its support for Bowes in December 1959, and a parliamentary announcement in August 1961 confirmed that the Broads would not become a national park (EDP 1961a; Nature Conservancy 1965: 96).

Opening waters

Post-war debate responded to diagnoses of Broadland waters threatened by vegetational encroachment (discussed in Chapter Two above), neglectful management, and private closure. The National Parks Committee noted: 'The most serious threat to the

Broads as a holiday area is the encroachment of aquatic vegetation on the open water ... The only remedy is in clearing and dredging' (National Parks Committee 1947: 117). Encroachment, accelerated by wartime closure, accompanied navigational restriction, Arrow offering 'A Study in Neglect': 'Remember, where our fathers sailed there are reed beds now. Unless we take action, our sons will find reed beds where we are sailing to-day' (Arrow 1948: 57). Photographic sequences showed 'Decay', with gated entrances, fallen trees, sunken hazards, and 'Neglect', with weed growth, channel choking, and an 'unfit for navigation' sign (Arrow 1949: 90). Diagnosis of neglect prompted calls to restore Broadland, River Board Chief Engineer SW Mobbs estimating in 1948 that £965,000 was needed to restore 1928 conditions (EDP 1948d; 1949a; George 1992). Mobbs' high figure, meant to spark action, contributed to concern over national park costs, the 'picture of a vast work of reclamation to be undertaken' having 'scared the Commission of the whole idea' (EDP 1958b).

Alongside general diagnosis came action over specific waters. In 1954–5 the British Transport Commission planned to close the New Cut due to maintenance costs after 1953 flood damage; its Private Bill was defeated in parliament, and campaigns to 'save the cut' brought a ten year reprieve in November 1956 (EDP 1956d). Individual waters were opened; in August 1952 the Norfolk and Suffolk Broads Yacht Owners' Association dredged Rockland Broad, part funded by brewers Bullards, owners of the New Inn, Rockland (EDP 1952a), while in 1958 the Chet navigation to Loddon was restored, the town back 'within the framework of the Broads' (EDP 1958a). Three boatyards opened there by 1960, 5000 cruisers coming up river in the first two years (EDP 1964c). The IWA unsuccessfully proposed similar navigation restoration to Aylsham, Bungay and North Walsham, to put 'three Broadland towns back on the waterway map' (Edwards 1955).

Open water argument also shaped the 1956 formation of the Broads Society, still a notable regional voice, prompted by an *EDP* letter from Leonard Ramuz of South Walsham: 'In response to approaches made to me by various people, I am drawn to the conclusion that there should be a society whose main object would be the preservation and expansion of our rivers and Broads in Norfolk and Suffolk, or alternatively that a local branch of the Inland Waterways Association should be formed with the same object' (Ramuz 1956). A meeting at the Norfolk Broads Yacht Club, Wroxham, on 4 June, established the Broads Society, IWA Secretary Louis Edwards' counter-proposal defeated by one vote. A steering committee included Ramuz as Chair, Ted Ellis and James Hoseason (EDP 1956b); Ramuz's successor, PV Daley, presented the Society as 'a sort of consumer council for the lovers and users of the Broads' (EDP 1957b).[32] The Society, with 750 members by 1970, articulated a genteel sailing culture at its heart, expanding waters yet controlling development (EDP 1956c). Navigational expansion was pursued through a 'Sub-Committee for More Water'. If the Society's symbol, designed by Hickling artist Roland Green in 1958, was a heron, and its magazine *The Harnser* (heron in dialect), naturalists played a minor role, although after retirement from the Nature Conservancy in 1990 Martin George became a key figure, serving as Chair from 1998, the Society increasingly acting as environmental watchdog. In early decades however boating interests dominated, though other leisure uses could be opposed, including 'Amphibious Caravans / Floating Mopeds' (1973), and Enfield Borough Council's plan for a children's field study centre at the Barton Angler Hotel.[33] Access to 'more water' was conditional on conduct.

'More Water' could however be interpreted as the Society straightforwardly calling for 'More Leisure', with Ramuz's initial *EDP* letter bringing a satirical reply from CA Boardman of How Hill: 'Always supposing I obtain enough support, I intend to found a "Society for the Preservation of the Broads against Broads Preservation Societies." ... These societies are really only supported by people who make a living out of desecrating the Broads with eye-sores with propellers' (Boardman 1956).[34] While William Mayes replied that Boardman, 'ridiculous rather than humorous', missed Ramuz's emphasis on quiet sailing (Mayes 1956), others (Betts 1956; Ruffell 1956; Perrin 1956) took Boardman at his word:

'As an angler and a lover of our Broads and "quiet" waters, I should like to associate myself with Mr. Boardman's proposal' (Arthur Betts, Norwich)

'As a visitor to Norfolk of many years' standing, and as a lover of the quieter rivers and Broads, I write to support Mr. Boardman's suggestion' (Jean Ruffell, Maidstone)

'I am sure his Society will find many members' (PM Perrin, Hickling).

The opening of waters also raised access rights. On 25 November 1947 Edward Evans, Labour MP for Lowestoft, tabled a Commons motion on Broads 'deterioration', attacking 'the most iniquitous and arbitrary enclosure of certain broads and waterways by private persons, riparian owners and dwellers by the waterways', with over 550 acres of open water lost since 1892: 'It is a very easy thing to do. A chain is thrown across a narrow waterway, or a heap of stones is gathered and weeds inserted.' National Liberal MP Brigadier Medlicott (Eastern Norfolk) cautioned against blaming private owners, with Minister Lewis Silkin concurring, though Silkin shared Evans's concern at 'wilful neglect'.[35]

In March 1949 *Picture Post* sent Fyfe Robertson to cover 'The Battle of Blackhorse Broad'. Herbert Woods and the Broadland Protection Society (Secretary, John Arrow) led the removal of a barrier to Blackhorse (Hoveton Little) Broad, closed for 40 years. Landowner TRC Blofeld had replaced an earlier chain barrier with ten foot wooden piles. Eight men laboured, 'Drawing the Teeth of Doubtful Privilege', restoring 'water liberties', Woods and company sailing around the opened Broad: 'It was all very British and very pleasant – and, a lot of law-abiding people think, very necessary' (Robertson 1949: 16–17). History was enlisted, Woods having issued a January 1949 pamphlet on 'The Fifth Freedom', for the return of the Broads from private 'water lords', citing the 400th anniversary of Kett's Rebellion, the 1549 Norfolk revolt against early land enclosures, and common reference point for assertions of Broads access and Norfolk independence (Woods 2002: 178–83; Dutt 1903: 128). Woods tapped an official post-war mood of Kett celebration, a plaque unveiled at Norwich Castle in 1949 asserting the justice of Robert Kett's cause and his unjust execution.[36] Woods and *Picture Post* quoted Kett: 'in his petition to the King he prayed "that ryvers may be ffre and common to all men for fyshinge and passage." What a king wouldn't do, a court might' (Robertson 1949: 17).

Kett also fitted the Black Horse 'battle' in having been a yeoman rebel; this was not a revolt of the waterless, and businessman Woods' trespass laid down a marker to broad owners, rather than challenging property. The case was heard in Chancery but settled out of court, Woods and Blofeld signing an agreement, witnessed by Anthony Buxton, giving Easter–September access and retaining owner's sporting rights (EDP 1949c; 1949d; Buxton 1950: 159; Woods 2002: 177–8). The agreement was later taken over by the Broads Authority, the Blofeld family still owning the Broad (George 1992: 53). May would reflect on Black Horse as a trespass producing inadequate compromise, the BPS a 'boat-owner's baby', with 'privilege sat tight': 'the wind had mysteriously gone from the sails of the invaders' (May 1952: 184–6).[37]

Planning a leisure society

In 1965 Michael Dower wrote of a 'Fourth Wave' of urban pressure breaking on the countryside, after industrial towns, rail development and car-based suburbs: leisure (Dower 1965). An assumed future of technological advance, full employment and increased leisure time generated 1960s debate over leisure landscapes, notably through the 'Countryside in 1970' conferences under the Duke of Edinburgh's patronage (Cosgrove and Jackson 1972; Matless, Watkins and Merchant 2010). The prospect of a leisure society sparked excitements and laments over the opportunities and constraints of a recreational mainstream. By 1971 David Bowie could place the regional name as a recognisable rhyme in 'Life on Mars': 'See the mice in their million hordes, from Ibiza to the Norfolk Broads'; in 1969 Tim O'Riordan could write of the region: 'This is an exciting time for amenity planners' (O'Riordan 1969: 40).[38] Dreams and fears for leisure Broadland can be tracked through unrealised 1965 plans, floated by developers testing profitable waters:

> Boat business Jenners of Thorpe planned an eight acre holiday development on Thorpe island between the Yare and its side channel, with seven storey flat and office blocks, boathouses, moorings and clubhouse. (EDP 1965m)

> Breydon Marine Developments (Holdings) proposed a £1million scheme for 86 acres of Crown Estates mudflats near Burgh Castle, with yacht harbour, chalets, restaurant, club-house and motel: 'Breydon Water, in this day and age, is tragically under-developed, both in terms of beauty and commerce'. The EDP commented that Breydon might be under-developed not 'tragically, but mercifully.' (EDP 1965n; 1965o)

> 'Rural Developments (Norfolk) Ltd' envisaged a 100 acre site between Acle Dyke and Acle New Road, with five storey modernist 'boatel' by the Bure, plus golf course, bowling alley, cafes, shops, caravan park and sailing school, bringing 'up-to-date facilities' to the region, with a 'motorway-type service area' on the A47. The EDP commented: 'In that never-never land where plans proliferate but fulfilment lags there must be growing up a city of unbuilt hotels and unconstructed yacht marinas.' (EDP 1965h)

What might Broadland have been like had such plans been realised?

Development plans were assessed from 1949 by the Broads Joint Advisory Planning Committee, criticised by conservationists for permitting upstream development on smaller rivers, notably the Ant, where increased Stalham hire traffic traversed Barton Broad (George 1992: 440–3). If the Nature Conservancy opposed such development, they also however became the conduit for a planned leisure future through their 1965 *Report on Broadland*, whose plans for new broads and a Bure–Yare cut were considered in Chapter Two. With the national park proposal dormant, and at a time of calls to planned order, as when Norfolk Deputy County Planning Officer RS Hookway called for a Tennessee Valley Authority for the Broads (EDP 1963h; 1963i), or the *EDP*, alluding to the influential 1963 Buchanan report on *Traffic in Towns*, called for a 'Buchanan afloat' (EDP 1963g), the NC was a national public body offering regional oversight.

The *Report*'s remit was to 'assess the ability of land and water to withstand different forms of use' (Nature Conservancy 1965: 7), with a Working Party representing local authorities, navigational, boating and agricultural interests, and including NC Regional Officer Bruce Forman and assistant Martin George. The NC extolled 'multi-purpose use' (24), the *Report*'s cover bringing blue, yellow and green cut out shapes of yachts, cruisers, fishermen, cars, windmills, eggs, birds and bridges into bright user harmony. An appendix 'Chart of Human Impacts on Broadland' proposed spatial or temporal zoning of incompatible activities (57–8). 'Piecemeal palliatives' were insufficient: 'To do nothing is to abandon the region to erosion, conflict and decay … East Anglia pioneered the agricultural revolution. There is now an opportunity for Broadland to be a pioneer in Great Britain in the multipurpose use of land and water on a large scale' (60; Forman 1966). A Norwich Castle exhibition, 'Focus on Broadland', conveyed the *Report*'s message for two months, its ethos receiving general welcome (EDP 1965i; 1965j; 1965k; 1965l; Mardle 1965).

Emphasis on 'multi-purpose use' was in part a response to controversy over the NC's March 1963 draft report, where an image of naturalists as restrictive had prevailed. The NC undoubtedly perceived nature to be threatened, former East Anglia Regional Officer Eric Duffey arguing in 1964 that: 'The stage has now been reached when the intensity and duration of the boat traffic, new types of craft, and new activities, including the use of speed boats and water skis, constitute a serious problem for conservation interests' (Duffey 1964: 292). The draft report's publication produced headlines of conflict, a 'Battle of Broads "getting worse"' (EDP 1963a), with reception turning on the NC's perceived connection to the region. If the *EDP* suggested the NC did not deserve 'to be convicted of scientific intolerance', and were 'reminded of Rachel Carson's "Silent Spring,"' (EDP 1963b), others were hostile, James Hoseason upholding the 'successful integration' of local interests: 'Time and time again the only odd man out has been the Nature Conservancy, who do not share the same interests as 95 per cent of river users … Presumably they prefer to see the Broads abandoned to uncontrolled wild life' (EDP 1963c). The Broads Society too saw conflict overemphasised: 'Mr. Nat Bircham, of Wroxham, said Norfolk people did not need outsiders coming in to tell them what was needed' (EDP 1963f).

The NC however already had local institutional connection, with Duffey, Regional Officer until 1962, serving as NNNS President in 1957–8, and presenting the draft report at the Castle as 1963 President of the Norfolk Research Committee (EDP

1963d). In January 1964 the NC met 12 local organisations to shape a Working Party, the national nature state making itself an appropriate vehicle for regional survey (EDP 1964a; 1964b). The EDP stated:

> from its contacts both official and informal, at local level as well as nationally, has come the looked-for break-through. From being 'those scientists,' little understood and often disgusted, it has now won widespread sympathy and respect, not just from a woolly and idealistic public, but from those who count: the farmers, the countrymen, the fishermen and wildfowlers. (EDP 1964e)

The 1965 *Report* led in early 1966 to a Broads Consortium of local authorities and river and navigation interests. Regional debate again however enacted tension between planning enthusiasm and the fending off of national authority. For Mardle the choice was 'between local co-operation and national intervention – possibly by some future Minister of Recreation, when it is too late for him to do anything but consummate the ruin of the Broads' (Mardle 1966).

The Consortium's 1971 *Study and Plan* included Norfolk County Planning Officer RI Maxwell's summary of regional leisure. Maxwell mapped 'Broadland in its Regional Setting', a south-east dotted with stars marking New Towns and overspill schemes, 1.25 million living within 50 miles, 12.5 million within 100, a crowded Broadland orbit. Forty-two per cent of hirers were from London and the south-east, with 'significantly large numbers of groups of "young people"' (Broads Consortium 1971: 23–5). A boat census revealed 'Acceptable Capacities' exceeded on a peak Sunday on the Bure between Wroxham and Thurne Mouth, on the Ant up to Barton Broad, and the Thurne through Martham to Somerton. To frame the crowded future, Maxwell established a landscape planning typology of 'landscape elements and areas'; 'enclosed valley', 'open valley, 'extensive open' and 'open' (4–5). A new 'Broads Authority' would shape the scene, with powers covering navigation above Yarmouth, drainage, planning and nature conservation, effectively anticipating the later BA (48–9). The Consortium itself however disagreed, suggesting navigation powers be invested in an amalgamated River Authority and River Commissioners. A dissenting memorandum, supporting Maxwell, was added by the GYPHC, ironically a body who would later obstruct the ambitions of a new BA, but whose powers the Consortium challenged. The GYPHC envisaged something 'similar to the Lee Valley Regional Park Authority', with navigation the priority, nature 'a secondary concern', and Broadland 'the boating and recreational "water park" of East Anglia' (GYPHC 1971: 12–13). Plans for planning authority articulate contrasting landscape visions.

Hullabaloo displacement

The Broads Authority was formed as a coordinating body in 1978, assuming statutory powers equivalent to a national park in 1989. BA formation and policy is considered in Chapter Six; this chapter concludes with the BA's narration of regional leisure history, questions of conduct continuing to shape policy.

In 1993 the BA stated: 'One of the fundamental principles of the Broads Authority is the promotion of quiet enjoyment of the area.' If the 1947 Broads Conference had planned to zone dance halls, the BA imagined quiet enjoyment blanketing the territory: 'large scale facilities and urban-style entertainments are difficult to imagine fitting in either physically or in character' (Broads Authority 1993: 100–1). A sense of post-war regional cultural decline underpinned this quiet mission, Chief Executive Aitken Clark writing in 1989:

> If on a dank and misty afternoon in autumn in the late 1970s, you had driven from Norwich along the 'Acle Straight', the road used by thousands of holidaymakers heading through the Broads towards Great Yarmouth, you might have seen little romance in the mist, and nothing but desolation in the flatness and in the derelict windpumps. If you had travelled to Wroxham and taken a diesel-powered launch along the River Bure, you might have wondered where you were supposed to find peace and tranquility.
>
> Indeed, in the sixties and seventies, the candyfloss and seaside atmosphere of chip shops, amusement arcades and souvenir shops selling trinkets and kiss-me-quick hats seemed suddenly to outweigh the quiet reedy reaches, where bitterns boomed unseen and the sails of a lone yacht drifted across the horizon. You could have been forgiven for believing the scare stories and blaming the heavy hand of commercialism that had taken its toll, exploited and spoilt for ever this fragile enchanting land. Once perhaps it may have made the grade as a national park – but alas, no longer.
>
> Perhaps it was too late for the Broads. (Clark 1989: 2)

The BA was to ride to regional rescue, saving (in its late 1980s slogan) a 'Last Enchanted Land'. The passage above is from Clark's contribution to a volume of Richard Denyer's photographs, itself a carefully pluralist aesthetic of Broadland life and landscape, with landscape quirks celebrated and Cantley factory on the front cover (Denyer 1989). After Clark's piece comes a critical essay on 'representation and reality' by historian Sarah Knights, hailing Denyer's 'varied and conflicting images' (Knights 1989: 5). Academic commentary on landscape ideology cohabits with Clark's landscape ideal.[39]

For Clark certain conduct went against the regional grain, and recent history needed displacement. Other historic leisure could however be gathered into Broadland heritage, the BA promoting electric boat tours at How Hill as echoing the Edwardian Thames, postcards reprinting pre-war rail posters, and Blakes (then still an independent company) reissuing early brochures and posters in the 1990s to lend themselves tradition. Regional discovery was rediscovered, and Arthur Ransome too revived, *Coot Club* and *The Big Six* on sale in BA visitor centres, Roger Wardle's 1988 *Arthur Ransome's East Anglia* praising an acute sense of Broadland place (Wardle 1988; Wardle 2013), and Norfolk County Council's 'Arthur Ransome Trail' tracking Broads locations.[40] Clark indeed tapped Ransome against post-war decline:

> It was against this background that the Broads Authority was set up in the late 1970s. Many people who remembered the idyll of Arthur Ransome, the crystal clear water, the days when sails outnumbered motor boats and when the bittern's boom was more often heard than that of the ghetto blaster were sceptical. The task facing the Authority was certainly a challenging one. (Clark 1989: 3)

The BA in effect revived the social aesthetic of *Coot Club*, with some success; from 2010 the region would be branded 'Britain's Magical Waterland', with tourism gathered into a language of sustainability (Broads Authority 2011: 51).[41] Wardle suggested that to find 'the world of the Coots' it was 'best to visit the area out of the holiday season' (Wardle 1988: 8), though one could of course argue that to find Ransome Country one should really visit in peak season, to relive the *Coot Club* action and sabotage some revellers. Or perhaps acts of cultural restoration, promoting quiet enjoyment, had helped see off the Hullabaloos, their noisy joys displaced.

Icon I
Wherry

Emblematic Sails

In 1954, outlining 'Problems of the Broads' in the *Journal of the Town Planning Institute*, Kenneth Grimes suggested: 'The gravest concern the preservation and perpetuation of the two most characteristic features of the district which are threatened with extinction: the sailing wherry and the drainage windmill' (Grimes 1954: 59). Grimes picked out regional icons, wherry and windmill, whose cultural status deserves close attention, whether as signs of a working region, or of regional loss.[1] Sails of mill and boat become emblematic regional shorthand, both taking the air, moving with wind, technologies working nature. Wherry and windmill stand out as verticals in flat terrain, novelist and poet Sylvia Townsend Warner writing in September 1931 from a houseboat near Thurne:

> Meanwhile I am very happy here on a sleek river with a flat marsh landscape all around in which only yachts, windmills and wherries stand perpendicular. Wherries are beyond words lovely – a very broad low build, with one enormous oblong sail, sometimes a black one. The most solemn gait you ever saw.[2]

Elsewhere in this book, chapters move across iconic Broadland features such as the bittern and the reedbed; between the substantive chapters, two essay miniatures will elaborate on the windmill and wherry.

The Museum of the Broads at Stalham has the corrugated Perspex end of a building painted with wherry and windmill, reedcutter and broad; inside, a 2011 exhibit

In the Nature of Landscape: Cultural Geography on the Norfolk Broads, First Edition. David Matless.
© 2014 David Matless. Published 2014 by John Wiley & Sons, Ltd.

showed children's miniature box-models of regional landscape, including one of wherry, mill, yacht, butterfly and reeds (see Figure 3). Visitors face on museum entry the rear portion of the trading wherry 'Volunteer', steps up to the deck and down to the tight hold, where audio describes working wherry life. A small replica wherry, made in 1934 by Norwich schoolmaster FC Lynes, stands on the deck. Elsewhere 16 card model Broadland boats, made by Rev Hugh Edgell, include 'Trading Wherry c 1800 AD' and 'Wherry Yacht c 1900 AD', black-sail furled on one, white-sail furled on the other, a two man working launch and a three person pleasure party.

A wherry photographed under white-sail made the cover of the 1965 *Broads* New Naturalist. Inside, Horace Bolingbroke described 'Native River Craft of the Broads' – wherries, keels, punts, reed lighters – with wherries transporting 'grain, coal, timber, bricks, manure, marsh litter, ice and more latterly, sugar-beet', heavy loads in shallow waters, 40 tonners sailing the Yare, 25 tonners the Bure (Ellis 1965: 260; Malster 1971; 2003).[3] Bolingbroke had run Norwich's Stranger's Hall Folk Museum, opened privately by his father in 1900 and presented to the city in 1921, 'native river craft' forming part of regional folk material culture. Three hundred wherries sailed Broadland in the mid nineteenth century, working numbers falling from rail competition, with some converted for pleasure use, and other pleasure wherries built from scratch, the wherry crossing Broadland work and play.

Present Norfolk visitors may also encounter 'wherry' in a glass. 'Wherry' bitter, brewed by Woodfordes brewery at Woodbastwick, was awarded the Campaign for Real Ale's Champion Bitter of Britain in 1996 and 2005, winning Supreme Champion Beer of Britain in 1996. Under Woodforde's logo of a wherry under sail, a Broadland boat travels cross country. The microbrewing pursuit of cultural geographical authenticity deploys names of regional connection, familiar for the county trade, translatable further afield. Visitors may also walk the Wherryman's Way, a 35 mile Yare valley footpath, established in 2005, commencing at the 'Rockland Trader' sculpture at the 'Three Ways Meeting Point' outside Yarmouth's Vauxhall Station, just above the Bure–Yare confluence.[4] A black metal sail a few feet high is set in stone, with poetic inscription on wooden seating ('A thousand miles of flight / Sanctuary here for the weary bird'), for rest before or after the five mile walk along Breydon's north shore to Berney Arms. Rail travellers will have arrived via the 'Wherry Lines', moniker for the Norwich–Yarmouth–Lowestoft network, revived as a label in the early twenty-first century after earlier deployment under public owner-ship, British Rail having promoted 'The Wherry', serving Norwich–Yarmouth–Lowestoft, from the mid 1970s to the late 1980s (Adderson and Kenworthy 2010). Take a wherry for a ride, a walk, a sail, a drink.

Norwich Vessels

The wherry achieved regional and national recognition in part through painting, featuring in early nineteenth century 'Norwich School' landscape art, including the Norfolk river work of Stark discussed in Chapter One. The 'Norwich School' encom-passes artists of national reputation such as John Crome and John Sell Cotman, and

Figure 16   'Rockland Trader' sculpture, at 'Three Ways Meeting Point', outside Vauxhall Station, Great Yarmouth. Source: photograph by the author, March 2011.

lesser art historical figures such as Stark, Henry Bright, Joseph Stannard (whose 'Thorpe Water Frolic' was discussed in Chapter Three) and John Thirtle. For Andrew Hemingway 'Norwich School' lacks purchase as an art historical 'hermeneutic tool'; if regional subject matter was shared, with Broadland prominent, painterly style and artistic biography varied. Hemingway however shows the power of the term as an 'ideological phenomenon', through which regional elites shaped civic and county identity, notably through urban bourgeois claims to county affinity and authority (Hemingway 1988: 36; Hemingway 1979; 1992; 2000). The wherry on the river here becomes a signature scene for such work, denoting operative trade or occasional pleasures; sailing with cargo, pausing for loading, moored for the evening, approaching a staithe, engaged in frolic.

   The 'Norwich School' negotiation of Norfolk city and country became central to late nineteenth and early twentieth century Norwich Liberal cultural philanthropy, notably through the Colman family, especially Jeremiah Colman and his son Russell.[5] Jeremiah gave Stannard's 'Thorpe Water Frolic' to the Castle on its opening as a museum in 1894, and bequeathed 20 Norwich School pictures to the city on his death in 1898. WF Dickes' 1905 *The Norwich School of Painting* gave art historical emphasis to 'the relationship between the Norfolk environment and Norwich painting' (Hemingway 1988: 24), while the predominant naturalistic aesthetic allowed for a 'common iconography of Norfolk scenery' to emerge (Hemingway 2000: 16); wherries, lanes, cottages, rivers, mills. The Castle's Crome Centenary Exhibition of 1921 included 123 exhibits, 41 from Jeremiah Colman's collection, presenting Crome as solid, English and showing 'the stable elements in a landscape' (Binyon 1921: 18). In 1946 Russell Colman bequeathed 228 oils and 985 watercolours to the city, special Colman Norwich School galleries opening in the Castle for the Festival of Britain in 1951, still the chief public display of such art. Broadland river craft adorn city gallery walls.

The 'Broadland' emerging from such painting is however a very particular one, with the 'Broads' as labelled region, or the broad as specific landscape feature, barely registered. This was painting before commercial boating leisure, and the emphasis is on working rivers, the Yare a 'flourishing commercial channel irrigating a fertile agricultural region' (Hemingway 1992: 273), with trade from the coast important for Norwich's prosperity. The wherry is the key vessel, the river in Norwich a common subject, and Yarmouth both port and emerging pleasure resort. The ruins of St Benet's, discussed in the 'Windmill' section later in this book, are a regular subject, but other northern river imagery is uncommon. Images of broads are rare, with few evident in Castle collections or elsewhere.[6] Paintings instead highlight rivers and riverside structures, navigation and movement in the scene, wherries under sail or quanted with masts lowered.[7] Crome's 'The Steam Packet' depicts wherries tied up while a passenger steam launch passes, likely travelling from Yarmouth to Norwich having just crossed Breydon Water, the wherry sails towering verticals over steam.[8]

The Colmans not only cultivated the wherry as regional icon in paint, but commissioned their own craft. The pleasure wherry 'Hathor' was built by Halls of Reedham for Ethel and Helen Colman in 1904–5, its Egyptian-style interior designed by their brother-in-law Edward Boardman, owner-architect of How Hill, hieroglyphic symbolism matching its naming after an ancient Egyptian goddess, also the name of a Nile boat on which the sisters had sailed with younger brother Alan, who had died in Egypt. Hathor remained in Colman and Boardman use until sold to Claud Hamilton (of Hamilton's Guides) in 1953, and was later restored for Wherry Yacht Charter, its 1989 relaunch restaging, in conjunction with the BA, the original 1905 event, and celebrating the award of national park status, an appropriate leisure heritage (Bower 1989).

Restorative

Richard Denyer's 1989 *Still Waters* includes, alongside photographs of Hathor, wherries in ruin; the 'Wherry Graveyard, Bargate', with skeletons of sunk craft, and wherries sunk to reinforce banks at Barton and Wroxham (Denyer 1989). In the mid 1960s the trading wherry Maud was sunk by the NNT at Ranworth as a bank bolster. Moved when the conservation centre was built in 1976, Maud was restored at Upton by Vincent and Linda Pargeter from 1981, sailing again by 1999 (Pargeter and Pargeter 1990). Maud is one of two trading wherries restored to sail; the other, Albion, owned by the Norfolk Wherry Trust, tours the broads each summer from its Womack Water base, carrying school parties and members' cruises, mooring for public open days, maintaining 'one of the most well known icons' of the Broads.[9]

Albion was built in 1898, the Trust formed in 1949, an early example of the enthusiast preservation culture examined by Raphael Samuel as 'resurrectionism' (Samuel 1995: 139–68), salvaging technologies no longer working, mixing technical enthusiasm and regard for cultural tradition, working at a tangent to the modern world. In 1949 a few wherries still operated as dredgers, and some were powered by motor, but none carried cargo under sail. Albion was another Colman craft, in this case a recently working wherry owned by the firm, which Lady Beryl Mayhew, daughter of Russell Colman, helped secure for restoration.[10] The Trust brought together the Broadland establishment (Humphrey Boardman was first Chair) and enthusiast activists, with bookshop proprietor Roy Clark (1911–2008) a driving force; his 1961 *Black-Sailed Traders* examined Norfolk and Suffolk wherries and keels (Clark 1961). On 14 January 1949 an *EDP* letter headed 'The Norfolk Wherry', with 12 signatories including Clark and Boardman, noted an 'informal decision' for 'the forming of a Trust to buy a Norfolk wherry, and put her back in commission', hoping for the 'good will of merchants' to provide 'occasional freight', with prospective use also by Sea Scouts (Benham et al 1949). A public meeting in Norwich's Stuart Hall followed on 23 February. The Trust's history highlights the 14 January letter as the first public move, though a week earlier the *EDP* had carried a similar letter from M Sawbridge of Newton Flotman, taking correspondence on saving marsh mills (to which Clark had contributed) to suggest such ideas 'be extended to save these vanishing craft' (Sawbridge 1949). Sawbridge disappears from subsequent discussion.

Albion was restored with volunteer labour in summer 1949 (Norfolk Wherry Trust 1952). Lead was sourced from sunken wherries, Clark rowing around the 'notorious graveyard on Rockland Broad' with Ted Ellis (Clark 1961: 80–1). Mardle covered Albion's tow to Yarmouth for refit and 'resuscitation', two 'gaudy placards' on the foredeck stating 'Albion will sail again'. Clark was a 'practical idealist', also arguing for windmill restoration, but for now concentrating on wherries, dreaming of 'a fleet sailing the Broads again' (Mardle 1949b). The first restored voyage from Yarmouth to Norwich on 13 October 1949 was featured in the national *News Chronicle*, Albion pictured passing Cantley on 'a triumphant voyage', skippered by Jack Cates, carrying 30 celebrity guests including the Mayor of Great Yarmouth, met by the Mayor of Norwich at Hardley Cross, historic limit of navigational jurisdiction between Norwich and Yarmouth Corporations: 'Next week, with her shining red and blue hatches, Albion will start work again in real earnest, carrying sugar beet to the Cantley factory' (News Chronicle 1949). Cates skippered until 1951, carrying over 1000 tons over 2000 miles in the first year, including sugar beet, timber, coke and grain, and hire parties of youth groups and naturalists. The Albion quickly became a regional sight, noted by visiting authors, John May meeting Cates on Lake Lothing (May 1952: 53–7), Grimes showing Albion ('Lone Survivor') leaving Yarmouth harbour (Grimes 1954: 60; Grimes 1953b), the 1965 New Naturalist picturing Albion mooring (Ellis 1965: 245).

Freight work fell off, and Albion turned to leisure use, from 1961 formally ceasing cargo operations. In 1957 Albion sank at Hardley Staithe with 40 tons of sugar beet, and sank again two years later at Berney Arms. A 1966 'Save the Wherry Week', with the Duke of Edinburgh as patron, helped raise £2000 for a refit (Malster 1971: 146–7).

Dreams of a new fleet stumble, with Albion quickly appearing the likely last restored specimen, itself needing salvage. As early as May 1953 the *EDP* had reported 'Wherry Albion to Cease Trading: Heavy Loss', the Trust annual meeting noting financial trouble, with fundraising activities misjudged. A club and museum in Norwich's historic cobbled Elm Hill had lost money, while a Theatre Royal concert lost £151, Clark as Secretary noting that 'the character had been changed after it was found that Mr. Benjamin Britten would not be able to appear after all' (EDP 1953i). The nature of Britten's proposed contribution is not recorded, but the possible connection of the Trust and a Suffolk coastal composer of nautical works – *Peter Grimes*, *Billy Budd* – carries a logic. In 1955 Ellis would be noted as having written a 'successful' 'Wherry Waltz', though whether for Trust theatricals is unclear (EDP 1955a).

Clark's *Black-Sailed Traders* set out social history, sailing technology and reasons for restoration, the opening chapter narrating a working March journey, Clark helping Cates carry logs from Buckenham to Beccles. Albion supports a cultural manifesto akin to that put forward by LTC Rolt for canal and steam rail (Rolt 1944; 1947; Matless 1998), rescued boats signalling cultural authenticity and lasting value in a world gone awry. For Clark, the 'age of atom bombs and moon rockets' divorced use and beauty:

> A time will doubtless come when all ships, and all sailors, will be as much museum pieces as bronze-age implements and the bones of mammoths; when nuclear-powered submarines, governed and directed by electronic devices, will move unseen and unmanned beneath the oceans; when men, women and children, fed on synthetic foods, will live all their lives in artificial light, deep down in the bowels of the earth, to escape the radioactive contamination of the air. When that time does come, people will never know sunlight nor the sparkle of waves upon the beach. For blind subservience to Progress, which never thinks to question whether, just because a thing can be done, it should be done, must surely, in the end, drive humanity to such straits.
>
> But until that day arrives, the sight of the *Albion*, sailing about our rivers, may help to remind us that utility and beauty were not always incompatible; that invention and sanity can go together; and that, as George Borrow put it, 'there is still the wind on the heath.' (Clark 1961: 51)

The 'wind on the heath' phrase, from Borrow's 1851 *Lavengro*, a geographically wide-ranging mix of autobiography and stories of gypsy life, was a common local literary reference, nature freedom evoked through an author born in Dereham, living in Norwich 1816–24, and at Oulton Broad from 1840–53, and 1874–81. Clark's Borrow citation lent Albion cultural values of nature freedom and regional pride.[11]

The windy 'heath' evoked by Borrow in *Lavengro* is Mousehold Heath, overlooking Norwich, Borrow's Romany acquaintance Jasper Petulengro stating: 'There's the wind on the heath, brother; if I could only feel that, I would gladly live for ever' (Borrow 1924: 166).[12] The passage served also for Clark and others as a specifically sensual atmospheric citation, conjuring value in the touch as well as sight of nature; the original comes from a conversation Borrow reports between himself as 'Lavengro' and the Romany Jasper, the latter taking sun, moon, stars and the wind on the heath as signs that 'life is sweet': 'who would wish to die?'. Lavengro then asks whether

Jasper would wish to live forever even 'in blindness'; the response is that the 'feel' of the wind would compensate. On their March journey, Clark briefly gets to sail Albion near Langley while Cates is relieving himself, and feels the tiller against his hip: 'In that contact was the feel of the very wind itself, that sense of exultation man gains from taming the elements to carry himself and his goods from place to place ... There was the added excitement, too, of knowing that a wherry's black sail had not been seen on the river for more than ten years' (Clark 1961: 14). The black-sailed trader touches and restores the scene, and its crew.

If Albion is restored as a working vessel, though, Clark is aware of its heritage spectacle, and recognises tension in the Trust's cultural work. The Albion is 'photo-graphed like a film-star', 'the last of her kind' (8), with Clark blunt about the wherry as a relic, playing at trade, its 'working span' being 'artificially extended' by 'a few years ... there was no economic need for it' (34). Albion is 'as much a relic of the past as the Tower of London' (109), something 'out of her time; an anachronism in this age of speed, space-rockets and ulcerated stomachs' (167); though survival means the wherry is not yet 'a thing of the past' (34). If whole fleets are not revived, restoration gives at least reproach to the present.

As in the period of regional discovery, the wherry allows claims to Broadland authority; appreciation of the peculiarities of native craft, the gathering of local testi-mony. Norwich School imagery, including from Stark's *Scenery of the Rivers of Norfolk*, illustrated *Black-Sailed Traders*, while Clark also talked to surviving wherrymen, oral enquiry marking him out as able to gain their respect, but nonetheless of a different social order. Albert Powley of Cobholm tells Clark: 'You know, you ought to write a book about these wherries of ours' (167). Clark produces a 'Glossary of Wherrymen's Terms' (240–1), and has Albion skipper Cates telling the Buckenham constable the names of river reaches, 40 between Norwich and Reedham; just as Rye noted the reaches in his *History of Norfolk* (Rye 1885; Malster 1971: 154–61). Clark himself looked back to discovery, finding Emerson a model explorer, one Broads authority to another, with Emerson 'no fair-weather sailor', and rising above the 'monotonous superficialities' of 'glib scribes' (Clark 1961: 93–5). Clark also echoed Emerson's masculine wherry world, women present only as wives or prostitutes, and pubs cen-tring wherry culture, commemorated in pub signs photographed by the author; Wherry Inn, Geldeston, Keel and Wherry, Norwich (128). Describing racing Albion at Oulton, Clark recalled: 'a real holiday treat ... That night ... we drank immense quantities of beer' (158). Clark however was sceptical over the drinking origins of the word 'wherry': 'And I feel it has no connection with a liquor, made in the West Country from crab apples, and called "wherry", from the Welsh meaning "bitter"!' (42).

Wherry resurrection prompted argument over sail colour. For Clark, sails were black, proofed with tar and herring oil, with no 'fancy colours': 'It has always aston-ished me that any doubt could possibly exist about the colour of a wherry's sail. It is just simply black.' (55). Others proposed an extended palette, G Colman Green's *The Norfolk Wherry* suggesting: 'The sails of the old trading wherries were originally tarred by boiling in herring oil with red lead, a few being tarred' (Green 1953: 17). If Green's own cover painting showed a black-sailed craft, a fold-out diagram noted for the sail: 'Tanned to various rich Reds and Purple Red for Trading Wherry / White Sail for

Pleasure Wherry' (20). Green's book, first published in 1937 by the Model Yachting Association, gave technical drawings to model 'the finest expression of a vessel ever evolved, for exposed estuaries and open marshland waterways' (3). Green offers a cultural contrast to Clark, the wherry as something delightful, for toy replication or historic reverie, but passing as a working vessel.[13] *The Norfolk Wherry*'s 'Supplement of Pictures' included Green's own pleasure wherry Zenobia alongside other paintings, while anecdotes and 'Broadland Lyrics' conveyed regional delight: 'Sailing far beyond the reed banks, / Vanished o'er some far horizon, / Majestic'ly they sailed, careening – / Unbelievably exciting! –' (supplement, 1). The wherry under trading sail was however gone: 'Never more shall mankind witness, / As he motors through the Broadlands, / Greater power, or more efficient' (iv).

Broadland, South Kensington

The making of boat models, whether from Green's template or in other form, could carry the Norfolk wherry far from Broadland. If model wherries feature in Norfolk museums, they also find a national stage. Within the Science Museum's displays of 'British Small Craft', installed in 1963 in the new 'Sailing Ships and Small Craft' gallery, stood a 'Norfolk Wherry'; Broadland in South Kensington. Regular Science Museum diorama artist Dunstan Mortimer painted a background of a broad and boathouse, a winter scene suggesting a vessel working all year. A wherryman was modelled at the tiller, 'The Trader Lowestoft' written in front of the hold, the sail pale, the model acquisition date and number on the rudder: '1927 – 822'. The display label was accompanied by a photograph of a black-sailed wherry, low in the water, passing a yacht, a capped man in silhouette at the tiller, the image underpinning modelling actuality.[14]

The British Small Craft displays emphasised technology shaped by local conditions, the wherry label beginning: 'This type of small cargo carrier was developed to meet the special conditions of the "Broads" districts of East Anglia.' Design details followed, although draft display labels included a feature not on the displayed model: 'A curious characteristic of these wherries was the quadrant of white paint on each side of the bow which, it has been suggested, bears a connection with the "eyes" or oculus so commonly found in some far eastern craft.' Such eyes were 'commonly depicted on the boats of the Ancient Egyptians and on the present-day boats of the Chinese and of several other eastern races'.[15] Attention to myth indicates further variation in wherry understanding; if the Science Museum noted such questions in connection with their global ship models, displayed elsewhere in the Shipping Gallery, and Colman Green delighted in reverie over white eyes, 'derived from ancient usage as far back as Egyptian civilisation of the XX Dynasty' (Green 1953: 29), for Clark 'eye' shapes on the bows were utilitarian, not symbolic (Clark 1961: 87), warnings for river users, not calls to myth.

If modelling took Broadland to the capital, London came to Broadland for craft skill. Museum files show the geographies of object acquisition, the cosmopolitan and provincial meeting, and attempting to converse.[16] Intelligence was gathered on local

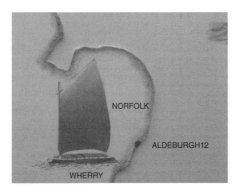

Figure 17   Model wherry and diorama, with location map, 'British Small Craft' displays, Science Museum, London. Source: photographs by the author, May 2010. Reproduced with permission of the Science Museum, London.

museum displays; a 1920 press article mentioning wherry models at Norwich Castle Museum, photographs showing models held in the Tollhouse Museum, Great Yarmouth.[17] In 1927 shipping curator Geoffrey Laird Clowes asked London stained glass maker Leonard Walker, who had Norfolk connections, for advice, Walker replying on 28 June: 'A Mr Darby of Oulton Broad who did my houseboat also builds models and he could get in touch with a certain Mr Hall at Reedham now elderly who used to build trading wherries, and who has a model and possibly drawings which Mr Darby could borrow to make a model wherry for you if you still require one.' Darby, Oulton Broad yacht owner and agent, might go 'to Reedham (a train journey) to negotiate the matter with Mr Hall'.[18] Clowes asked Walker to ask Darby to ask Hall: 'What we want is a model of a trading wherry of as early a type as possible, untouched by later outside influences … it would be kind of you if you would impress on Mr. Darby that accurate scaling in all proportions is of the first importance to us.' Clowes was going to Norfolk, and could see Darby from 28 July.[19]

Intermediaries communicated for Clowes across class and geography, Darby translating between Museum and Hall. Darby wrote to Clowes on 22 July: 'In my search for plans of Trading Wherry of 50 years back I have found a model just completed and

I have asked the man to bring it here for you to see.'[20] On 31 July: 'I have seen the builder of the Wherry Model this week end, and he can arrange to have it here and meet Tuesday or Wednesday next.'[21] After meeting, Clowes and Hall corresponded direct, Hall accepting Clowes' £20 price, setting himself as craftsman-servant: 'I can assure you that the work will be completed to your satisfaction as well as mine my motto is and always was that anything thats' worth doing is worth doing well. I am very pleased to know that you appreciate the work already done.'[22] Hall had a museum track record, making a model wherry in 1912 and a keel in 1923 for Norwich's Bridewell Museum, the wherry black-sailed and still displayed (Ellis 1965; Malster 2003). Clowes collected Hall's model on 22 August, Hall writing on 7 September: 'I should be very pleased if they would send payment for the Wherry as I am rather short of money. I should be pleased if you can get it through this week.'[23] Clowes was happy with the transaction, also commissioning a Norfolk Keel from Hall, completed in 1928 for £20; Clowes would comment to the Museum: 'The model is very well made and is decidedly cheap.'[24]

Wherry sail colour is again however uncertain, indeed the Science Museum model itself is marked with doubt. On 16 August 1927 Hall asked Clowes: 'do you prefer a black sail or white?'[25] Clowes' reply is not recorded, but his colour views are indicated in annotations to tourist postcards collected for Museum files, where sails appear white or red-orange: 'A wherry's sails are really a very definite deep dull black.'[26] Yet the model displayed in British Small Craft had white sails, as did Hall's Beach Yawl and Norfolk Keel for the same galleries. Are we to assume a white-sailed commission, against Clowes' postcard judgement? Or did Hall make sails against Clowes' colour choice?

Hall's sails turn out not to be those in the Museum display case, at least not for the wherry. Colman Green's *The Norfolk Wherry* includes an undated photograph: 'Model of a Wherry. This beautiful model was made by W. Hall of Oulton Broad, and is in the So. Ken. Museum, London' (Green 1953: supplement, 6). The model stands in isolation, its sail dark, suggesting Hall received a black-sailed commission, sails changed between 1927 and 1963, though when is unclear. An authoritative national museum model holds uncertainty, wherry sail colour contentious.

Chapter Four
Animal Landscapes

The Norfolk Room

In the Norwich Castle Museum, the Norfolk Room presents six dioramas, three along each side wall, around six feet high: Broadland, the Yare Valley, Breydon Water, Breckland, the North Norfolk coast, a Norfolk Loke. Broadland and the north coast are central and double-sized, 26 feet long. Shaped in the 1930s by Curator Frank Leney and Keeper in Natural History Ted Ellis (and renamed the Ted Ellis Norfolk Room in 1987), the Norfolk Room replaced the Gurney collection of birds of prey, a consciously modern shift from private trophy taxidermy to the publically commissioned display of regional habitat, encased open air, on view to all. The dioramas landscape the animal, stuffed specimens posed in action, living dead in their local habitation. The county is shown to the city, the Broads region making half the scene; Yare, Breydon, Broadland.[1]

Leney, inspired by American natural history dioramas, claimed Norwich as a British pioneer: 'By means of these dioramas it is hoped that many visitors will become Nature observers and field naturalists' (Leney 1935: 185; Wonders 1993; 2003; Nyhart 2004; Insley 2008; Nahum 2010). For Ellis the dioramas provided the 'ecological section' of the Castle's displays, habitat in 'panorama'.[2] An upstairs mezzanine gallery held, for the 'serious student', examples of 'every living organism found in Norfolk', with a reference library, card index, hand-lists and staffed Naturalists' Desk (Castle Museum 1949: 16). The Yarmouth Naturalists' Society donated the desk, marking the centenary of CJ and James Paget's natural history of Yarmouth (Paget and Paget 1834); the NNNS granted 100 guineas towards the displays. Seasonal mezzanine

In the Nature of Landscape: Cultural Geography on the Norfolk Broads, First Edition. David Matless.
© 2014 David Matless. Published 2014 by John Wiley & Sons, Ltd.

Figure 18   The Norfolk Room, Castle Museum, Norwich: Top: Broadland diorama; Bottom left: Breydon diorama; Bottom right: Yare Valley diorama. © Norfolk Museums & Archaeology Service (Norwich Castle Museum & Art Gallery), photographs by David M Waterhouse.

exhibits included: 'May: Life-history Swallowtail butterfly & a Dragonfly' / 'Nov.: Winter buds of broads water-plants'. September 1935 brought 'Parasites of the reed'.[3] For Ellis the Norfolk Room offered 'county naturalists' a 'clearly conceived and boldly executed memorial' to their efforts, by which 'the public will be aroused': 'Surely, this Norfolk Room is not a thing to trifle with'.[4]

The Norfolk Room registers Broadland as nature region. Chapter Five will attend to cultures of Broadland plant life; here Broadland is considered as animal landscape. The ecological Norfolk Room of course highlights the artifice of such a separation, but plant and animal nevertheless shape distinct cultures of natural history, and reward specific chapter attention. The Breydon and Yare Valley dioramas (and Breckland and the north coast) were designed by Ernest Whatley of the pioneering Imperial Institute Studios, London. 'Yare Valley' shows a river and pools, snow, a reedbed behind, 'a few miles east of Norwich' (Castle Museum 1968: 12). 'Breydon in October' foregrounds birds on mud, with others suspended landing, the open estuary behind, a line of posts into left distance, a flint estuary wall to the right, a distant mill. Norfolk naturalist BB Riviere arranged birds 'in natural positions': 'By means of an electric light switch visitors are able in a moment to bring before their vision a characteristic scene of wild life within twenty miles of Norwich. A second switch illuminates a map of Norfolk' (Leney 1933: 362–3). A key identified birds, photographs showed Breydon in different seasons.

The Norfolk Room carried theatrical quality, Norfolk nature performed. 'Broadland' was designed by Owen P Smyth, scene painter and actor at the Maddermarket Theatre, Norwich, and later wartime camouflage artist. The 'Broadland' label indicates that within the Norfolk Broads region, some scenes may be deemed more essentially Broadland than others: 'Lord Desborough's sanctuary at Hickling has formed the basis of the scene.' The painted backdrop shows a wide broad, with distant wherry and yacht sails. Regional realism is the aim: 'It was tempting to put in many other characteristic Broadland Birds but the Committee adhered to their scheme of having a few objects well shown instead of, as in the old days, exhibiting a crowd of birds with little or no resemblance to nature' (Leney 1933: 364). Reeds fringe the front, inviting the viewer to look within and through for birds, plants, nests, eggs, small mammals, insects. Species regionally iconic come into view; swallowtail butterfly, Montagu's harrier, bittern. In recent decades sound has supplemented the view, with buttons pushed for Broadland bird song. The viewer walks along the field of vision to take things in. Here is what you might see, if you went to look, though only a dedicated naturalist might see such detail in the field.

The Castle dioramas capture prominent themes of this chapter: animal landscapes surveyed and displayed, iconic sites and species, county nature on urban display, the institutional shaping of nature knowledge. The Norfolk Room as animal (and plant) landscape shows Broadland performed as nature region.

Fauna Surveyed

Nineteenth century Norfolk naturalism traced ancestry to Sir Thomas Browne, seventeenth century scientist and physician, codified for the county in Thomas Southwell's edition of *Notes and Letters on the Natural History of Norfolk More Especially on the Birds and Fishes* (Browne 1902), Broadland creatures prominent. A statue of Browne was erected on Norwich's Haymarket for his 1905 tercentenary; his skull, held in the Norfolk and Norwich Hospital from 1840, was reunited with his body in St Peter Mancroft church in 1922. Browne's skull would later open WG Sebald's East Anglian meditation *The Rings of Saturn*, his enquiring spirit a touchstone (Sebald 1998: 22). Southwell presented one of the 'pioneers of Natural Science' in 'the originality which pervades all his observations' (Browne 1902: xvi). Browne, 'proud of his adopted county' (xviii), becomes patron of Norfolk natural history.

If Browne appears a polymathic genius, with Norfolk natural history one of many talents, Rev Richard Lubbock (1798–1876) is claimed as Norfolk's Gilbert White, a lifelong local observer.[5] Lubbock's 1845 *Observations on the Fauna of Norfolk, and More Particularly on the District of the Broads*, noted in Chapter Three for its human fauna, was revised in 1879, with essays and notes by Southwell and Henry Stevenson, founder NNNS members in 1869, Lubbock's *Fauna* 'regarded almost as a sacred book' (Lubbock 1879: xi). *Fauna of Norfolk* was derived from 'lectures' at Norwich Museum between 1835 and 1844, presenting within the 'conversazione' format common to Victorian science, mixing talks, debates and displays (Alberti 2003). The Norwich conversaziones had begun in January 1835 with a scientific celebrity, Adam Sedgwick, professor of geology at Cambridge, who gave further lectures in January 1836 (Norwich Mercury 1835a; 1835b; 1836b). Lubbock addressed the third meeting, in March 1835, on 'Zoology', presenting 'Natural History as a rational source of amusement and consolation', and emphasising 'the effect of civilisation on the habits of animals' (Norwich Mercury 1835c), and returned in November 1836 to consider 'British Birds of Prey': 'The room was comfortably filled, and the audience highly gratified' (Norwich Mercury 1836a).

Lubbock's reputation was ornithological, half the *Fauna* addressing 'Water Birds', with 'reference rather to the region of the broads' (Lubbock 1879: 175). The Broads were Norfolk's special district, the word broad 'entirely provincial', though 'the waters of Norfolk have little to interest the seeker after picturesque beauty' (79–80), Lubbock writing before Broadland leisure. Sport and bird appreciation join, as in the Pagets' more localised 1834 *Sketch of the Natural History of Yarmouth and its Neighbourhood*, dominated by Breydon birds shot and sold on Yarmouth market (Paget and Paget 1834). Lubbock highlighted decoys, with their reed screens, pipes, decoy ducks and luring dogs, as 'perhaps the best place in which to speculate and gain knowledge on the habits of various birds … in the view given through the reed screen of a decoy-pond, you see nature as she really is' (Lubbock 1879: 134–5). The decoy becomes a proto-diorama, artifice heightening actuality. Place names such as Decoy Carr or Decoy Marsh indicated lost sites: 'the days of decoys will soon be past' (224). Lubbock's reflections 'On the Remains of Falconry in Norfolk' likewise saw faunal arts decayed: 'The sight of a falcon is somewhat like that of the rusty mail or the monument of a departed hero' (32; Macdonald 2006). Bird life was also diminished by drainage, shooting and egg collecting for the London market, but Lubbock held to a general 'compensating principle' in nature: 'By reclaiming waste lands and draining marshes, we gradually lose certain species; but by cultivation and planting, we either encourage or actually gain others' (Lubbock 1879: 64).

Lubbock's survey was supplemented by NNNS naturalists, Stevenson's 1866 *The Birds of Norfolk* (with a second volume in 1870) similarly emphasising the broads' 'peculiar' conditions, 'possessing the greatest amount of interest for the naturalist and sportsman' (Stevenson 1866: xvi; 1870). Stevenson could state positively: 'Probably more rare birds have been killed on Breydon than in any other part of the United Kingdom' (Stevenson 1866: xviii). Stevenson set his Broadland scene with a 'literally "bird's eye" view of this singularly level district' from Great Yarmouth's Nelson monument, 'say from the summit of the Nelson Column, if twice its present height', tracing each river valley in turn; the 'trimness' of the drained lower Bure and Yare, the 'more natural and unrestrained fertility' upstream in Bure and Thurne (xviii–xx). A third volume of *The Birds of Norfolk*, including a memoir of Stevenson, was completed by Southwell in 1890 (Stevenson 1890). Southwell sought an 'Ornithological Archaeology of Norfolk', arguing that, except for spoonbills and cormorants, all species nesting in 1671 were still nesting in 1800, but that many had since disappeared from drainage, shooting and railway development: 'By improved communication the London market was opened to the remote wilds of Norfolk, and the London market was insatiable' (Lubbock 1879: vii; Southwell 1871).[6] Southwell's socio-economic analysis of the lost geographies of Norfolk nature implied the need for sporting restraint, and a language of nature protection emerges, to which we return in the 'Broadland Preserve' section below.

Murder Most Fowl

In Arthur Conan Doyle's 1893 short story 'The "Gloria Scott"', Sherlock Holmes recalls an estate 'in the country of the Broads': 'There was excellent wild-duck shooting in the fens, remarkably good fishing, … he would be a fastidious man whom could not put in a pleasant month there' (Conan Doyle 1988: 374). The reference indicates national regional recognition, and sporting reputation. If tensions of conduct were emerging over shooting, it is important to register the celebratory shooting culture generating Southwell's naturalist unease, an unproblematic joy in killing, whether in estate game shooting or wildfowling, with a concomitant sense of ranging masculinity.

Before his bestselling *Handbook* guide, Davies published *The Swan and Her Crew*, subtitled 'The Adventures of Three Young Naturalists and Sportsmen on The Broads and Rivers of Norfolk' (Davies 1876). Boyish natural history accompanies joyous sport: 'Now, Jimmy, we have got a prize. Crossbills are not seen every day. Let us go to the boat-house and skin them, and read something about them in our books' (Davies 1876: 9). In Davies's story of nature study, manliness, class and carnage, akin to imperial adventure tales (Mackenzie 1988), delicate Dick Carleton turns to vigour under the care of Hickling boys Frank Merrivale and Jimmy Brett: 'I did not know that life could possibly be so jolly, until I learnt something of natural history' (Davies 1876: 114). Proto-scouting 'habits of self-reliance' (134) accompany shooting, hawking, egg collecting, fishing, botany, entomology. Poachers receive a 'drubbing' (172), the deserving poor surplus fish, and egg-smashing Yarmouth boys are dispatched, Frank taking up an 'excellent boxing position, his blue eyes gleaming with such a Berserker rage' (85).

Natural history, masculinity and the gun also aligned for Emerson. While criticising the 'extreme game-culture' of 'the worshippers of St. Partridge' (Emerson 1888: 86), Emerson photographed Broadland wildfowling (Emerson and Goodall 1886: 31–2/49–51), and *Birds, Beasts and Fishes of the Norfolk Broadland* conveyed his relish of bird taste, including the golden plover ('no better bird on toast' (Emerson 1895b: 270)) and waterhen: 'Beautiful as the waterhen is on the lilied lagoons, he is equally attractive in a dumpling' (259).[7] Emerson used anthropomorphic animal species' 'artistic biography' to enact human class judgement, with supposedly authentic upper and working class lives contrasted to an inauthentic middle class concern for domesticity and propriety. The rook was thus bourgeois and cowardly, coyly mating in early morning, while the swan was rapaciously aristocratic, an aggressive masculinity rendered through a detailed account of adultery and rape at Buxton. *Nature*'s reviewer commented: 'if the book is intended for the eyes of ladies and young people, why are we treated … to a very unnecessary anecdote concerning the amours of swans' (Lydekker 1895: 196).

The Broads as masculine sporting ground was surveyed in Nicholas Everitt's 1902 *Broadland Sport*; solicitor and businessman Everitt would donate Everitt Park by Oulton Broad for public use. Everitt's 'Sport' also includes yachting, the effect (as in Oliver Ready's 1910 *Life and Sport on the Norfolk Broads* (Ready 1910)) being to set sailing, shooting, fishing and hunting within an elite Broadland leisure round. Broadland sportsmen, marked out from reckless 'crowds of armed Cockneys' (Everitt 1902: 154–5), carry their own style, Everitt showing 'wildfowling costume' modelled by Richard Fielding Harmer, aged 70, including leather boots to the hip, and 'a peculiar loose pair of dressed waterproof knickerbockers' (110); Emerson's *Wild Life on a Tidal Water* had included a Breydon appendix by Harmer (Emerson 1890a: 125–45). *Broadland Sport* presents a heterosexual bachelor domain, in part via literary allusion.

A woman is sketched peering through reeds: 'Molly, The Broadland Trilby' (Everitt 1902: 187), an upper Bure dairy girl known to Everitt's friend: 'there Molly, a modern Trilby, domineered as the Svengali of his existence' (186). In George Du Maurier's 1894 Parisian novel *Trilby*, Svengali had mesmerised artist's model and milk girl Trilby, though Everitt stresses his is 'not *the* Trilby … we are far away from the Latin quarter of gay Paris' (181; Du Maurier 1995). Elizabeth Wilson terms *Trilby* a 'picturesque version of bohemian life', featuring 'decent Anglo-Saxons on temporary leave from the professional middle class' (Wilson 2000: 223–4). Everitt domesticates an already reduced Bohemia, offering mild Broadland eroticism: 'No sooner has she left us than one by one we silently and sorrowfully crawl back to our blankets, each to enjoy his own reverie of the bright, bonny little English maiden' (Everitt 1902: 182–3).[8]

Sporting masculine adventure shaped Broadland well into the twentieth century. Thus Alan Savory's *Norfolk Fowler* (1953) and *Lazy Rivers* (1956) eulogised wild-fowling and fishing against 'bird protectionists and anti-sports cranks' (Savory 1953: vii; 1956); Savory's *Thunder in the Air* (1960) recounted South African sport. A parallel imperious masculinity inhabits Broadland through James Wentworth Day (1899–1983), sometime editor of *The Field* and *Illustrated Sporting and Dramatic News*, and author of many books on shooting, agriculture, wildlife, ghosts, gambling, speed, royalty and more. Day, resident in Essex, presented the Broads as part of his East Anglian writing patch, claiming Broadland notables including Savory as friends. Roland Green, 'hermit artist of the Broads' (Day 1950: 48), illustrated two Day books (Day 1948b; 1960), while Day reported encountering Ted Ellis, 'that excellent naturalist' and 'King of Wheatfen Broad', convincing Ellis that he was a 'harmless' visitor: 'Since then we have been good friends' (Day 1967: 60); Phyllis Ellis recalled Day as an 'awful man, awful man … he claimed us as his great friends'.[9]

Broadland Adventure's dust jacket showed Day, gun shouldered; the frontispiece photographed 'The author in a flight pit on the Waxham marshes' (Day 1951), Day in trilby, jacket and tie, gun ready, dog ready, a serving man holding already dead ducks: 'I may be prejudiced, but somehow I think that wildfowling is one of the last forms of strenuous relaxation left to the man whose heart and soul are so essentially masculine that he must by necessity escape from the shams, conventions and ortho-doxies of modern life' (Day 1935: 25). Day eulogised estate and free shooting, setting both above supposedly detached modern leisure. Science too could be out of place, whether from ornithologist television expert 'birdy-boys' (Day 1967: 100), or agricul-tural science harming land. Day could eulogise agricultural improvement (Day 1943; 1946), but in 1957, six years before Rachel Carson's *Silent Spring*, he condemned the pesticide 'war on wildlife' in *Poison on the Land*: 'Who wants the Britain of broad bright fields, of tall woods brilliant against gentle hills, of water-meadows and shining rivers where birds sing and animals gladden the eye to become a silent land where no wild life stirs? It is a nightmare vision' (Day 1957: 3).

Day was active in hard Right politics, assisting maverick Lady Houston in the 1930s (Day 1958), and as agent provocateur provoking Labour's Harold Laski into a notable failed libel case in 1945 (Daily Express 1947).[10] Day's politics informed his

Figure 19   *Broadland Adventure*. Source: Day 1951, front cover dust jacket.

mid-market topographic commentary, the sporting stress on action a common Right theme, crossing interest in nature and technology. In the 1930s Day produced eulogies of motor racing speed (Day 1931), while *Gamblers Gallery* likewise lauded well-heeled risky thrills (Day 1948a).[11] Active masculinity moves from casino to track and fen. Day's 'Gunroom Philosophy' envisaged 'a snug house-corner forever masculine, a room whose four walls are an enclave of enchanted memories of marsh and mountain, river and tidal mudflat … nothing keeps the domestic female more effectively at bay than a half dozen stuffed birds' (Day 1948b: 211). Day was married three times, twice divorced; his second wife (1936–43), bisexual writer and journalist Nerina Shute, left him after a year (Hillier 2004). *Coastal Adventure* (Day 1949) was dedicated 'To That Good Sportswoman My Wife', third wife Marion McLean. Day's public persona of commanding male promoted, as with Everitt and Emerson, a Broadland nature nurturing aggressive masculinity, full of itself outdoors.

Broadland Preserve

Public rights and regulations

Concern emerged in the late nineteenth century that Broadland fauna might need protection. Plans for public regulation ran alongside models of the private estate as protective nature preserve, while private ownership also clashed with public rights to fish and fowl. Animal landscapes entailed judgements of human conduct, shaping the Broads as public and private space.

Broadland's animal life had long been subject to public authority, including through custom noted by those discovering the region. Thus Dutt described swan upping, the annual August Yare tally by Norwich Corporation, the swanherds 'almost unobserved' at Buckenham Ferry, marking the cygnets and taking some for fattening in Norwich Great Hospital's swan pit (Dutt 1906: 252; George 2000a: 9–11).[12] Swan custom received attention in *TNNNS* from NF Ticehurst, whose 1957 *The Mute Swan in England* would offer a definitive history of law, swan-rolls (including the Cantley Roll) and swan-marks, bills knife-engraved with dark shapes, crosses, triangles, dots (Ticehurst 1921; 1925; 1928; 1929; 1937; 1957; Long 1925a).

If marked swans signalled archaic authority, others sought new public authority to protect birds. From the 1869 Sea Birds Preservation Act, legislation mixed species protection with shooting management, the 1880 Wild Birds Protection Act establishing a March–July close season (Sheail 1976; George 2000a). Stevenson's 1872 NNNS Presidential Address supported legislation, condemning '"excursionist" gunners', 'indiscriminate egging' and 'the plume mania of the last few years' (Stevenson 1872: 13/18). NNNS lobbying secured egg protection on the upper Thurne and Trinity broads in 1895, while from 1888 the Breydon Wild Birds Protection Society appointed watchers and monitored Yarmouth game dealers (George 1992; 2000a). From 1921 the NNNS Norfolk Wild Birds Protection Committee assumed the duties of the Breydon Society, its remit extended across the county, *TNNNS* publishing annual reports (TNNNS 1922).[13]

Broadland's animal landscape also however saw contention over public shooting and fishing rights. In the 1892 Hickling Broad Case, plaintiff HSN Micklethwait claimed ownership of Hickling Broad under an 1801 Enclosure Act, and sought to restrain defendant Robert Vincent, local marshman, from shooting, fishing or sailing, other than within a marked channel. On sailing the case failed, the GYPHC held to exercise navigational jurisdiction, but on fishing and shooting Micklethwait won, with Hickling judged to be non-tidal and therefore liable to private control. Local commentary suggested freedoms infringed, via judgement deficient in tidal knowledge (Dutt 1903; Maxwell 1925). Walter Rye represented Vincent as Honorary Solicitor of the Norfolk Broads Protection Society: 'after a long and obstinate struggle – in which with hardly an exception the local fishermen and boatmen swore to a tide, and were supported by such experienced local yachtsmen as Messrs. G. C. Davies and Bolingbroke – the Judge found on expert evidence, that it was not tidal' (Rye 1893: 5). Rye suggested the Plaintiff's superior 'heavy metal' informed 'first-class expert witnesses' (19–20).

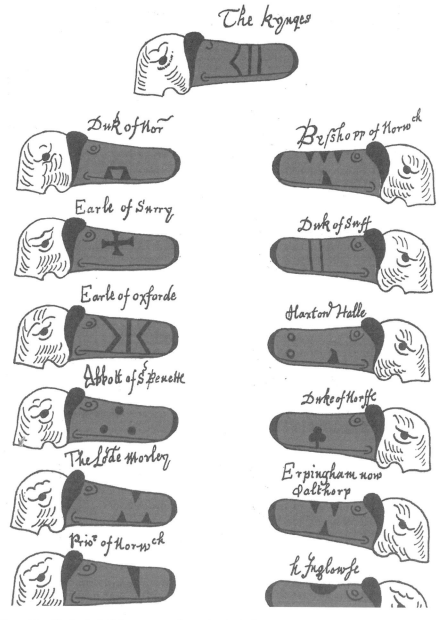

Figure 20   'The Cantley Roll'. Source: swan-roll reproduced in Ticehurst 1957.

The 1949 Blackhorse battle was anticipated, with Kett's rebellion and its 'prayers' that 'the rivers may be free and common to all men for fishing and passage' cited (29), and Rye noting: 'The first aggressors had been Blofeld of Hoveton, who chained up the two entrances of Hoveton' (72; Rye 1916). George notes continuing Hickling sensitivities, as when 1985 NNT/NC proposals to close parts of the broad proved 'anathema' to Hickling Parish Council, Hickling Sailing Club and the Broads Society (George 1992: 52), the 1892 case a sore memory.

Hickling also framed Emerson's 1895 story 'The Tide Pulse' (Emerson 1898b: 144–9). Emerson had criticised broad closures as 'land thefts' (Emerson 1888: 81), and gave evidence in the Hickling case (Wilkins-Jones 1986). 'The Tide Pulse' began: 'It was my fortune once to have to watch the ebb and flow of a tide on a wide lagoon, and delicate work too, for a sixteenth of an inch rise or fall was invaluable' (Emerson 1898b: 144). Emerson, 'feeling the pulse of Nature's arteries as they softly ebb and flow up and down the face of a deal stick' (144), finds all Broadland variably tidal, water flowing free across human claims: 'And when my task was finished I learnt, after all, that the Geographies were correct: the tide ebbed and flowed in even my far-away capillary' (149).

Nature's private domain

If public regulation, navigating public rights, offered one route to nature protection, others proposed the private estate as sanctuary. From the early twentieth century Hickling and Horsey in particular became demonstration spaces of sporting reserve, nature protected as private domain.

Hickling: kept sanctuary

At Hickling a metropolitan social elite acquired land, employed local labour, regulated communal action, remade the built environment, and gave the place national reputation. The Whiteslea estate was acquired by Liberal peer Lord Lucas after visiting in 1908 for shooting, birdwatching and egg collecting with Edwin Montagu (Liberal MP, and Secretary of State for India 1917–28) and Sir Edward Grey (Liberal Foreign Secretary and later author of *The Charm of Birds* (Grey 1927)). In 1916 Lucas died on active air service, his 97 hectares passing to Ivor Grenfell, then a minor, whose father Lord Desborough managed the estate, assuming ownership on Ivor's death in 1926.[14] Desborough's main residence was Taplow House in Buckinghamshire; Lady 'Ettie' Desborough was a leading hostess and socialite (Davenport-Hines 2008). Desborough expanded Whiteslea to 256 ha by his death in 1945, including the southern half of Hickling Broad, with the northern half leased from Col John Mills, who had inherited in 1908 from his father, who had in turn inherited from Micklethwait, the Hickling plaintiff, in 1901. Desborough consolidated Hickling as elite nature space. Whiteslea Lodge was refurbished in the late 1920s as a comfortable wooden thatched shooting lodge. A private road gave access from the village and estate, while underlining exclusivity. Five hundred prints from Gould's *British Birds* decorated the interior, while in 1930 Roland Green painted four friezes for the sitting room; marsh landscapes, churches and mills, harriers and herons, duck and waders (Cable 1991: 144–5). Before the Castle dioramas, Desborough made a private Norfolk Room.

George V visited twice as Desborough's guest, while George VI shot at Hickling, Woodbastwick and Ranworth before and after becoming King (Buxton (Aubrey)

1955; Linsell 1990). Hickling fostered elite regional-national sporting networks, shooting syndicate members including Ranworth estate owner Harry Cator and Dereham farmer and wildfowler Colin McLean. Common birds were shot and rare birds protected in a sanctuary of discrimination. McLean also recalled interwar Ranworth as seeing 'the greatest wildfowler team of all time' (McLean 1954: 51), members including Cator and racing driver Tim Birkin, who had met helping organise supplies to break the 1926 General Strike; the kind of men Day might truly have liked to call friends. From 1895–1951 McLean shot 19,147 birds, including 9128 duck and 2042 coot, his record underpinning rather than undermining his 1935 Presidency of the NNNS (McLean 1936).

Permanent Whiteslea care for birds, owners and guests was vested in keeper Jim Vincent (1884–1944). Son of the Hickling case defendant, Vincent crossed a social boundary, first engaged as assistant by Montagu in 1900, and again when Lucas acquired Whiteslea. Vincent recorded 247 bird species at Hickling; Montagu commissioned George Lodge to illustrate Vincent's 1911 bird diary, published in 1980 as *A Season of Birds* (Vincent 1980). Elected to NNNS in 1917, Vincent contributed 'Notes from Hickling' to protection reports, and would have been NNNS President in 1945. Anthony Buxton's *Country Life* obituary praised 'A Man of the Broads'; *The Field* put 'Jim Vincent and Friends' (two young marsh harriers) on a memorial cover (Buxton 1944; Field 1944). Service of wildlife, place and employers conjoin, captured when Vincent wrote in the Colman's works magazine on the death of shooting syndicate member Geoffrey Colman: 'As I write, there lies before me a silver match-box inscribed thus: J.V. from G.R.R.C., Jan. 22–23, 1932. / 10W.F. Geese, 1 Mallard, 1 Teal, 1 Pochard, 3 Tufted, 1 Shoveler, 9 Geese in 10 Shots' (McLean 1954: 104–5).

Vincent also organised the Hickling Broad coot shoots, run in part to preserve aquatic flora from wintering birds, visiting gunners punted from Whiteslea, the record day bag 1213: 'All the coot shot are given away to the people of the village … after a shoot the scent of coots cooking can be detected here and there' (Vincent 1935). Aubrey Buxton suggested that if on a pond a coot had 'a *bourgeois* appearance' and 'cringing servility' ('And on a slippery surface there is grotesque humiliation; Norman Wisdom has much to learn from coots on ice'), on a broad, en masse, the coot was 'a different creature' (Buxton (Aubrey) 1955: 121), baffling the gunner, though the King naturally mastered things, with 961 killed in January 1951 on 'His Majesty's last day in Broadland' (131). Vincent noted the shoot held from 'time immemorial', though with records only kept since 1894 (Vincent 1935: 191). Vincent would seem however to have himself reformulated communal practice. In 1895 Emerson described the former coot shoot as a communal working people's event, now lost, a '"learned"' judge having ruled the broad not tidal (Emerson 1895b: 264–7). In 1902 Everitt noted public shoots discontinued: 'The annual coot shoot at Hickling was an event eagerly looked forward to by everyone who owned a gun within ten miles. The slaughter was great and the shooting wild, reckless and dangerous.' Everitt added: 'Had the Broads Protection Society of London won their case concerning the public rights on Hickling Broad, the shooting … would have been free to all comers, as they doubtless were some years ago' (Everitt 1902: 190–1;

Everitt 1903). From Vincent father to Vincent son, coot shoots shift from communal to newly feudal nature management.

Coot shoot conflict revived after 1945, with Hickling under NNT ownership, discussed below. The coot as object of communal sustenance resurfaced however only as quaint memory. In June 1955 Ellis noted: 'The other day I came upon a reference to a peculiar spring frolic, known as the Coot Custard Fair, said to have taken place annually at Horsey a century and more ago. For the celebration sweet foods were prepared from an abundance of eggs collected.' Egg numbers however declined through 'reckless exploitation': 'The Coot Custard Fair had perforce to die out' (Ellis 1955b).

Horsey: estate curiosity

Lord Lucas purchased Horsey Hall around 1912, the estate passing to his sister Lady Lucas (Nan Herbert) on his death in 1916. Anthony Buxton (1882–1970) purchased hall and estate in 1930, becoming a key figure in Norfolk naturalism (Seago 2000). Family wealth and nature interest shaped his life; Buxton was the youngest son of Edward North Buxton, who had helped preserve Epping Forest and in 1903 found the Society for the Preservation of the Wild Fauna of the Empire. Educated at Harrow and Cambridge, Buxton worked in London for the family brewery, before being commissioned in World War I. From 1919–31 Buxton served on the League of Nations Secretariat in Geneva, establishing a beagle pack, with wartime servant and future Horsey keeper George Crees as kennelman (Mardle 1970). Buxton's first book was *Sport in Peace and War* (Buxton 1920), *Sporting Interludes at Geneva* following in 1932 (Buxton 1932). Neither of his first two *TNNNS* contributions, 'Birds of the Western Front' (Buxton 1916) and 'Spring Birds at Geneva' (Buxton 1921), had Norfolk reference; Buxton was elected to the NNNS in 1921, his membership address near Epping. Moving to Horsey in 1932, Buxton immediately became central to county nature institutions, serving as NNNS President 1932–4 and *TNNNS* editor 1936–50, contributing to the wild bird protection notes on Horsey and Broadland, and serving on the NNT Council.

Buxton maintained Horsey as shooting estate and bird sanctuary. If Hickling saw highly organised sporting management, Horsey became an experiment in curiosity. Buxton's three post-war books, *Fisherman Naturalist* (1946), *Travelling Naturalist* (1948), and *Happy Year* (1950), recorded nature knowledge acquired in Britain and overseas, Buxton a worldly traveller rooted at Horsey as owner-observer, shooting, photographing and filming birds, 'Stalking with a Camera', photography carrying 'that supreme difficulty of achievement, which constitutes real sport' (Buxton 1946: 143; Ryan 2000). Buxton exhibited his 'Birds of the Broads' films in Norwich on 9 June 1933, addressing the South-East Union of Scientific Societies (Buxton 1933; 1946). Buxton displays proprietorial curiosity, as when studying otters' snow slides: 'One of them had made six slides, with only one galloping stride in between each slide. He had apparently gone at full gallop for the performance.' Buxton's 'Diagram of Slide and How I Think it is Done', first published in *TNNNS*, remains on display in Norwich Castle Museum (Buxton 1946: 120–1; Buxton 1939a; Matless, Watkins and Merchant 2005).

Horsey allowed Buxton nature experiment, management relaxed (apart from sea defence) to see what followed: 'The fun of the thing to me, and I believe to many naturalists, has been to see the struggle for life.' Horsey predators had 'at least as good and very much longer rights than I to the sport which both of us enjoy', though once 'I lost my temper and overlooked my principles over some rogue owls' (Buxton 1946: 189). Buxton set himself apart from the 'dull form of natural history', from 'Ecology and other weird 'ologys' (Buxton 1948: 194–5), retorting with idiosyncrasy, including a *Happy Year* chapter as written by his terrier, 'Jane (By Herself)', where birdwatching bores: 'He would go looking at those great silly birds that float about over the marshes. They bore me absolutely stiff … I got sick of it and wandered off' (Buxton 1950: 78). Idiosyncratic narrative underwrites estate curiosity, Buxton projecting a singular ownership.

Species of Writing

Varieties of writing convey Broadland's animal landscape; Buxton's author-owner accounting, Day's ranging gunnery. Writing enacts modes of cultural authority, authors gaining a public for their Broadland works, becoming voices of the region. Two early twentieth century figures, Emma Turner and Arthur Patterson, exhibit contrasting species of writing, in literary style, economic rationale, social operation, scientific standing, geographical focus and mode of engagement with the animal. Just as contemporary nature writers such as Mark Cocker work a particular country, in Cocker's case the 'crow country' of the Yare valley (Cocker 2007; Matless 2009a), so Turner and Patterson voice nature through landscaped performance.

Emma Turner

In 1922 Emma Turner (1867–1940) addressed the NNNS as President:

> It is twenty years since I first explored the Broadland, and seventeen years since my houseboat was brought by land from Sutton Staithe, and launched on Hickling Broad … carrying as her crew Mr. and Mrs. Bird and myself; steered and propelled by Alfred Nudd she came to anchor in the quiet haven, where some of the happiest days of my life have been spent. (Turner 1922: 228)

Until 1912 Turner lived near Tunbridge Wells, restricted by 'the circumstances of my life' (Turner 1924: vii). Broadland offered release, and would dominate Turner's public persona. NNNS membership lists from 1926 give a permanent Cambridge address, but her second dwelling was the Hickling houseboat 'Water Rail', built at Sutton. Turner moored against Booth's Hill island, named after ET Booth, whose stuffed specimens, many shot there, fill Brighton's Booth Museum;

Mardle would later term it 'Emma's Island' (Mardle 1955c; 1963; 1970). Turner stayed by permission of successive Hickling owners, despite her opposing shooting. Of one who knew birds only through shooting Turner commented: 'we were, and should always remain strangers to each other' (Turner 1929: 44).

Journalists reported Turner as 'the Loneliest Woman in England', a reputation furthered as National Trust watcher on Scolt Head Island in north Norfolk from April 1924–November 1925 (Turner 1928). Turner described isolation, being viewed by locals as a 'harmless lunatic', as '"That ther mad nat-turialist over the water"', though later becoming 'one of their chief shows' (Turner 1924: ix). Turner's local assistants were Jim Vincent and the Nudd brothers, Alfred and Cubit, the latter 'my most faithful henchman' (viii), the former a bird guide, first met in April 1902, when Turner asked a man where Alfred Nudd might be: '"I am Alfred Nudd. Yon goes a Hen Harrier". And that was my first introduction to Nudd, and to the Broadland birds' (vii). Knowledge, phlegm and deference combine in a photograph captioned: 'Alfred Nudd awaits his turn of the Stereoscope'. Turner also presents Nudd deploying knowledge for gain, sending redshank eggs to market as lapwings': '"But Alfred," I said, "they are quite different from Lapwings' eggs." "They're all the same to them Londoners; *they* don't know what they eat."' (56).

Turner carried national scientific status, one of the first ten women fellows of the Linnean Society elected in 1904. On the standard Linnean Fellowship recommendation form, the printed word 'gentleman' is crossed out, with 'lady' handwritten above: 'a lady attached to the study of Natural History, especially Ornithology'.[15] Turner's head and shoulders creep into James Sant's painting of the ten women,

Figure 21   'Alfred Nudd Awaits his Turn of the Stereoscope'; Alfred Nudd and Emma Turner. Source: Turner 1924.

now on the Linnean Society staircase (Gage and Stearn 1988). Turner was Fellow of the Zoological Society, Vice-president of the RSPB, and Gold Medallist of the Royal Photographic Society, co-authoring technical photographic works (Smith and Turner 1927; Smith, Turner and Hallam 1932). Broadland featured from Turner's first book, *Home Life of Some Marsh Birds*, a 1907 photographic number of *British Birds* (Turner and Bahr 1907). *Nature* termed Turner 'the pioneer of women bird photographers', influenced by Richard Kearton in technique (Nature 1940). Turner compared her method to a sessile oyster, waiting under leaf litter: 'The rubbish-heap method of photography was absolutely exhausting, but it had lively compensations. The hiding-tent has wholly done away with the old intimacy which so often existed between the photographer and stray birds' (Turner 1924: 52). If Turner could downplay science, presenting *Broadland Birds* as 'personal observations' for the 'bird lover' (v), *Nature* highlighted the work's 'considerable scientific interest' (Nature 1925).

Turner engaged with Norfolk natural history through houseboat-maiden-voyager MCH Bird, rector of nearby Brunstead and 1908–9 NNNS President, and naturalist Robert Gurney of Ingham Hall, her co-author for *A Book About Birds* for scouts and guides (Turner and Gurney 1925). Turner served as NNNS President in 1921–2, before publishing her chief Norfolk works, *Broadland Birds* (1924) and *Bird Watching on Scolt Head Island* (1928). Her Presidential Address on 'The Status of Birds in Broadland' discussed shooting, collecting and drainage, with the petrol engine shaping bird life, reducing hay demand for London horse buses and altering marsh management (Turner 1922). The substantial *Broadland Birds* would describe Turner observing amid windmills, regattas and night mists, with 'Autumn and Winter in Broadland' conveying out-of-season authority (Turner 1924: 159–70). Close photographic work on individual species presented watcher and birds in intimate space, images seldom showing a wider scene.

Nature praised *Broadland Birds* for avoiding 'anthropomorphic interpretation' (Nature 1925), though Turner's later children's writing, *My Swans, The Wylly-Wyllys* (Turner 1932a) and *Togo, My Squirrel* (Turner 1932b), embraced anthropomorphism. *My Swans* projected swan family values far from Emerson's amorality tales, with 'the true story of a family of swans which lived near my house-boat on Hickling Broad'. A frontispiece showed 'Mr and Mrs Wylly-Wylly and Family', two adults and three cygnets, their name from the fenmen's 'wulla, wulla' swan call. The Wyllys feed off Turner, with bread an attraction, the whole family visiting three days after hatching, receiving half a loaf , though Mrs Wylly is 'jealous of the time Wylly spent philandering with us' (Turner 1932a: 34); Mr Wylly reaches in the houseboat, seizing a cigarette from Turner's mouth. Turner described swan history and myth, with incidents from *Broadland Birds* reprised, though there the swans were unnamed. The adult Wyllys are pinioned, the cygnets not, Turner welcoming the end of pinioning and bill marking as bringing 'the priceless gift of freedom' (126).

If Day worked masculinity around Broadland, Turner too worked gender, satisfying publishing expectations, and conveying nature domesticity to counter gunning values. An upper middle class femininity styles the Broads with eccentric stability, skilled work practised within estate parameters. Another species of writing, contrasting to Turner in gendered performance and regional geography, is found in Arthur Patterson.

Arthur Patterson

In October 1957 a Portland stone tablet was unveiled near the new Patterson Close, ordinary cul-de-sac housing in Yarmouth. Road and tablet marked the centenary of Arthur Patterson (1857–1935), born nearby. A Town Hall Yarmouth Naturalists Society meeting followed, speakers including Ted Ellis (who also produced an *EDP* tribute) and Roland Green. Janet Spruce, aged 13, received 'The Patterson Shield' for best school nature notebook in the Borough (Ellis 1957; EDP 1957a).

Patterson grew up in the back-to-back Yarmouth Rows, last of nine children, the only to survive to 21 (TNNNS 1935b; Manning 1948; Tooley 1985; 2004). Patterson's *Nature in Eastern Norfolk* included an 'Autobiographical' opening, born in 'one of the poorest streets', enthused by cemetery bird life and allotment insects, shooting and watching on Breydon, finding fish specimens through quay contacts, writing manuscripts on waste paper (Patterson 1905: 1–13). Patterson's Yarmouth life became central to his reputation, natural history pursued between long working hours. Cocker's 2008 NNNS Presidential Address highlighted his rare working class presence in the natural history of the time (Cocker 2008). Patterson's voice is as carefully honed as a Turner or Buxton, a literary autodidact performing his self. Patterson cultivated a Thoreauian Broadland presence ('I see myself not far removed from him' (Patterson 1923: 29)) on the houseboats Moorhen (from 1894), Moorhen II (1913) and Moorhen III (1930), and the punt Yarwhelp (dialect for black-tailed godwit) after World War I (Manning 1948: 65). *Wild Life on a Norfolk Estuary* (1907) presented 'The Skipper of the Moorhen' at his 'Breydon observatory' (Patterson 1907: 94), while later works ranged around Broadland; *Through Broadland in a Breydon Punt* (Patterson 1920), *The Cruise of the 'Walrus' on the Broads* (Patterson 1923), *Through Broadland by Sail and Motor* (published by Blakes) (Patterson 1930a). Patterson also looked to Samuel Smiles' self-improving shoemaker-naturalist Thomas Edward, writing in 1897: 'I am a bit of a field naturalist, a sort of Bloaterland (Great Yarmouth) Tom Edward, I believe, they look upon me here' (Manning 1948: 70; on Smiles see Finnegan 2009: 106–11).

From his first book, *Seaside Scribblings for Visitors* (Patterson 1887), Patterson wrote for local and national publishers. Patterson's Breydon chapter in Dutt's *The Norfolk Broads* (Dutt 1903: 200–14) led Methuen to publish six books, some, like Dutt's, illustrated by Southgate (Patterson 1904; 1905; 1907; 1909; 1929; 1930b). William Beach Thomas's 1907 *Times Literary Supplement* review found Patterson's life adversely manifest in style, 'the only leisure hours of a busy and anxious life' producing 'a strange ill-assorted hotch-potch, sometimes verbose, sometimes too compact, and always ill-arranged', though 'raw material of a quality very rare in natural history books' (Beach Thomas 1907). Patterson was a school attendance officer (1892–1926, his only salaried post), postman, warehouseman, taxidermist, shadow puppeteer, Primitive Methodist preacher, travelling showman exhibiting

Figure 22   'The Skipper of the "Moorhen"'; Arthur Patterson. Source: frontispiece of Patterson 1907.

exotic animals, zoo keeper (in Preston and Dublin, with a book giving *Notes on Pet Monkeys and How to Manage Them* (Patterson 1888)), cartoonist, lecturer drawing 'lightning sketches' of fauna, local newspaper columnist as 'John Knowlittle' (from 1896), and dialect columnist of 'Melinda Twaddle's Notions' in the weekly *Yarmouth Mercury* (1893–1931). Patterson's work exhibited curiosity, *Notes of an East Coast Naturalist* headings including 'Strange Companionships' and 'Curious Manoeuvres'. 'Strange Fatalities' include a kingfisher killed by the vibration of the punt gun it perched on. 'Fish Notes' include deformed codfish and hermaphrodite herring, Patterson haunting Yarmouth quay, curiosity after curiosity (Patterson 1904).

Patterson received varying institutional recognition. Southwell proposed him for NNNS election in 1889, paying his subscription until Honorary Membership was granted in 1910. Patterson never held county office, refusing the 1921–2 Presidency (held instead by Emma Turner): 'I am not the sort of man for the job' (Manning 1948: 107). In 1902 the Duchess of Bedford wrote to Patterson for advice, and became his patron, writing the preface for *Wild Life on a Norfolk Estuary*. In a sketchbook Patterson offered himself as a human-headed, bird-bodied 'Rare Hybrid', stuffed and mounted, 'Lordium Mudflattus Pattersonii / from Breydon Gt Yarmouth / Discovered 1902 by Her Grace the Duchess of Bedford' (Tooley 1985: 67). Paterson's parody registers the class predicament of the patronised, though also alludes to Bedford's scientific standing as, with Turner, one of the first Linnean Society women fellows. From 1906 Bedford sought to nominate Patterson for Linnean Associate status, 'an honour restricted to those financially unable to offer themselves as candidates for Fellowship' (Gage and Stearn 1988: 198); Smiles' shoemaker naturalist was 'Thomas Edward, A.L.S.'. Only 25 Associates were permitted, Patterson eventually successful in 1935, the Duchess writing to the

Society: 'He is an old man now, over 70 and it has been his great wish that he might one day be considered worthy of the honour.'[16] Patterson was elected on 9 May 1935; he died that October (Long 1936).

Patterson offered a very human natural history, covering 'The birds and insects, plants and fishes, and a host of other natural objects – even Broadland Man himself' (Paterson 1892: preface). 'Broadland Man' indicates a social anthropological vision rendering an individually eccentric Broadland type, 'so characteristic of the locality as to seem almost part and parcel of its indigenous *Fauna!*' (Patterson 1905: 37). Breydon was Patterson's focus, with 'Lady Breydoners' (Patterson 1907: 172–4; 1929: 33–9) rarely figuring in an outdoor male world very different to stratified Hickling, fowlers independent, rough, killing for income and survival, transgressing poaching laws: '"Now, young Patt'son, I never stole from a man of my own class, … whatever I *did* nab come from a richer one, and worn't missed"' (Patterson 1929: 253; Hopkins 1985; Howkins 1985). Unowned Breydon shapes a different animal landscape. Patterson documented a 'disappearing class', clinging to Breydon though gone elsewhere (Patterson 1929: xi). Patterson set field-lore of 'Men and Manners' against Emerson's portrayal of Breydoners as drunken and vicious, dialect transcriptions of 'Native's Opinions' giving a hearing to Breydon watcher Jary, 'Snicker' Larn, 'Short 'un' Page and others, with Patterson's the only non-dialect voice (Patterson 1907: 95–100). Manning notes Breydoners regarding Patterson as 'a sort of camp follower and an enigma' (Manning 1948: 22). If, for Patterson, 'Like Lubbock's interesting "Broadmen," they seldom strayed far from their favourite resort' (Patterson 1905: 34), the world came to Breydoners, and they sold to the world. When, in *Wild Life on a Norfolk Estuary*, Patterson describes passenger steamers disrupting Breydoners' net fishing, the fishing is for the London market (Patterson 1929: 180).

Patterson differed from Breydoners on the gun: 'Since 1891, I have entirely discarded the gun as a "help" to observation, and have derived comparably more pleasure and interest in the pursuit of wild-life with a field-glass' (Patterson 1904: vii). Patterson's 1896 Society for the Protection of Birds pamphlet, *A Protest by a Masculine Naturalist*, found fashion absurdity grounded in cruelty, highlighting 'The Goura Mount', a whole New Guinea Crowned Pigeon worn as headgear, feminine vanity feeding barbarous trade just as masculine vanity shaped shooting trophy culture (Patterson 1896). Patterson criticised the 'sanctuary humbug' whereby bird protection and shooting cohabited (Manning 1948: 122), and mocked Fielding Harmer, Everitt's model fowler, whose Breydon contribution to Emerson's *Wild Life on a Tidal Water* was 'more of a butcher's record than a bird lover's diary': 'He rowed and shot over the Estuary with the airs and imperiousness of a Tudor lord; … slavishly dressed as recommended by Colonel Hawker. The rank and file of the illiterate, unrefined horde of fishermen and fowlers were, I think, more amused than awed by his assumption and self-assertion' (Patterson 1929: 140). Patterson juxtaposed photographs of 'R. Fielding Harmer, Gentleman-Wild-Fowler', posed in a studio, and 'Little Pintail Thomas, Wild-Fowler', sat by his shed (142–3). Patterson continued to support fowlers' shooting rights, including through the revival of Breydoners' 'swan suppers' at his Yarmouth home, the last in 1934 (Tooley 2004: 125–6). Class judgements shape Patterson's toleration of shooting.

That Patterson might not be an eccentric one-off, but part of an autodidact culture, is indicated by his co-authored 1903 memoir of *Charles H. Harrison, Broadland Artist*.

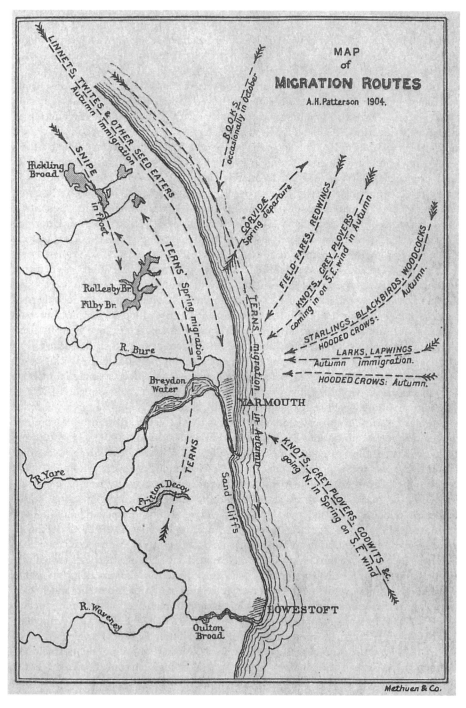

Figure 23   'Map of Migration Routes', by Arthur Patterson. Source: Patterson 1905.

Harrison (1842–1902) appears a painterly Patterson, of humble Yarmouth origin, learning from nature ('To him landscape was rightly the beginning and the end' (Patterson and Smith 1903: 29)), facing class barriers, aided by aristocratic patronage (from Lady Crossley of Somerleyton), devoted to his region. Broadland holidays were taken with fellow painter Stephen Batchelder, and Yarmouth telescope maker Calver: 'It was fitting that the Yarmouth self-taught scientist, and the self-taught artist, should be life-long friends' (27). Patterson campaigned to restore Harrison's reputation, organising a memorial 1902 Tollhouse Museum exhibition. At Harrison's funeral Patterson placed Filby bulrushes on the coffin, naturalist to artist: 'From one of his well-beloved Broads – to be buried with him' (54).

Yarmouth and Breydon were Patterson's distinctive Broadland territory, Patterson describing himself on Breydon watching Yarmouth, and in Yarmouth looking to Breydon, the town as port and wildfowling centre, the seaside incidental. Patterson's 'Map of Migration Routes' showed knots and godwits going north in spring, larks and lapwings from the sea in autumn, Yarmouth the fulcrum of arrowing flight, Patterson's locally set eyes noting all (Patterson 1905: 58). Patterson's Breydon curiosity also involved pastiche accounts of 'Breydon in 1755', and 'In Norfolk Bird Haunts in A.D. 1755': 'How the bundle of torn and almost illegible manuscript which purports to be the diary of one Sylas Hardley came into my possession matters not; for its authenticity I am not prepared to argue' (Patterson 1923: 133; Patterson 1930c).[17] Patterson presents Hardley shooting with his 'philosophic henchman' (Patterson 1930c: 27), birdwatching through his 'perspective glass' (Patterson 1923: 139), listing 'strange local names of birds' (Patterson 1930c: 70–2). Hardley concludes: 'And then we made for the Lady Haven once more, much pleased to have seen this remarkable estuary.' Patterson adds: 'Here the entries finished, for quite a number of leaves had been badly torn; and the entries that followed referred not to Breydon' (Patterson 1923: 145). Hardley leaves Patterson's Breydon in self-invention.

Reserving Nature

On 29 October 1957, under the headline 'Cheap Peeps', the *EDP* reported a Broads Society meeting:

> Put a charge on bird watching in Broadland. This was the suggestion made last night by Mr L. Ramuz, retiring chairman of the Broads Society. Tolls for boat users were high and were going higher to pay for Haddiscoe Cut, he said. 'Bird watchers are allowed to watch absolutely free. I don't think it would be a bad idea if they were charged half-a-crown a peep.' (EDP 1957c)

By the mid twentieth century, structures of conservation and rituals of seeing had developed around Broadland bird life such that Leonard Ramuz's half-joke could make sense. Ornithological culture had shifted from Turner and Patterson's time, with the management and seeing of birds a matter of public interest and policy. By 2000 Broadland would include seven National Nature Reserves (NNR) and 29 Sites of

Special Scientific Interest (SSSI), while 4623ha of Broadland and 1203ha of Breydon Water were designated as Wetlands of International Importance under the Ramsar Convention (George 2000c). This section considers the legislative and institutional processes reserving post-war nature.

The 1948 establishment of the Nature Conservancy, and the 1949 National Parks and Access to the Countryside Act, registered nature as object for public regulation. The Wild Life Conservation Special Committee, chaired by Julian Huxley, had considered the Broads as of 'outstanding scientific value', recommending two Broadland national nature reserves, Barton and Hickling/Horsey (Wild Life Conservation Special Committee 1947: 107). The 1954 Wild Birds Protection Act established protection for all birds, nests and eggs, except for game and a few 'black list' and 'quarry species', including 8 duck, 4 geese, 8 waders, coot and moorhen. Sites could be subject to sanctuary orders, and punt guns were effectively outlawed (George 2000a: 25–30). Wildfowler concern was mitigated through meetings organised by NC Director-General Max Nicholson, with the Wildfowlers' Association of Great Britain and Ireland promoting a modern 'New Wildfowler' in alliance with conservation (Matless, Watkins and Merchant 2005). Tensions could however persist, including at Hickling, discussed below.

The NNT and NNNS worked closely together, shaping wartime planning via a local Nature Reserve Investigation Committee (Riviere 1943). The NNT had been formed in 1926 by Norwich doctor Sydney Long, with its first reserve at Cley on the north coast. Small Broadland reserves were acquired before the war, at Starch Grass, Martham (1928) and Alderfen Broad (1930), isolated and without water access. Post-war reserves were of different scale and complexity; Barton Broad (1945), Hickling Broad (1945), Ranworth Broad (1948), Surlingham Broad (1948, discussed in Chapter Five) (NNT 1976a). Resources came from membership, donation, public appeal, share dividends and property income. Until the mid 1960s this was a small body with membership in the hundreds, headed by sporting landowner-naturalists.[18] Links and differences between the landholding NNT and scientific NNNS were evident when in 1950 the NNT failed in a rate exemption appeal, their Articles of Association not deemed to constitute a scientific society. Ellis and Lambert spoke in court on their behalf, their case won in 1952, with a new Clause having stated the Trust's first object as 'To encourage and facilitate the advancement of the natural sciences' (EDP 1951b).[19] In 1963 the NNT established a Scientific Advisory Committee, with Ellis as Secretary and Cambridge ecologist AS Watt as Chair, and representatives from the NC and new UEA, science moving into trust.

NNT reserves, and those managed by other bodies, extend a dual sense of 'reserve'; of nature enclosed, under reserved human conduct, with tensions of access and regulation following. Even open Breydon could be gathered, made Norfolk's first Local Nature Reserve in 1967, under local authority control, in part to regulate shooting, voluntary supervision made statutory 'to put a stop to indiscriminate shooting by "marsh cowboys or teddies"' (EDP 1963e; George 1992: 473–5). Unruly shooting and aggressive working class youth culture are aligned, neither belonging.[20] Horsey was conveyed to the National Trust in 1945 by Anthony Buxton, the family continuing to manage the estate on a 99 year lease; his son John would take over in 1958. Buxton had proposed to the NNT in February 1944 that National Trust rather than NNT ownership was best for both Horsey and Hickling. A joint NNT–NT appeal was explored, but the NNT refused to accept NT ownership.[21] From 1979 Horsey would see the return of cranes to the UK after 400 years, their breeding presence carefully nurtured by John Buxton (Buxton and Durdin 2011). A linocut of cranes by Robert Gillmor takes the cover of the BA's *Broads Plan 2011*, symbol of a region renewed.

NNT reserves were themselves properties turned to Trust ownership. Barton Broad was acquired in 1945 through donation and purchase of different sections.[22] Ranworth was offered in 1948 by Henry Cator, Ranworth Inner Broad and Cockshoot Broad to be 'a permanent sanctuary for ducks' with no shooting, except for the 'right of H.M. the King to shoot over these Broads on one day per annum selected by him'.[23] February–September public access to Ranworth Inner Broad (the adjoining Malthouse Broad was already public) was withheld after tern egg thefts in 1949, the *EDP* reporting 'a melancholy confirmation of the worst fears of those who prefer to see nature reserves closed to the general public' (EDP 1949e). Ranworth and Cockshoot came within the Bure Marshes NNR in 1958, management agreements negotiated with the NC, with Hoveton Great Broad added by agreement with owner TRC Blofeld (George 1992: 457).

The NNT Hickling reserve was acquired in 1945, becoming a major showpiece, preoccupation and burden. Lord Desborough died in January 1945, Lady Desborough seeking to sell Whiteslea; a public NNT appeal for £10,000 supplemented donations. Key to negotiation and finance was Christopher Cadbury, of the chocolate manufacturing family, NNT Council member from 1945, and President of the Royal Society for Nature Conservation 1962–88. Cadbury first attended NNT Council in March 1945, having discussed purchase with Desborough before his death (George 2000b).[24] Resident in Worcestershire, Cadbury illustrates the political plurality in local nature elites, from Liberal Lucas through Conservative Desborough to Cadbury, leader of the Labour group on Worcestershire County Council. Cadbury effectively took Whiteslea Lodge as a second home, with annual shooting holidays in April/May and August/September, and responsibility taken for repairs. With further royal visits (the Duke of Edinburgh brought Prince Charles twice, staying at the Pleasure Boat Inn in January 1959 due to flooding), Whiteslea retained social exclusivity.

Labour and leisure preoccupied the Hickling Broad Management Committee. Cadbury helped fund wages of keepers and marshmen, and purchased a reed cutting machine in 1962. Vincent's former assistant Ted Piggin acted as Warden until 1965, contributing 'Notes From Hickling' to *TNNS* reports. The Trust found Piggin less knowledgeable and dutiful than Vincent, with occasional 'incivility' to visitors.[25] An unsigned memo suggested Piggin nevertheless looked after things well: 'he does not know a great deal about birds, but I cannot see that this matters either. His job is to keep the sanctuary quiet, and if he says he has seen an orphean warbler when in fact he has seen a hedge sparrow, it is surely immaterial as long as the bird, whatever it is, is protected.'[26] Hickling was the NNT showpiece, Eric Fowler suggesting wardens might be model guides, 'wise in the ways of the Broads and the wild creatures that inhabit them, happy to instruct and encourage the curious, but standing no nonsense from the "cocky" and the destructive' (Fowler 1947: 326). Visits were controlled, with hides provided from the late 1950s. The NNNS, and Lambert's ecology students, were welcome, though other leisure less so; the NNT opposed 1959 proposals for a water skiing club and 1962 plans for 'flat-a-floats' (permanently moored floating chalets) at Catfield Staithe (EDP 1962c).[27] Piggin showed Fowler-as-Mardle visitor vandalism on 'Miss Turner's Island': 'The hut where she developed her photographs stands yet ... Piggin ... pointed, with a grunt of disgust, to smashed window panes ... his contempt for some of the species holidaymaker is abysmal' (Mardle 1963). Visitors received guided welcome from the late 1960s, Martin George of the NC devising an innovative Water Trail, inaugurated in May 1970 by naturalist James Fisher, with tours on a Norfolk reed lighter, a tree top observation post and thatched viewing hut near the Lodge. Over 800 took the trail by August.[28]

The Water Trail occupied a space recently vacated by shooting, and signified new nature values coming to regional prominence. In 1946 the RSPB as major Hickling appeal donor had hoped that 'all shooting on the estate will cease as it cannot be a real sanctuary with that going on',[29] and tensions continued over the coot shoot, and shooting rights held by NNT Council members. From 1952 the NC sought management agreements over Hickling, but encountered stubborn NNT independence over shooting. If scientific NNT figures such as Watt urged agreement, others including Anthony Buxton resisted. Shooting provided Trust income, Council member Aubrey Buxton (Anthony's cousin) making a seven year syndicate subscription in 1952. Buxton personifies the complex relations of tradition and modernity shaping the NNT and conservation, upholding traditional field sports while founding Anglia Television and instigating the influential *Survival* wildlife series, producing a special on the NNT in 1972, *Norfolk in Trust* (EDP 1972). Like his friend the Duke of Edinburgh, Buxton proffered dynamic engagement with nature built on sporting natural history (Matless, Watkins and Merchant 2005). John Buxton likewise combined shooting and film, producing *Broadland Summer* and *Broadland Winter* as RSPB films in 1965, technically impressive hymns to the 'balance of Broadland conservation', noting 'sport' and 'bird protection' in 'paradoxical' alliance.[30]

Hickling was declared a National Nature Reserve in June 1958 despite ongoing tension, with public complaint building over the coot shoot, Cadbury regretting that publicity was antagonising 'public sympathy'.[31] Letters of complaint cited 'wanton slaughter' in a 'bird sanctuary' (J Hoddinott, Suffolk, 29 September 1956), and questioned 'how your members square their consciences about the killing of birds' (Miss Peart, Manchester, 25 June 1956). Newspaper reports of Prince Philip and son prompted anger: 'the poor Prince Charles has probably been bullied into taking part in this terrible pastime' (A Hind, London, 5 January 1957).[32] When Aubrey Buxton asked to see the correspondence, Constance Gay, NNT Secretary 1938–58 (and NNNS President 1943–4) replied:

I have been told many times that the Trust's membership would be increased considerably if shooting were stopped and I believe this to be true. The Norfolk and Norwich Naturalists' Society is against us almost to a man ... If I may be allowed to say so I think it must be very difficult for you and other members of the Council who have shot since boyhood to appreciate the feelings of quite 90% of the Trust's members who have probably never owned a gun in their lives – except possibly in the army.[33]

Buxton replied forcefully, arguing against views 'based on sentiment, and not on reason': 'Unless you are prepared to be a vegetarian (perhaps you are!), it is illogical to object to a bird being <u>shot</u>.' Coot shooting controlled excessive numbers, as would shooting 'of all those frightful swans'.[34]

In March 1958 NNT Council resolved to end shooting on all reserves in five years, as 'financial and other considerations permit'.[35] Internal debate continued, Aubrey Buxton, Timothy Colman and John Buxton resisting cessation. Shooting ended in 1965, the reserve soon to welcome Water Trailers, another culture of reserved nature inhabiting Hickling.

The Public Animal

Feathersight, feathersound

The post-war public value of nature was shaped by the animal in broadcast and print media, generating public understanding and affection, with shooting critiqued or downplayed. Conservation and communication were two sides of a post-war 'new naturalism', fostering popular appreciation as part of a broadly social-democratic citizenship, with the Collins 'New Naturalist' book series one manifestation, and sound a notable theme, whether as the medium to relay nature, or as an aspect for understanding (Marren 1995; Matless 1998; Macdonald 2002).

In 1955 Tyler Whittle's *Spades and Feathers* extended Ransome's children adventure landscape, excavating treasure at St Benet's and caring for avocets near Barton. The avocet was a contemporary cause célèbre, re-established in Suffolk in 1947, though only nesting in Norfolk in 1977 (Taylor et al 2000; Davis 2011). Young enthusiasts accompany eccentric older ornithologists, plus local 'expert naturalist' Osbert Parsley.[36] Uncle Bertie, Parsley and the children record sound, cables, microphones and signal systems capturing the bird:

> just as Sarah and Kate and Paul were giving up hope, they all heard a bird's cry, as loudly and as clearly as if it were sitting on the collapsible table and giving them a special performance.
> 'NGUH, NGUH, NGUH, CLEWT,' cried the bird.
> 'Great Scott,' said Uncle Bertie. ...
> 'NGUH, NGUH, NGUH, CLEWT.'
> 'He's suspicious of something,' said Uncle Bertie excitedly. 'You've caught him properly.' ...
> 'It's the best recording I've ever heard,' whispered Uncle Bertie in awe.
> 'NGUH, CLEWT,' murmured the avocet. (Whittle 1955: 186–7)

Through sound, children enter a new naturalism, the avocet's attitude caught 'properly', their earlier archaeological finds surpassed by ornithological treasure.

Whittle plays on sonic topicality. If John Betjeman's 1937 poem 'Slough' had mocked suburban dwellers who 'do not know / The bird song from the radio' (Betjeman 1974: 23), new naturalism inverted Betjeman via bird song *on* the radio.

Sound pioneer Ludwig Koch, refugee from Nazi Germany, who in the late 1930s had produced wildlife sound-books with leading naturalists (Nicholson and Koch 1936; 1937; Huxley and Koch 1938), took aural new naturalism to Broadland, his *Memoirs* showing him at Horsey in 1945, punted by Crees to record the bittern (discussed below) (Koch 1955). Alongside broadcast recording came imitation, with Percy Edwards impersonating over 150 birds, a field naturalist variety turn whose light Suffolk voice called the radio listener to a wonderland of accuracy (Edwards 1948). On a 1959 *Woman's Hour* from Norwich, Edwards conveyed 'A Fairyland' at Black Horse Broad: 'Those rocket winged little ducks, the teal, they flash past, but no protests, just dainty call notes. (Example)'.[37]

Children were a special radio nature target. In 1955 the *Country Schools* slot dramatised a Broads story by naturalist 'BB' (Denys Watkins-Pitchford), with children boating and camping, investigating insects and birds, and catching an egg collector.[38] *Children's Hour* in June 1952 featured Wynford Vaughan Thomas and Leslie Anderson's 'Broadlands Journey', a regional tour including Roland Green at Hickling, Crees at Horsey, and the Norfolk Room, the presenters discussing the Yare diorama with Curator Ellis: 'It's the nearest thing we have in England to a jungle.'[39]

EA Ellis: voicing the region

By 1952 Ellis was the most prominent Broadland media voice; his career requires brief introduction here (Stone 1988). Ellis (1909–86) was born in Guernsey, moving to Gorleston in 1920; his parents came from Great Yarmouth. From 1946 until his death Ellis lived at Wheatfen in the Yare Valley, the property now owned by the Ted Ellis Trust; Wheatfen and Ellis's plant studies are discussed in Chapter Five. Wheatfen has since overshadowed Ellis's earlier Gorleston life. Ellis only moved to Norwich after marriage in 1938, living in the south-eastern suburbs until 1942, then at Rockland and Brundall before Wheatfen. Ellis's mentor from 1923 was Patterson, shaping his curiosity, mode of nature articulation (notes, talks, sketches) and sense of nature as socially open. Ellis's first public talk inspired the 1927 re-formation of Yarmouth Naturalists Society. By Patterson's 1957 centenary, Ellis would be the public figure commemorating his mentor.

Ellis worked at the Castle Museum from 1928–56, as assistant, then Keeper in Natural History. Tubercular diagnosis in 1940 prevented war service. Elected to the NNNS in 1929, Ellis served as President 1953–5 and 1969, *TNNNS* Editor 1951–82, and was NNT Secretary 1958–62 and Vice-president 1979–83. At the Castle too Ellis performed a public nature role, answering queries, identifying specimens. His 1956 departure brought an *EDP* editorial:

> To amateurs of natural history in East Anglia … the news of the impending resignation of Mr EA Ellis … will convey the same sort of shock as an announcement to Norwich City supporters of the impending transfer of a greatly gifted centre-forward to a First Division Club.

Fortunately, ... he is not being sold down the river to the British Museum. His work has contributed to keep Norfolk and Norwich very near the top of the First Division as regards natural history, and he has no intention of deserting the team. (EDP 1956a)[40]

Newspaper columns and radio talks made Ellis a public regional voice (Ellis 1982; 1998). Ellis contributed to the *EDP* and *EEN* from 1928 under pennames including 'Chirp' and 'Moldwarp', and from 1 October 1946 produced a daily *EDP* 'In the Countryside' column for 40 years, along with 'Young Naturalist' and gardening columns. A fortnightly national *Guardian* 'Country Diary' contribution began in 1964. Journalism honed Ellis's voice; observational, reflective, sometimes moral though rarely moralising, alert to curiosities. Columns often conveyed night or early morning at Wheatfen, sensory country experience beyond town readers, Ellis becoming in this sense, rather than via agricultural subject matter, a country writer.

EDP columns were reshaped in Ellis's 'Naturalist's Diary' contribution to Mottram's 1952 *The Broads*, making a composite year, beginning:

You may be a summer voyager, ... You may be a week-end angler and dreamer, ... Perhaps you journey in search of the fantastic bittern, to catch a glimpse of the swallowtail butterfly in all its glory, ... Belonging, you may keep cattle on our vast green levels ... Whether you find yourself here for a day or a lifetime, mysteries will crowd upon you and the peace that is felt in solitude will be yours in the midst of this land of bright waters, jungle wildernesses and twinkling facets of life on every side.

Ellis addresses, almost blesses, different modes of being in Broadland, with shooting a notable absence. If farming denotes 'belonging', that does not give it cultural priority, Broadland implied as a place of common cultural ownership. Animal and human life co-occupy, co-existently territorial. Under overall 'enchantments', the 'twinkling facets of life' warrant close observation (Ellis 1952f: 185).

Ellis first broadcast in 1946, becoming a regular national and regional radio voice. Ellis's summer 1957 talks, *Through East Anglian Eyes*, began with 'Night Excursions and Alarms on the Broads', Ellis inviting the listener into curiosity: 'For many years I was puzzled by loud smacking noises, coming from the water's edge, particularly on moonlight nights in July'; eels suck in air after eating snails on water-logged grass.[41] Ellis was also national broadcasting's regional contact, as when Philippa Pearce wrote in preparation for a 1946 Home Service Rural Schools programme, *Norfolk Broads*. Ellis suggested staying at Wheatfen, visiting Hickling, even sleeping in Turner's island hut: 'you would need a sleeping-bag, some mosquito-lotion and gum boots'.[42] Pearce stayed at Wheatfen, though preferred Hickling's Pleasure Boat Inn. Ellis briefed Piggin: 'Miss Pearce, of the B.B.C. is coming down to Norfolk this week ... we are anxious that she should see Hickling and have a chat with you ... We feel that if anything is to be broadcast about the Broads, the right things ought to be said.'[43] The eventual broadcast featured an unnamed 'Bird Warden': 'I wished I was like the Bird Warden – able to recognise bird after bird at a distance with the naked eye.'[44]

From the 1960s Ellis also became a BBC television personality, receiving a Royal Television Society regional award in 1982. Eugene Stone would entitle his 1988 Ellis biography *The People's Naturalist*, though the phrase contains complexity, the public new naturalist emerging through a dialectic of detachment and connection. Detachment from society enables nature pursuit, institutional and media attachment enables nature communication, the latter connection resting on the former solitude, making the perfect cultural space for eccentricity, which in turn carries popular appeal, the naturalist a possible species put up for entertainment. On camera and air, Ellis walked a fine line of erudite curiosity.

Ellis's scientific and popular reputations combined in editing the 1965 Collins New Naturalist volume on *The Broads*, a book initially listed as forthcoming in 1948 (Ellis 1965). Robert Gurney was first commissioned, but editor James Fisher deemed his material unsuitable; Gurney died in 1950, Ellis taking over as editor and chief author. Editorial disorganisation, and loss of colour plates, delayed completion.[45] The new origin theories necessitated revisions, Lambert and Jennings, with Smith and Green, making substantial contributions; other authors were Gurney (aquatic fauna), AE Ellis (no relation, on molluscs, woodlice and harvestmen), Duffey of the NC (spiders), Horace Bolingbroke (native river craft), BB Riviere (birds) and Rainbird Clarke (archaeology and history). Three authors (Gurney, Clarke and Bolingbroke) were dead by publication. The book's cover departed from the New Naturalist series' usual distinctive designs by Clifford and Rosemary Ellis, showing instead a colour photograph of a wherry on a river.[46] *The Broads* sold 8500 hardback copies, 5000 by June 1966.

Ellis continues to be evoked as a regional voice. The BA have adopted Ellis's phrase, 'a breathing space for the cure of souls', as capturing a Broadland essence, his words becoming a regional slogan, as in the *Broads Plan 2011* (Broads Authority 2011: 9/16). The BA's claim to a Broadland phrase however belies an original complexity, Ellis's April 1964 column 'Let Us Keep Our Serenity' evoking 'the pastoral peace of the East Anglian countryside' in general, including agricultural and forest country: 'Let us remain a breathing space for the cure of souls rejoicing in honest agriculture, forestry and the like and cherishing the serene beauty of our Broads and coast. I am sure we can still do this and live' (Ellis 1982: 67).

Feathercraft

If authors and broadcasters evoked an enchanted Broadland, another regional institution offered animals in different fashion, dealing in game and adornment. From 1921 Pettitts of Reedham processed 'Food, Feather and Fur' (Pettitts 1949: 10), animals shot for food and plucked for ornament. Masculine shooting culture meets a feminised culture of decoration. Peacocks attracted visitors to the Reedham showroom, another public space of Broadland birdlife.

In the Valley of the Yare, published around 1949, set out Pettitts' commercial territory, a colour illustration promoting the 'Pettitt Pack', pheasants hanging, rabbits grounded, and (with flying duck behind) duck lying dead with a goose, heron, oyster

Figure 24   Advertisement for the 'Pettitt Pack'. Source: Pettitts of Reedham 1949, image courtesy of Norfolk County Council Library and Information Service.

catcher and other quarry. Daily rounds gathered in wildfowl, poultry and rabbits; the north side of the Yare in the Monday Round, the south side on Friday, the Ant, Bure and Thurne villages on Tuesday. 'The Sportsman's Guide' stressed Pettitts would not receive protected or out-of-season birds, while Anthony Buxton contributed on 'If you don't know, don't shoot' (Pettitts 1949). When the NNT Council discussed Pettitts in May 1949, Buxton suggested it was 'genuine in its wish to cooperate in regard to bird protection'.[47]

In wartime Pettitts developed their 'Arts and Crafts Department', making something of feather and fur by-products, female staff turning feathers to flowers, birds to bouquets. A feather commemorative wedding 'Anemone Posy' for Princess Elizabeth gave a 'best selling' publicity coup (Pettitts 1949: 24), while the victory dinner of the National Country Packers' Association received 'A six-foot model of "Big Ben" in Feathers' (26). Designer K Cross recalled: 'I was introduced to Dr. Edith Summerskill, and was congratulated on the beauty and colour of the feathercraft.' Socialist feminist Summerskill, attending as Parliamentary Secretary in the Ministry of Food, met Broadland ornament: 'We left our model of Big Ben for a week in London in the window of Galleries Lafayette, where it attracted great attention' (24). National publicity supplemented Pettitts' regional presence: 'From the Broads each season we have many visitors to our Showroom at Reedham, and they are pleasantly surprised at the variety of fur and feather novelties which can be purchased as gifts, suitable for all ages' (26). The feather exoticism of Victorian millinery is turned to workaday wonders, a food sideline becoming profitable decorative beauty; palace posies, feather tablemats. Broadland birds find their set place.

Icon and Intruder

If individual species such as the swallowtail butterfly and marsh harrier have become symbolic of the region, others have been deemed invasive and out of place. Consideration of the bittern as Broadland icon, and the coypu as alien rodent, shows animal species as gathering points of regional cultural landscape.

Bittern

Broadland bittern numbers shift from early nineteenth century abundance (*'Botaurus stellaris*, common bittern – not uncommon; breeds at Ranworth' (Paget and Paget 1834: 7)), late nineteenth century loss, early twentieth century restoration, mid twentieth century abundance: 'Bittern. Resident; about 60 booming males recorded in 1954 ... the population has probably reached saturation point' (Ellis 1965: 207). After saturation, loss again; 28 Norfolk boomers by 1970, nine in Broadland in 1976, just three in Norfolk by 1997 (Taylor et al 2000: 120). Conservation works to re-restore, notably at Hickling, and the RSPB Strumpshaw reserve, encouraging migrants to breed in residence.

Four takes on the bittern indicate varieties of bird culture; the contemporary regional symbol, the awkward historic curiosity, the irresistible plot device, the national concern:

1. The bittern serves as regional brand. The railway from Norwich to Sheringham, skirting Broadland, becomes 'The Bittern Line'. In the 1990s the Bure broads were served by a floating take away, 'Once Bittern'; in March 2011 Woodfordes brewery, based in Woodbastwick, issued a beer supporting Norfolk Wildlife Trust, 'Once Bittern'. Puns trade on cultural value.

2. In *The Dovecote and the Aviary* (1851) Rev ES Dixon recounted an unfledged bittern, kept in confinement for five years, becoming pugnacious, booming all night, dying unlamented: 'Those who believe in the transmigration of souls ... regard the Bittern as having owed the spark of life to an idle servant, named Ocnos, who was punished for his laziness by this metamorphosis. Ocnos continues to be of but little better domestic use or profit in his second than in his first character, except when artistically roasted, or skilfully mounted in a glass case' (Dixon 1851: 326; Patterson 1905: 174).

3. The spring sound of the booming male could serve as novelistic plot device. In Sylvia Townsend Warner's *The Corner That Held Them*, the fourteenth century Broadland nuns of Oby convent ponder worldly economic engagement: 'we are also told not to hide our light under a bushel. We cannot forever go on in the old way, booming in our swamp like so many bitterns' (Warner 1948: 53). The bittern suggests the insular and remote. Bittern as mystery bird also served Keble Howard's comic *Fast Gentleman*: '"What on earth's that row?" asked Muriel. / "The bittern, darling," ... "It is a bird allied to the heron, my sweet, once abundant in Britain, but now only an occasional visitor"'. Leonard is later tempted by

Olga's moonlit deck advances: 'Boom on, O bittern, boom on! What do *you* know about life on the Broads?' (Howard 1928: 187–90).

4. The bittern also registers as national Broadland concern. At 12.07 am on 25 November 1947 in the House of Commons, during a Broads national park debate, Mr Maclay (Montrose Burghs) interjected: 'Will the Minister remember that there also arises the important question of the preservation of wild bird life to which the hon. Member for Lowestoft (Mr. Edward Evans) referred? There is the case of the bittern, and I think something ought to be done in this matter.'[48]

Bittern numbers shift with reed management and shooting, Lubbock noting the bittern 'decreased much in number in the last twenty years', Southwell adding that in 1819, 'Mr. Lubbock ... killed eleven Bitterns without searching particularly for them' (Lubbock 1879: 87). The nineteenth century bittern received visceral attention, traced back to Browne, the bittern 'esteemed the better dish' over a heron (Browne 1902: 17), having 'more of the true wildfowl flavour' (Stevenson 1870: 173). The bittern was by 1870 largely a migrant, Stevenson asking: 'Can we wonder if the Bittern's nest is robbed, when we consider the sum which a recently laid Norfolk example of its egg would command, as compared with foreign specimens' (Stevenson 1872: 18). Southwell noted a last breeding boom in May 1886 at Sutton: 'and on August 10th a young female, with down still adhering to some of its feathers, was killed at Ludham, probably the offspring of the bird heard in the spring' (Southwell 1901: 236).

Sutton Broad also saw bittern restoration, the 1911 reedbed nest discovery 'one of the great ornithological romances of our times' (Vincent 1926: 303; Clarke 1921). The romantic leads were a bittern, Jim Vincent and Emma Turner, the latter addressing the NNNS in April 1912, and *Broadland Birds'* opening chapter telling of 'the never-to-be-forgotten July 8' (Turner 1924: 2; 1912; 1918). Naturalists and unidentified locals guided the find; an anonymous 'watcher who took an interest in these Bitterns from the time of their arrival' (Turner 1924: 4), an angler friend reporting possible breeding, Robert Gurney as unnamed landowner, Rev Bird helping the search. Turner and Vincent's 'discovery' was the finding of bittern young: 'How we gloated over our prize as he stood there, transformed into the resemblance of a bunch of reeds!' (3). The '1911 Bittern' was removed for third person authorisation, returned next morning and photographed by Turner (4). The bird appeared on lantern slide at the Linnean Society in January 1912.[49]

In the 1920s Vincent as observant protector usurps Turner as lead human character, narrative geared to Hickling's presentation as estate sanctuary. In 1926 Vincent conveyed 'The Romance of the Bittern' in *Country Life*, with photographs by Humphrey Boardman, son of the How Hill owner (Vincent 1926; Boardman 1926; Patterson 1923). Vincent's 1929 *TNNNS* Hickling notes included a photograph of keeper Jim Linkhorn holding a bittern by the wing, the bittern a live trophy, displaying reserve triumph (Vincent 1929). The mid twentieth century bittern becomes an icon under estate care, as in Anthony Buxton's Horsey photographic studies, and Walter Higham's 1931 'Secrets of Nature' film *The Bittern* (Buxton 1946; 1948).[50] The estate bittern

The 1911 Bittern: The crouch before
the upward spring

Camouflage: Turning itself into the
semblance of a reed

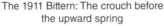

Figure 25   Bittern, photographed in 1911 by Emma Turner. Source: Turner 1924.

receives fullest elaboration, alongside herons and water-rails, in Lord William Percy's 1951 *Three Studies in Bird Character*. Percy, youngest son of the 7th Duke of Northumberland, elected to the NNNS 1926, owned the Ant valley Catfield Hall estate from 1925–45, then moving to Horstead House near Coltishall (Percy 1951; Murphy 1963). *Three Studies* scrutinised the 1934 Catfield bitterns, birds and Percy co-existing, the estate naturalised. A young bittern, late fledged, is photographed on a doorstep, alongside a young girl. Experiment aids photographic character study, Percy placing a stuffed tawny owl alongside eels and fish near the nest, the bittern feeding unperturbed.

If camouflage makes the bittern an elusive (and so still more deeply embedded) visual Broadland symbol, the distinctive boom of the spring male stands out. In *Coot Club*, Dorothea's projected adventure book includes a chapter on 'The Bittern's Warning' (Ransome 1934: 126), while after the Hullabaloos sink: '"Boom … boom … boom … boom …."' The call of the bittern sounded over the marshes in the quiet May evening' (349). Bittern boom trumps Hullabaloo racket. For Patterson the sound was 'weird, eerie, rare, unique. To me it was music sublime and appropriate, as are most of the cries of the Open' (Patterson 1923: 69). For HJ Massingham: 'Whoever has heard the occult, resonant gong of this baroque fowl … will realize that a new mastery of interpretation has been given to the watery landscape of Broadland' (Massingham, 1924: 116; Massingham 1923: 43–51). Strangely-in-place sound shaped vernacular bittern names; bottle-bump, bog bumper (Dutt 1906; Clarke 1921).

Sound also fitted the bittern for radio nature. From Hickling in 1947 Anthony Buxton described the process: 'three noises, clicks, … intake of air, a sort of gasp, and expulsion of air which makes the boom. The boom can be heard on a still day inside my house with all the windows closed, and the nearest booming place is 700 yards away. And yet it is not a very loud sound even at 10 yards.'[51] Pearce's 1946 *Norfolk Broads* programme included:

(Boom of bittern, distant)

BIRD WARDEN: Listen!

(Boom of bittern again)

LEWIS: (doubtfully) It sounded like a fog horn, but.....
BIRD WARDEN: It's the bittern booming – that's its song – it only calls like that in the spring.
LEWIS: (excited) A Norfolk bittern! I'm glad to have heard that!
BIRD WARDEN: Yes, real Norfolk birds they are – stay here summer and winter. But we lost them altogether for a time.[52]

In summer 1945 Buxton invited Koch to make a first boom recording, the expedition reported in *The Times*, Koch punting a floating studio around Horsey Mere: 'To get ten to fifteen seconds of booming we had had to be on the alert for about a hundred hours and have discs running for about 130 minutes altogether' (Koch 1955: 77–8; Koch 1945). On '*The Naturalist' in East Anglia*, chaired by Ellis, Bagnall-Oakeley introduced 12 seconds of boom, the bittern having 'had this delightful North American name introduced – the thunder pumper ... Well, there it was "bog-bumping" or "thunder-pumping", have it how you like.'[53] After six seconds of bittern noise on a 1952 *Saturday Review*, presenter Doreen Pownall interjected: 'Don't be alarmed – it isn't your radio set that's gone wrong – that's the booming of a bittern.'[54]

The boom also allowed ludicrous satire, Rooke offering a didgeridoo skit of 'Bird-boom at Twilight':

> The story put about in Norfolk is that it comes from a small bird called the bittern. But to the thinking man, the idea of such a large deep-chested boom coming from such a small narrow-chested bird scarcely bears scrutiny. Personally I lean to the view expressed by an Australian cruiser-hirer we met in Wroxham whose home town is Alice Springs. He thought the booming came from a hollowed-out tree, blown through at night by members of a Secret Norfolk Booming Society. He admitted that he hadn't actually seen any members of Secret Norfolk Booming Societies blowing through hollowed-out trees at night. But if it comes to that, how many people do you know who've seen a bittern boom?

A sketch showed blazered trilbyed men, one holding, the other shouting into, the trunk, with 'Boom!' in trembling fount (Rooke 1964: 92–3). The bittern moves from restoration romance to leisure farce, Patterson's music sublime sounding something ridiculous.

Coypu

In 1934 Robert Gurney warned *EDP* readers of 'a new pest to beware of', the 'hairy-fisted crab' (now termed the Chinese mitten crab), spreading across Europe since 1912: 'have we got this pest in Norfolk rivers already, or is it likely to invade us?'

Figure 26   Coypu, 'kept as a mascot by students of Joyce Lambert', Museum of the Broads, Stalham. Source: photograph by the author, August 2011. Reproduced by permission of the Museum of the Broads.

Vigilance was necessary: 'Actual specimens may be seen in the entrance hall of the Norwich Castle Museum' (Gurney 1934). Vigilance continues, the Norfolk Non-native Species Initiative, launched in 2008, seeking 'prevention, control and eradication of invasive, alien species' in Broadland's 'priority England Wetland Vision area' (Broads Authority 2011: 33; Landscape Research 2003). For a 'heritage of pro-active management' the Initiative highlights the 'successfully eradicated' coypu.[55] If the bittern symbolises belonging restored, the coypu sees invasion defeated, although, like the bittern, its story carries cultural complexity.

A large South American rodent, the coypu was brought to Britain for 'nutria' fur farming in 1929, with farms in the Yare and Wensum valleys. PET Carill-Worsley described 'A Fur Farm in Norfolk' for *TNNNS*, his breeders £50 a pair from France, 'delivered by air mail at Croydon' (Carill-Worsley 1931: 107). Coypu were however 'great escapers' (112), and a group of 'outlaws' would 'soon become a nuisance' (114). Most farms closed by 1939, with several escapes, the coypu moving into the Yare. The thriving Broadland coypu received scientific and political attention, the county War Agricultural Committee organising trapping, and inviting EMO Laurie of Oxford University to survey distribution (Laurie 1946). Concern over crop and riverbank damage was accompanied by zoological fascination, and welcome for coypu eating encroaching vegetation, Cator making 'The Case for Coypu': 'I feel the slogan for the time being should be "The coypu will help to keep the Broads open free, gratis and for nothing"' (Cator 1952; Nicholson 1957). Coypu had extended Ranworth Inner Broad by six acres in four years; Cator also acted as receiver of skins for sale to a London furrier, raising £56 12/- for NNT in 1955.[56] The controlled coypu offered good harvest.

In 1956 the Ministry of Agriculture's RA Davis concluded the coypu formed 'an interesting addition to our fauna' which should not be 'proscribed as a pest' (Davis 1956). He later changed his view: 'By 1957 it had become apparent that the feral coypus were not an unmixed blessing' (Davis 1963: 346). The market for pelts collapsed in 1958, population increasing to 200,000, and a state-sponsored eradication campaign began (Fitter 1959; Ellis 1960a; Norris 1963; 1967; Sheail 1988; Gosling and Baker 1989; George 1992: 231–4; Sheail 2003). It is striking however that the campaign did not play on alienness. If a native/alien polarity was then strong in the politics of human immigration, the language did not transfer, the coypu damned as intrusive pest rather than non-native. Farmers at Horsey and Hickling fired the marshes to drive coypu out (EDP 1958d; 1959c), rabbit-clearance societies were authorised to kill coypus, 100,000 were killed in the year to April 1962 by Norfolk River Board employees (EDP 1962b), and in 1962 coypu joined the proscribed list under the 1932 Destructive Imported Animals Act, and the Ministry established the Coypu Research Laboratory in Norwich. The severe winter of 1962–3 killed 90 per cent, John Buxton's *Broadland Winter* film showing the dying and dead, 'his domain of marshy jungle having now turned solid'.[57] A military style 'Coypu Campaign' co-ordinated further 'Operations', Ministry posters urging vigilance (Times 1962). In February 1965 'Operation "Broadland"' began with a 'Blitz' on Wroxham Broad, the coypu in their 'last redoubt' (EDP 1965b). The cull was scaled down, with science helping monitor movements via individual coypu fitted with radio receivers: 'Lines can then be drawn on a map to show his whereabouts' (EDP 1966). Numbers recovered in mild winters, but a further eradication campaign from 1981 was deemed successful, terminated in January 1989 (Gosling and Baker 1989).

If alienness does register in the coypu, it is for novelty and charm. Though conceding the need for control to prevent the 'ruin' of areas 'cherished by naturalists' (Ellis 1960c), Ellis suggested 'moderate numbers' would be fine (Ellis 1960a: 1591), the live coypu signifying happy landscaped foreignness: 'A warm summer day will sometimes bring them forth to frolic in quiet waters. On such occasions they may be seen floating with their tails stiffly erect, or wallowing for pure pleasure, like hippopotamuses in an East African river' (Ellis 1965: 226). Ellis told radio listeners of night sounds:

> We have South American music on the Broads now: genuine jungle stuff --- not the sort they imported from Spain. It echoes across the valley when the moon is full. If you've ever heard a cow in trouble at any time, imagine listening to one lamenting the loss of a calf in the middle of every reed-swamp. That's the nearest description I can give of a coypu's love-call ... I wouldn't say they make the night hideous with their cries, but they're very mournful under the moon's influence. (Give the coypu's cry here – MNAAHW – MNAAHW —).[58]

Ellis wore a hat of coypu fur (now in the Broads Museum) and ate coypu regularly, secretly serving it to a British Association party in 1961: 'The meat of coypu has long been appreciated abroad and, incidentally, consumed under various pseudonyms in this country. A few Norfolk people are beginning to overcome their scruples and are eating casseroled coypu regularly now, but there is no organised demand for the local product so far' (Ellis 1960a: 1591; 1958b; Stone 1988: 102). Overcoming culinary prejudice might help coypu control.

It was sporting writers, often the most reactionary cultural voices, who most consistently favoured the coypu. Former Brundall coypu farmer Savory appreciated escapees clearing 'Coypu Marsh', making 'the most delightful little duck swamp ... the perfect teal shoot', the coypu as sporting ally (Savory 1956: 83–4). Day also welcomed the 'Great South American Marsh Rat' (Day 1967: 181) as potential 'permanent and truly wild Broadland animal' (Day 1953: 154), adding 'a primeval echo of an older and larger, but vanished, fauna' (Day 1951: 181). The coypu restores rather than erodes regional spirit, Day making alliance in a roguish ecology. If the government cull was 'a wonderful excuse for orders, instructions and memos in quadruplicate', then 'the coypu beat them every time. He bred, increased and travelled!' The coypu, like Day, knew Broadland best, and Day knew the coypu: 'He is probably here for good. He is harmless to man and will only attack you if you attack him. Incidentally he is good to eat and, when roasted, very much like roast pork. If you attempt to skin one, cut the skin down the backbone, not on the belly. Thus you will obtain the best fur' (Day 1967: 182–3).

Wings Conflicting

In 1954 Broadland reached the big screen, in an episode which can serve as an epilogue for this chapter. *Conflict of Wings*, directed by Australian Don Sharp from his novel of the same name, set birds against planes, filming at Norfolk sites including Hickling and Cley (Sharp 1954). Australian Sharp would go on to direct Hammer horror, and the cult 1972 *Psychomania*; *Conflict of Wings* would gain US release as *Fuss Over Feathers*. In 1952 Ellis related to the NNT that a company were proposing to film 'The Norfolk Story' at Hickling and Alderfen: 'After consideration it was agreed that, provided neither the Trust nor its nature reserves are mentioned by name in connection with a foolish plot, no harm would be done, but Mr Ellis was asked to ensure this.'[59] Early in the film a local antiquarian ex-solicitor (a latter-day Walter Rye) nicknamed 'Bookie' is noted as writing 'The Norfolk Story'; Ellis's film would seem to have been *Conflict of Wings*.

Conflict centres on Hickling Broad, site for a projected air firing range. The film's opening shot (from the top of Whiteslea Lodge) has a jet roaring over the broad, startled swans lifting off. Bird life is at risk, including nesting gulls on the 'Island of Children', acquired by the Ministry of Land Acquisition. Love twists loyalties, a 'Saltingsby' (Hickling) girl falling for a pilot yet campaigning against his work. Sharp delights in nature and technology; exquisite Broadland bird life, thrilling jet speed. Legal research helps fight the range, Bookie's antiquarian enquiry showing that the tidal waters mean that fishing and eel rights granted by Henry VIII still hold, as if Rye's Hickling Broad Case was refought and won: 'No power, military or otherwise, can stop the people from exercising that right' (Sharp 1954: 140). Locals take boats across the broad (Kett's Rebellion is evoked), lining up in defiance of descending planes, waving as the jets near. Air training is stopped, but a public enquiry defers decision, as the squadron head off to late imperial conflict in Malaya. Local patriotism and national patriotism meet in tension, as in other stories of military land acquisition (Wright 1995), Sharp exploring post-war British identity through a Norfolk story. Should Saltingsby feel pride at saving the island, or guilt at leaving pilots unprepared for war? And are wars against a spirit of independence, whether in Norfolk or overseas, worth training for?

Film sound marks conflict; jets and bitterns booming, birdsong interrupted. Sharp's novel begins by setting silence, a 'silence' akin to John Cage's contemporaneous evocation of everyday sounds, whose musicality might be evident to those pausing to hear (Cage 1968). Sharp drives from Norwich to 'that flat region of fen and water and sky known as Broadland' (Sharp 1954: 10), to a Saltingsby shifted slightly coastal from Hickling. Broadland's ambient silence shows an animal and plant landscape to cherish, worth defence, though Sharp ends at something also signalling threat:

> You are entering a wild, enchanted place, a place of lonely beauty. You stop for a moment and hear the sounds that will never leave you again. The first sound is all about you, for even on the stillest day when no breath of air stirs them, the reeds whisper ceaselessly across the marshes. And then you are aware of that other sound, low and persistent, a steady murmur broken only now and then by a slap. Is there a note of menace in it? You can't place the unseen sound and so you go slowly on. On under that high, wide sky where the small white clouds sail steadily and where the air is never void of birds. Away in the reed beds a pheasant crows. You cross yet another dyke and a snipe gets up in erratic flight. But there are only the birds – you see no more of man. On you go through the silence ... yes, silence it is, for even the soughing of the reeds and the sudden lonely call of a wild goose do not break the silence – they increase it. Where is the road going? The other sound is coming nearer now and as the road completes its wide sweep you seem to be running beside the sound ...
>
> And then you know what that other sound is – it is the sea. For just behind the thin sandhills which shelter Saltingsby is the grey North Sea.
>
> This is the end of the road – there is no other place for it to reach. (11–12)

Chapter Five
Plant Landscapes

Plant Cultures

Writing on Wheatfen in 1939, Ellis noted 'the card index kept at the fen', a record system in the field, nature indexed on the spot (Ellis 1939: 115). Ellis's chief research preoccupation was fungi, and Wheatfen as plant landscape, sheltering common and rare botany. Ellis's ecological outlook of course highlights the interrelationship of plant and animal, but his work also demonstrates specific scientific traditions and popular cultures pertaining to plant life. This chapter examines (with occasional faunal digression) Broadland as plant landscape, in terms of regional scientific cultures, strategies of proprietorship, economic resource and cultural icon.

In 1901 Dutt described 'A Broadland Botanist', finding a landscape curiosity itself curious. Dutt's account opens up key aspects of Broadland's plant landscape:

> This same botanist was a queer character in his way ... From a distance I took him to be a marshman, for he wore a broad-brimmed Tyrolean hat, such as the dyke-drawers often wear in all weathers ... He believed he had found specimens of all the flowers native to his district, and not a few interesting aliens: now he was devoting his time to the compilation of a list of the local grasses and sedges. When he had done so he would have completed a manuscript *Flora* of ten parishes; but he doubted whether any publisher would issue it at his own expense. 'If he did,' he remarked, 'he'd sell perhaps fifty copies in as many years.'

The botanist, carrying specimens, often from 'the least accessible spots', in a rush 'frail' basket, reflects: 'Some of my friends think that a man who troubles himself so

In the Nature of Landscape: Cultural Geography on the Norfolk Broads, First Edition. David Matless.
© 2014 David Matless. Published 2014 by John Wiley & Sons, Ltd.

much as I do about wild flowers must be a little queer in his head; but I tell them that he is a lot better occupied in finding and classifying plants than in collecting birds or birds' eggs or sticking pins through butterflies and dragon-flies' (Dutt 1901: 141–3).

Dutt's botanist is an eccentric specimen, self-aware of such characterisation. If flowers are a delight for all, botanical knowledge grabs only a few, inaccessible in site and interest. Botany is curious, and harmless, odd people doing what they will but harming none. Devoted to locality, the botanist gathers aliens alongside the native, flora registered before it might be lost. Dutt's 'Broadland Botanist' characterises a field culture central to Broadland as plant landscape.

Becoming a Scientific Region

County natural science

The sense that Broadland might be a region for scientific study was shaped by the NNNS, founded in 1869, mentioned through earlier chapters but, as a key botanical body, requiring detailed discussion here. As in Naylor's Cornish 'biography of a scientific region' (Naylor 2010: 10), and Withers and Finnegan's account of Scottish natural history and 'civic science' (Withers and Finnegan 2003; Finnegan 2009), the NNNS informed civic, regional and county identity, science documenting flora and fauna to match, with the field 'not just a collecting spot, but the object of study in itself' (Naylor 2010: 82; Lowe 1976; Naylor 2002). The NNNS's 'objects', which remained identical until after World War I, included scientific study, knowledge dissemination, flora and fauna protection, species record, and 'friendly intercourse between local Naturalists' via 'Field-Meetings and Excursions'. As Alberti (2001) notes, county naturalism brought together science, aesthetics, utility and recreation.

While early published membership lists do not give addresses, by 1909 the Society had 110 members in Norwich, 106 elsewhere in the county (with 22 of these in Broadland), 71 beyond the county (of whom 19 were in London), and four beyond the UK. Eighteen of 291 were female, the Society having admitted women from the outset, with the first female presidents botanist Alice Geldart (1913–14, and 1930–1) and ornithologist Emma Turner (1921–2). The NNNS connected Norwich and Norfolk, with city meetings and county excursions, and Broadland a key site, excursions in the early years including Ranworth, Horning, Surlingham, South Walsham and St Benet's. The first 1869 meeting adopted George Munford's 1864 botanical division of Norfolk into eastern (containing 'the whole of the broad country' (Geldart 1870: 21)), north-central, south-central and western, perpendicular map lines 'of permanent value to Norfolk botanists in all time to come' (Crompton 1870: 6). Watershed boundaries were 'unfortunately impossible in our own county' (Geldart 1870: 20; 1901). Geographical record was supplemented by analysis, in Broadland attending, in Stevenson's terms, to human management alongside the 'natural causes at work, which, unchecked by man, must eventually close up a great many of these Broads':

we must hope that the marketable value of reeds and rushes will henceforth increase, and the area of demand be extended far beyond our own borders. Thus by a yearly harvesting of such marsh produce, the slow processes of nature might be effectually checked, and the majority of our Broads preserved to us for many years to come, to afford sport and pastime to the gunner and angler, and hours of recreation to the scientific collector of birds, plants, and insects. (Stevenson 1866: xxiv–xxv)

Alberti notes the concern of late nineteenth century amateur scientific bodies to 'replace the image of the lone naturalist collecting for aesthetic or other unsuitable ends with a new, rigorous, collective identity', thereby cementing status alongside a professionalizing academic science (Alberti 2001: 133). Tensions arose when naturalist curiosity risked straying beyond science. The territorial holdings of NNNS county members could foster curious private enquiry, with, for example, Rev MCH Bird attempting to grow wild rice, 'my father then owning Somerton Broad' (Bird 1913: 603). Swans and voles intervened: 'Thus ended my attempts at introducing a very desirable alien ... some of my readers may like to make further attempts at its cultivation in Broadland' (605–6). Speculative enquiry could provoke internal NNNS argument, and reputational policing, as when RJW Purdy's 1908 'The Occasional Luminosity of the White Owl (*Strix Flammea*)' suggested owl luminosity as accounting for field lights near his Twyford home.[1] Purdy hoped for a 'scientific' explanation, against 'insinuations' of a 'practical joke' (Purdy 1908: 547), but insinuators came from within the NNNS, including Patterson, who addressed 'unnatural history' in the *EDP*, Purdy's vision perhaps impaired after 'a good dinner'. Patterson considered resignation, refusing to attend Purdy's NNNS lecture, being 'none too proud of a Society which encourages spurious natural history' (quoted in Manning 1948: 90; cf Norgate 1918).

Uncontentious study gave groundwork for systematic science, including from the leisure time of working members. Yarmouth naturalist HE Hurrell's micro-faunal Polyzoa studies showed the 'storehouse of wonders' (Hurrell 1911: 197) in Yare valley ponds, ditches and rivers: 'A great deal of my own collecting has been done in the early morning hours snatched from sleep and spread over some years; but I trust that some one with more time at their disposal may carry on the quest' (205). The NNNS also cherished Norfolk rarity, as when Arthur Bennett discovered the only British specimen of *Najas marina* in Hickling Broad in 1883. Bennett, based in Croydon, took the train from Liverpool Street with his daughter, hiring a 'lad' and 'his boat' at Potter Heigham:

we took the channel near the keeper's house to Hickling Broad, and exactly at the spot marked 'B M 4-6' on the 6-inch Ordnance Map, my daughter at the bow of the boat brought up a lot of aquatics in the 'drag' ... I at once saw we had *Naias*! [sic] Giving a good 'Hallo,' and making the boat rock considerably, I knew we had the new British plant. Curiously enough three days after Mr. H. Groves went over the same ground and found the *Naias*, being accompanied by the same lad. He said to Mr. Groves: 'Be ye from Lon'on after weeds? Ah, yer too late.'
    He knew the plant again as I had pointed it out to him.

Bennett offers a scientific image echoing Emerson's contemporaneous 'Gathering water-lilies', plants hunted for science rather than art and bait. The plant was later raided, and for protection 'Mr Cotton' bought the adjoining rond, after a request from Bennett at the Linnean Society (Bennett 1910: 47–9). Scientific transactions stake Broadland.

The Sutton Broad Laboratory

From 1902 Norfolk naturalist science had a private institutional focus, the Sutton Broad Fresh-Water Laboratory (SBL), recording freshwater plant and animal biology. Brothers Eustace Gurney (NNNS President 1905–6), of Sprowston Hall near Norwich, and Robert Gurney (NNNS President 1912–13), of Ingham Old Hall, near Stalham, established the private laboratory, the first of its kind in Britain, in an isolated newly built house at Longmoor Point, overlooking the former Sutton Broad, 'on a spit of land running out from the uplands into the marshy region'. A motor-boat allowed collecting excursions, while the wherry 'Cyclops', bought from travel author HM Doughty in 1902 and renamed from 'Gipsy', was converted 'into a floating residence and improvised Laboratory' (Gurney and Gurney 1908: 1; Malster 1971: 142). F Balfour-Browne was appointed SBL Director in 1903, Robert Gurney taking over in 1907 after Balfour-Browne's move to Belfast.[2]

Eustace and Robert were educated at Eton, both studying zoology at Oxford. Eustace abandoned 'a scientific career' on becoming responsible for the family Sprowston estate in 1902 (TNNNS 1928). Robert's freshwater crustacean research brought an Oxford DSc in 1927, and election as Fellow of the Linnean Society in 1922. The Ray Society published his standard three volume *British Fresh-Water Copepoda* (Gurney 1931–3); Balfour-Browne produced the equivalent *British Water Beetles* (Balfour-Browne 1940–58). Gurney also published for Scouts and Guides on plants and trees (Gurney and Gurney 1920), and, with Turner, on birds (Turner and Gurney 1925). AC Hardy's Linnean Society obituary praised a man 'before his time', without an official position, working 'in his own private laboratory at home' (Hardy 1950: 118–19; Calman 1950). *TNNNS* published SBL papers by Gurney (1905; 1907; 1913), Balfour-Browne (1905; 1906) and others (Freeman 1905; Soar 1905), Gurney also researching Broadland salinity, discussed in Chapter Six below. SBL papers demonstrated NNNS science, without a luminous owl in sight.

The SBL was also an NNNS excursion site:

> On 21st June an excursion of the Society was made to Sutton Laboratory. The party drove to Stalham, lunched there, and proceeded in boats to the Laboratory, from Stalham Staithe. The director of the Laboratory, Mr. F. Balfour Browne [sic], had prepared an interesting exhibition of objects and specimens illustrating the biology, etc., of fresh waters. Some botanising was done on the marshes, great interest being displayed in a dyke which contained a quantity of *Stratiotes aloides*, L., in flower. (Gurney 1906: 157)

Alice Geldart's *TNNNS* paper on '*Stratiotes Aloides*, L.', prompted by the excursion, reviewed plant biology (Geldart 1906). Robert Gurney also offered Calthorpe Broad, part of his Ingham property, as polite field site, the NNNS visiting in 1910: 'July 7th. By the kind invitation of Mr. and Mrs. Robert Gurney, an excursion was arranged to Calthorpe Broad, Ingham. Owing to the bad weather and other causes the attendance was unfortunately small, but the Broad and the plantations with their fine clumps of Osmunda were much admired by those members who joined the party' (Crowfoot 1911: 145). The NNNS returned in 1931, an unsigned *EDP* report, probably by Ellis, describing 'Dr. R. Gurney's story of plant life', with 'botanical and nature treasures' including osmunda fern and yellow loosestrife. The broads were 'peculiarly transitory', with weed growth after wartime suspension of cutting only recently cleared. Only 'man's interference' maintained open water. After tea, pressed plants were examined in the summerhouse: 'On leaving hearty thanks were accorded to Dr. and Mrs. Gurney for the delightful afternoon that had been spent under such able tutelage.'[3]

Botanic survey

The NNNS encouraged systematic county botany, registered in the 1901 Victoria County History volume on *Norfolk*, edited by Rye, where over 250 pages of natural history approached a county flora and fauna (Rye 1901). Beckett and Watkins (2011) note the early VCH volumes' remarkably detailed natural histories, national experts working with local authorities. Five former and future NNNS presidents contributed, Herbert Geldart dominating botanical sections (with CB Plowright on fungi). The '"Norfolk Broad" country' held many rarities, two unique in Britain: '*Naias marina*, discovered in 1883 by Mr. Arthur Bennett, and *Carex trinervis*, first found in 1884 by the late Hampden G. Glasspoole' (Geldart 1901: 46).

In 1908 Robert Gurney proposed a botanical Broads survey, the northern valleys covered in 1908–9 (Nicholson 1908; Driscoll and Parmenter 1994). Gurney built on SBL research by WA Nicholson, whose 'A Preliminary Sketch of the Bionomical Botany of Sutton and the Ant District' highlighted plants 'most characteristic of the marshy and aquatic regions' (Nicholson 1906: 266). Botanical richness included 11 species found in one Sutton patch: 'a square yard was marked off in the Middle marsh …, and a rough census taken of its vegetable inhabitants' (284). Nicholson ended with a manifesto for a geographic botany:

> the ideal to be aimed at is the preparation of a map exhibiting the vegetation of the area, as minutely as is possible, with a separate description of all the conditions under which plants grow, such as the plant-associations, chemical nature of soils and waters, meteorological phenomena, action of insects and animals, etc. In short, the help of all branches of natural science is required to thoroughly elucidate the plant-phenomena of even a comparatively small area. (288)

Nicholson (1858–1935), born in Birkenhead, moved to Norfolk in 1873 when his father became curate of Gislingham, worked for Gurney's Bank in Norwich

1875–1922, was elected to the NNNS 1889, and became Secretary and *TNNNS* Editor 1891–1912. The NNNS published Nicholson's *A Flora of Norfolk* in 1914. The only previous *Flora of Norfolk*, by Rev Kirby Trimmer (1866), was out of print, its coverage skewed to Trimmer's residence in Stanhoe and Crostwick: 'Scattered though the pages of this County-Flora, two Parochial-Floras will be found' (Trimmer 1866: v). As with Lubbock's *Fauna*, Trimmer's *Flora* had presented human action destroying plant life yet opening new ground, new colonists appearing in 'places unintentionally provided for them by man' (xvi). From 1909 Nicholson guided an NNNS committee, the eventual 150 page *Flora* published in 300 copies, with introductory essays on climate (AW Preston), soils (LF Newman of Cambridge School of Agriculture) and physiography and plant distribution (by WH Burrell, also author of the 'Mosses and Liverworts' section). Nicholson's *Flora* accumulated and systematised observation, giving Latin and common names, with occasional notes on vernacular lore standing out in their peculiarity. In 1933 Nicholson donated his herbarium to the Castle Museum, 'for the Norfolk plants to be selected for the Norfolk Room herbarium'.[4]

Nicholson listed 'Plants of Special Interest in Norfolk' (Nicholson 1914: 36–7), but this was not a nativist botanics, indeed the NNNS also delighted in alien species. The water fern *Azolla Filiculoides* was noted by Burrell, drawing upon FH Barclay's 1909 *TNNNS* misidentification as '*Azolla Caroliniana*' (Nicholson 1914: 24). Barclay, staying at Woodbastwick Old Hall, and finding 'great quantities' in a dyke, decided to experiment: 'Last September (1908), I brought a small bag full of *Azolla* from the dyke near Horning Ferry, and introduced them into a small pond in the garden at Woodbastwick Old Hall.' By May 1909 there were thousands of plants, a 'wonderful' increase, though showing 'how careful one ought to be in introducing such rapidly increasing plants into new localities' (Barclay 1909: 856–8). River floods in 1912 dispersed *Azolla* along Bure and Ant: 'It was reported from St Benet's Abbey in August by Miss E L Turner, and near the bridge at Potter Heigham in October by Miss A M Geldart' (Burrell 1914: 741). If however *Azolla*'s density risked blocking navigation, botanical fascination overcame qualms, WG Clarke noting: 'A ditch in autumn or winter entirely covered with this brick-red vegetation is one of the most remarkable sights of East Norfolk' (Clarke 1921: 45).[5]

After war

Clarke had been 1918 NNNS President, a serving Lance-Corporal: 'I am the first – and, I hope the last, to deliver my Presidential Address in khaki' (Clarke 1918: 294). No wartime meetings were held, though *TNNNS* continued publication. NNNS Broadland coverage began however to shift post-war. In 1928 Robert Gurney, Treasurer since 1915 and *TNNNS* Editor from 1926, moved to Boars Hill outside Oxford, his December 1928 diary giving social comment: 'We rejoice too in the Change of Society. Here we have nice and congenial people all round us, so different from the Norfolk people, and O. has parties of all sorts to go to – almost too many.'[6] As Gurney's voice receded, a different scientific presence dominated, shifting NNNS focus to the north coast, the NT reserves at Blakeney (1912) and Scolt Head (1924) fostering research by UCL ecologist FW Oliver (from 1913) and Cambridge geographer Steers (Oliver and Salisbury 1913; Steers 1934), Broadland only returning as significant subject with Ellis's mid 1930s Wheatfen publications, discussed below.

In 1927 a *TNNNS* forum on 'The Functions of a Local Natural History Society' considered the amateur and academic, with contributions by Gurney, Oliver, UCL botanist and future Kew Director EJ Salisbury, and Hoveton landowner TC Blofeld. Gurney submitted suggestions to Oliver and Salisbury 'for criticism' (Gurney et al 1927: 308). For Salisbury natural history societies accumulated knowledge for subsequent scientific analysis: 'Each observation is, moreover, complete in itself, and therefore entails no liability on the further leisure of the individual' (313). Local societies could keep their place, with 'useful work' possible for the British Ecological Society's new national flora: 'By allocating the Norfolk plants among the membership everyone taking part would have a clearly defined task' (314). Edwardian hopes for Norfolk naturalist science fade to contained enrolment.

The date of SBL closure is uncertain; the Nature Conservancy (1965) would state 1914, while 'Sutton Broad Laboratory, Director of' disappears from NNNS membership lists in 1918. Visitor book entries cease in 1912, but appear again in 1923, with further holiday visitors for 1937–45.[7] Long refers to 'a visit to the Sutton Broad Laboratory' in 1925, though this may simply be premises rather than a functioning laboratory (Long 1925b). A 'Laboratory' remains however on the 1932 OS Tourist Map, and 'Sutton Broad Laboratory' lingers on holiday maps issued by Esso in 1960 and BP in 1963. Gurney died in 1950, and Longmoor Point and Sutton Broad were sold in 1954, and again in 1964, the YHA seeking but failing to purchase through 'lack of capital' (EDP 1965a; EDP 1964d). Gurney published into late life (Gurney 1947; 1949a; 1958), in August 1949 reflecting on 'Vegetational Changes' for the *EDP*: 'This year I have revisited several places in East Norfolk in which certain rare or interesting plants used to grow, and I have been very much struck, or shocked, by the changes which have taken place.' A new botanical survey was needed: 'Looking through Nicholson's "Flora of Norfolk" one wonders how many of the rarities there recorded could be found today!' Cessation of mowing had released natural succession, birch wood over marsh, and at Sutton 'this summer I walked dry-shod where 50 years ago I could easily push a boat'. Examples might 'have been multiplied if the petrol shortage did not so greatly curtail an old man's range of observation' (Gurney 1949b).

Science on Tour

Visitors from another scale: the 1911 International Phytogeographical Excursion

If the NNNS included occasional scientists from beyond the county, in August 1911 another scale of scientific enquiry came to Broadland. The first International Phytogeographical Excursion, organised by Arthur Tansley of Cambridge University, saw British, European and North American scientists exploring British Isles vegetation, formulating common language for the emerging science of ecology. The IPE's four week tour began in Cambridge and ended at the British Association conference in Portsmouth. Broadland was the first stop (Cameron and Matless 2011).

The Broads tour was led by Marietta Pallis, then at Newnham College, working alongside Tansley at Cambridge University. In a letter home to America, IPE member Edith Clements wrote:

> We had three delightful days on the Norfolk Broads ... When we reached our starting point the first day, I called Fred's attention to a striking-looking person coming toward us. It had all the appearance of a lad & yet was most undoubtedly a woman! This was Miss Pallis: ... with knickerbockers & Norfolk coat of tan cravanette, leggings of the same color, hat of soft felt to match drawn closely down over her head - straying wisps of curly brown hair, framing a nut-brown face with rosy cheeks & red mouth. Moreover she was jauntily swinging a cane, & later, as jauntily smoked various and sundry cigarettes!![8]

Photographs show Pallis and Tansley, fieldworker and mentor, in strikingly similar attire, Pallis's field coat, breeches, tie and hat, very different from the dresses of other female party members, Pallis a scientific new woman in the field.

Figure 27   Marietta Pallis and Arthur Tansley at Barton Broad, International Phytogeographical Excursion, 1911. Reproduced by permission of the British Ecological Society.

Pallis had surveyed Broadland vegetation with Jean Shaw in 1908–9, the two elected the first Associate Members of the British Vegetation Committee (BVC; later the British Ecological Society), chaired by Tansley, in April 1910. Nothing more is known of Shaw. In 1911 Tansley edited *Types of British Vegetation*, Pallis contributing on 'The River-valleys of East Norfolk: Their Aquatic and Fen Formations', the first academic scientific survey of Broadland, discussing river valleys, broads origins, aquatic and fen formations, and plant associations of water, swamp, fen and carr woodland (Pallis 1911a; Tansley 1911a). Pallis's fold-out transect diagram of Barton Broad explained succession, open water to reed swamp to carr: 'The consequent change in the boundary of land and water is so rapid that a mapped boundary cannot long remain correct' (Pallis 1911a: 230). Frederic Clements would reproduce the diagram in his key 1916 text *Plant Succession* (Clements 1916: 256), by which time Pallis's scientific reputation had been furthered by research on plav, the floating fen of the Danube delta, including comparative Broadland analysis (Pallis 1916; Cameron and Matless 2003).

Tansley presented *Types* to international IPE members, participants also receiving 'Descriptive Notes on the Topographical and Geological Features and on the Vegetation of the Route' (Tansley 1911b). The party travelled by train from Cambridge to Brundall, met by Pallis. Train travel gave general valley views, boat excursions explored intimate waters. Pallis's regional route demonstrated differences in fen association between Yare and Bure, and in aquatic association between the brackish Thurne, strongly tidal Yare, and fresh, less tidal Ant and upper Bure. Pallis guided the IPE by boat from Brundall to Surlingham Broad, on by train from Brundall to Buckenham, by boat to Rockland Broad, on to Yarmouth by train for supper, before another train north to Sutton. Day 2 took boats to Sutton and Barton Broads, visiting the SBL (discussed below). Day 3 saw a morning motor-boat from Potter Heigham to Hickling and Horsey, the afternoon back by river to Horning, disembarking into fen carr, taking tea at Woodbastwick Old Hall, exploring Hoveton Great Broad, and to Wroxham station, Broadland quickly left behind as the train pulled north into agricultural country.

The IPE moved among other peak season parties, Edith Clements noting: 'the Broads were full of white & black-sailed yachts, launches, steamboats, rowboats & house-boats with everyone out for a good time'.[9] The IPE also explored closed waters, entering private landscapes at Hoveton and Woodbastwick Old Hall. Little attention however was given to local naturalists. In 1904 the BVC had resolved not to rely on 'Local Workers' from 'local natural history societies etc': 'members of the committee shall do what they can to obtain suitable workers'.[10] Pallis was the perfect locally knowledgeable non-local guide, knowing the Broads but in no sense (by origin, residence or class) a 'local worker'. NNNS vegetational studies were bypassed, the IPE styling itself an international elite, forging a new ecology. Conversely the IPE barely registered in NNNS public statements, with no taking of pride in esteemed visitors. Nicholson's *Flora* cited *Types*, and IPE-derived accounts in the *New Phytologist*, but the IPE was never named. Pallis was likewise made minor, noted only as a contributor to *Types*, for three observational records, and as named in ES Gregory's *British Violets* for 'Miss Pallis's Violet' between Waxham and Palling (Gregory 1912: 82–3; Nicholson

1914; Bennett 1916).When Nicholson had announced Gurney's Broadland botanical survey in 1908, he added: 'Unfortunately, some weeks after our proposal was brought before this Society, it was ascertained by Mr. Gurney that two ladies, the Misses Shaw and Pallis of Liverpool, had already commenced a survey of the Broads District.' Nicholson commented:

> I do not think this fact ought to deter us from prosecuting our Survey. A local Society must have many members who possess a considerable amount of knowledge of the natural phenomena of their district, a knowledge which cannot be acquired by those non-resident in the district, in the short space of time which they could necessarily devote to its acquirement. (Nicholson 1908: 621)

The BVC's 'local workers' judgement is reversed.

Pallis and Gurney would share a stage at the RGS in December 1910 addressing salinity, discussed in Chapter Six below (Pallis 1911b; Gurney 1911a), and in later life would cite each other approvingly, Gurney in *Our Trees and Woodlands* (1947), Pallis in 1950s pamphlets, by which time she was friendly with Gurney's naturalist sister Catherine, resident in Ingham. Their next formal encounter after the RGS however was within the IPE, with differences in scientific scale telling. On 4 August the IPE visited the SBL, some staying overnight at Longmoor Point, others in Sutton village, and 'the whole party … entertained to tea' (Tansley 1911b: 277). The SBL Visitors' Book shows the 'International Excursion of Ecologists', names and place of residence: Drude (Dresden), Rubel (Zurich), Elizabeth Cowles and Henry Cowles (Chicago), Edith Clements and Frederic Clements (Minneapolis), Ostenfeld (Copenhagen), Oliver (London), Lindmann (Stockholm), Hill (London), Pallis (Liverpool), Yapp (Aberystwyth).[11] International excursionists meet Norfolk science. While much of the IPE highlighted national differences between participants, Sutton foregrounded another geographical distinction, between the international and the regional. If several IPE members were British, all register here as international, landing in, and apart from, regional science. Sutton signatures denote visitors to Broadland from another scale.

Science escorted

The Broads were a regular showpiece for scientific meetings in Norwich, with itineraries often paralleling tourist excursions. If the IPE had toured at a tangent to local nature institutions, for later scientific groups Norfolk naturalists played a key role in representing county nature, regional standing gained. Thus in June 1933 the South-East Union of Scientific Societies were escorted from Wroxham by Yarmouth naturalist Hurrell and City Librarian Stephen, down river to Barton, St Benet's and Ranworth, meeting Rev Everard (who would greet the IBG 26 years later), and ascending for the view (South-Eastern Naturalist 1933). British Association for the Advancement of Science conference delegates in Norwich in September 1935 could likewise take 'General Excursions' of sightseeing science; a whole day from Wroxham to How Hill for tea and gardens, and to Stalham for the return coach (BAAS 1935: 6–11; Withers 2010). Humphrey Boardman was co-leader, How Hill his family home. The BAAS's *A Scientific Survey of Norwich and its District*, included Broads geography and zoology, noting the former SBL and the Norfolk Room's 'fine panoramic displays' (Mottram 1935: 49). Nicholson and Ellis covered botany, Wheatfen noted for lessons in succession management: 'It is more than ever necessary for an authoritative conclave of botanists to keep watch over the county's flora at vulnerable spots, and be recognised in its efforts to avert botanical calamities' (32).

The BAAS returned in August–September 1961, the *EDP* noting: 'The streets of Norwich are already becoming gay with flowers in preparation for the coming of the British Association at the end of this month' (EDP 1961b). The programme booklet noted: 'Special provision has, of course, been made for trips on the Norfolk Broads' (BAAS 1961b: 130). General excursions included a 'Coach and Launch Tour' from Wroxham to Ranworth for the church and view, while specialist sectional trips also toured Broadland. If Economics were content with a dinner at the Horning Ferry Inn (dress informal), Zoology visited Horsey and Hickling, and Geography were led by Lambert, Smith and Green on the Ant and Bure. Botany went to Wheatfen, with 'Tea as guests of the Local Secretary', namely Ellis, who reportedly served coypu:

> Wheatfen Broad, Surlingham, is a private nature reserve of 160 acres, comprising small broads, reedswamps, fens, swamp and fen carrs and mixed woodland. The flora and fauna have been studied intensively for the past thirty years and the flowering plants, fungi, plankton, mollusc, woodlice, harvest-spiders and Heteroptera have been the subjects of numerous published papers. Many new species of agarics and micro-fungi have been described from the fen and wood in recent years.
>
> On arrival, it is suggested that some members shall be shown the types of vegetation present in the area, while others shall take part in a fungus foray.

Ellis offered a scientific treat: 'Gum boots are essential' (BAAS 1961c: 57).

In 1950 Sainty reviewed the *New Naturalist Journal*'s special East Anglia issue: 'Norfolk men in particular will feel pride in recognising that so many of the authors … are members of the Norfolk Naturalists' Society' (Sainty 1950). The NNNS had worked to increase its regional role, enhancing its public profile and asserting scientific status. In June 1951 the Castle Museum hosted a Festival of Britain NNNS exhibition on 'Norfolk Naturalists at Work', with members' photographs, watercolours of Norfolk fungi, and displays of 'penetrating research on the Norfolk Broads', conveying naturalism to a Norfolk public (EDP 1951a; EEN 1951). If such events gave formal publicity, *TNNNS* under Ellis's editorship from 1951 became more consciously scientific, ceasing reviews of Society meetings and publication of Presidential addresses, and in the mid 1950s experimenting with specialist theme issues on botany and mycology (the latter including three articles authored, and one co-authored, by Ellis). Lambert recalled arguments with Anthony Buxton, Editor from 1945–50, over the rejection of scientific pieces, including one of her own. Lambert, 'diametrically opposed', saw Buxton devaluing a journal of national standing.[12] Buxton resigned and Ellis took over, reasserting science.

The BAAS's 1961 *Norwich and its Region* showcased the NNNS, with Lambert, Ellis and west Norfolk botanist EL Swann contributing as NNNS Joint Botanical Recorders, Seago as Recorder of Birds, ET Daniels and KC Durrant as Joint Recorders of Insects, and FJ Taylor Page as Recorder of Mammals. Lambert wrote on 'The Chief Norfolk Habitats', Ellis on fungi, the NC's Duffey on nature conservation (BAAS 1961a). Ellis, Lambert and Swann acted as BAAS sectional officers for Botany, and Ellis also for Zoology. In 1968 Swann and CP Petch edited a new *Flora of Norfolk* for the NNNS 1969 centenary, recapping botanical history: 'The 1914 *Flora* dealt mainly with the plants of East Norfolk, and in our *West Norfolk Plants Today*, 1962, we attempted to redress the balance of "the neglected half"' (Petch and Swann 1968: 11). The 1968 *Flora* was by implication the first true county survey.

Ellis and Lambert were key in recasting the NNNS, their individual work discussed in the next section. As a bridge, discussion of the July 1951 British Ecological Society Summer Meeting, written up by Lambert as Acting Local Secretary, can demonstrate their collaborative contributions. Sixty-six attended, accommodated at Wymondham College, south of Norwich. An opening visit to the NNNS Castle exhibition allowed viewing of 'the large-scale dioramas of typical Norfolk habitats, which are a well-known feature of the Museum' (Lambert 1952b: 415). Three of six daily expeditions visited Broadland, while the final afternoon gave a Bure tour by 'hired launch ... as somewhat lighter relief' (419). Lambert had earlier led the party to Ranworth and Upton, the group following one of her published transects, and viewing 'an area of fen occupying the site of a former peat cutting' (416). Lambert and Ellis led to Barton, while Saturday 7 July saw 'Surlingham, Rockland and Wheatfen Broads (Yare Valley)', guided by Lambert, Ellis and Prof Roy Clapham (who would be BES President 1954–5). The party 'walked (or rather, scrambled) through rough fen and carr to Wheatfen, where a halt was made for lunch'. Some examined Yare valley fen plant communities, others 'took advantage of small punts, kindly lent by Mr Ellis, to make a tour of the waterways under the guidance of a local marshman'. Gathered again, 'Tea was provided at Wheatfen through the kindness of Mr and Mrs Ellis, and the party returned to Wymondham about 7 p.m.' (415).

Two Yare Valley Scientists

Lambert and Ellis have featured throughout this book, but their scientific identity requires further discussion, their public roles resting on Yare valley plant life expertise. Lambert's Yare field studies and later Brundall residence, and Ellis's Wheatfen life, set both at remove from showpiece Thurne and Bure bird reserves, shaping a different culture of Broadland nature via Yare valley science.

EA Ellis

Ellis visited Wheatfen from 1933, having met owner Captain Cockle at the Castle Museum. Cockle died in 1945, the Ellises offered the lease on the cottages and 150 acres, moving in January 1946, with no mains services, water from a well, the house

Figure 28   Wheatfen nature reserve, August 2010. Source: photographs by the author.

half a mile down a boggy track. They would purchase outright in 1960. Wheatfen became Ellis's nature experiment, monitored from his chaotic study and well ordered library. Reeds were cut, a meadow mowed, fires put out, a garden maintained, but general fen management was relaxed. Ted and Phyllis (1913–2004) married in 1938, with five children born between 1941 and 1953, plus one foster child from 1960 (Stone 1988; Kelley 2011).[13] Phyllis, herself elected to the NNNS in 1938, worked as a teacher and organised home life. Russell Sewell, born at Wheatfen, gave the Ellises part-time help, living in from 1949, initially while Ellis was treated in Kelling tubercular hospital. Sewell was an unnamed character in Philip Wayre's 1961 film *Wind in the Reeds*, filmed at Wheatfen, with script commentary written by Ellis (Wayre 1965).

Wheatfen was a cherished experiment. At a 1967 Nature Conservancy symposium on 'The Biotic Effects of Public Pressures on the Environment', Ellis asked: 'Are Research Activities Always Compatible With Conservation?', noting researchers potentially damaging habitat, bringing invasive species, or making experimental introductions: 'Every nature reserve is a "gene bank" and it may be storing some rare treasure' (Ellis 1967: 35). Scientists were however welcome on good behaviour, Wheatfen becoming a showpiece under Ellis's care; the BAAS in 1961, the 1951 BES, NNNS visits including 1943 'Fungus Forays' with the British Mycological

Society (BMS) (Mason 1944; Ellis and Cooke 1943). Student parties visited from 1950, eight universities making annual visits, including Lambert's Southampton botanists. Ellis hosted scientists as a respected scientific observer, conducting specialist study of rusts and smuts from the 1930s, and elected Fellow of the Linnean Society in May 1948 for mycological research.[14] Ellis published in the BMS *Transactions*, with his Kew mycologist brother Martin, and with Prof CT Ingold of Birkbeck College: 'In December 1950 one of us (E.A.E.) encountered a collection of fungus spores in the scum on the surface of a fresh-water tidal ditch bordering a wood at Wheatfen Broad near Norwich' (Ingold and Ellis 1952: 158; Ellis 1934a; Ellis, Ellis and Ellis 1951). Ellis would serve on the BMS Council, received an honorary UEA doctorate in 1970, and had two fungi named after him, *Nectria ellisii* (1959) and *Microscypha ellisii* (1971).

Wheatfen was Ellis's core scientific site, with a *TNNNS* series on 'Wheatfen Broad, Surlingham' issued between 1934 and 1943. Ellis began: 'The area here dealt with … is the Yare Valley swamp covering some hundreds of acres to the north of Rockland Broad … The owner, Mr. M. J. D. Cockle, possesses a land of fascinating desolation, known to few but the marsh folk' (Ellis 1934b: 422). Wheatfen Broad was the '"home" water', its 'traditional name' given to 'the whole estate', with species mapped within 13 sub-areas:

The Pool / Middle Marsh / Carr South-East of the Pool / Shores of Deep Waters / Marsh, Deep Waters to Island Reach / Broads Marsh / Home Marsh / Two-Acre Marsh and Osier Carr / Old Mill Marsh / Pool Marsh / Thack Marsh / The Carr / Surlingham Wood. (425)

Richness and diversity followed internal variation: 'Atmospheric humidity is greatest in the carrs, where lichens and mosses grow luxuriantly on the trees' (423); dense aquatic plants were 'swarming with mollusca and other animal life' (424). Invertebrate notes gave 'a "taster" of good things in store for naturalists who wish to work in special directions, helped by Mr. Cockle, in years to come' (448). Ellis concluded: 'All students of natural history are invited to visit Wheatfen Broad for scientific purposes' (451). Further *TNNNS* studies built up 'The Natural History of Wheatfen Broad Surlingham' in six parts, Ellis authoring parts 2, 3 (on micro-fungi) and 5 (on bugs) (Ellis 1939; 1940, 1941). Part 4, on woodlice and harvestmen, was written by AE Ellis, authority on British snails; his 1941 Conchological Society Presidential Address discussed 'The Mollusca of a Norfolk Broad', namely Wheatfen (Ellis 1926; 1941a; 1941b). Stanley Manning, later Patterson biographer, authored part 6, on lichens (Manning 1943). Ellis noted his 1934 article prompting naturalists to 'pursue their special studies there in collaboration with Captain Cockle and the writer, with the result that numerous records have accumulated in the card index kept at the fen' (Ellis 1939: 115). Wheatfen studies surpassed, in 'friendly rivalry', the 1925–32 Cambridge University-led *Natural History of Wicken Fen*: 'Wheatfen has proved far richer in plant life than Wicken and … there seems little reason why the insect fauna should not prove correspondingly superior' (Ellis 1939: 117; Gardiner 1932; Friday 1997).

Ellis disseminated a scientific plant poetics in print and on air:

> A warm summer night has a special music. The thing to listen for is the singing water-weed. You almost have to hold your breath to hear it, as I've done once or twice when conditions have been perfect. It's a queer experience; a sort of fairy music rises from the water. If your ear is not attuned, you might compare it with the 'muzz' in sound broadcasting before VHF came along; but it's really rather like a faint jiffling of cymbals, or the hiss of rain on water. Only one sort of weed performs in this way; the hornwort.

Hornwort leaves, moved by oxygen surfacing after a sunny day, 'skiffle as they dance. Very much the same sounds are made by musical water-bugs in our marsh ditches.'[15] For the 1965 New Naturalist volume Ellis described 'Some Flowering Plants of the Broads', a glossary giving Latin, standard English and local names, collected from marshmen over 20 years ('Their authenticity has been checked in every case by the examination of specimens' (Ellis 1965: 115)), including:

> Apple-pie: *Eupatorium cannabinum* (hemp agronomy)
> Bolder: *Scriptus lacustris* (bulrush)
> Bulrush: *Typha latifolia* (greater reed-mace)
> Bunk: *Angelica sylvestris* (marsh angelica) (113)

Ellis also catalogued ferns, mosses, liverworts, algae, stoneworts, lichens, fungi, mycetozoa: 'Even if the broads were not famous for their birds and the beauty of their scenery, British mycologists could make out a very strong case for preserving them for posterity' (116).

Ellis extended popular plant science in 1970s colour guides for Jarrold, on countryside seasons, fruits and berries, rare plants and wild flowers, *Wild Flowers of the Waterways and Marshes* presenting the broads as a site where 'human interference has put the clock back to the beginning' (Ellis 1973: 1). Colour photographs, often taken by Ellis, accompanied species description. Two volumes on *British Fungi* conveyed 'diversity and beauty', with 'bizarre' products including hallucinogens and luminescent luciferin (Ellis 1976a: 1–2; 1976b). Ellis's descriptions could themselves carry hallucinogenic quality, as when, 'Probing the Undergrowth' on radio in 1957, Ellis stated: 'Now I'll try to tell the story of a sort of dream fungus. One I imagined might exist before it was ever discovered in the world and which eventually appeared before me.' Ellis's plant poetics is one of luminosity through scientific precision, taking listeners into scientific wonderland:

> one day I happened to stoop rather lower than usual to search for something … I parted the undergrowth and found I'd taken the lid off a strange new world. It was rather like looking into Aladdin's cave. There were rows of little white and grey and yellow fungi shaped like wine glasses: these were sprouting from dead stalks. There were tiny flask-shaped things glowing like rubies and orbs of crystal. I saw chocolate coloured moulds like miniature ninepins and salmon pink moulds in tufts with stiff black bristles round them. This was my first glimpse. Since then, I must have crawled many miles on hands and knees through the reed beds, the sedge marshes and willowy jungles, searching for further wonders of the fungus underworld. Moreover, I go on doing it all the year round.[16]

JM Lambert

In an unpublished poem written late in life, entitled 'Broadland', Joyce Lambert (1916–2005) evoked Broadland botany:

> Swaying reed and nodding sedge, whispering at the water's edge;
> A hinterland of tangled marsh, with alder carr and sallow bush,
> Purple loosestrife, yellow flags, sturdy rushes, tufted sags,
> Valerian, sweetgrass, meadow-rue, scabious, skull-cap, orchids too;
> A treasury on deep, rich peat, tanged by the scent of meadowsweet.
>
> Joyce Lambert, 'Broadland', 2000[17]

If she became a public figure for her origins work, conducted across the region, Lambert's core science was botanical. After early research on Welsh salt marshes (Lambert and Davies 1940), Lambert concentrated on Yare valley plant life, mentored by Ellis. Lambert was elected to the NNNS in 1943, and joined the committee in 1949, becoming 1951–2 NNNS President for her studies of Yare fen ecology; she was President again in 1961–2. Following her origins work, Lambert would promote statistical botanical analysis (Lambert and Dale 1964), research *Spartina Townsendii* (Lambert 1964a) and support the teaching of ecology (Lambert 1964b; 1967).

In 1961 Lambert noted: 'Most of the detailed work on the composition of broadland vegetation was carried out between 1947 and 1953, at a time when interference with the marginal vegetation was at a minimum and many of the broads were becoming rapidly overgrown' (BAAS 1961a: 79). The Yare valley, before significant coypu effect and after suspension of human management, retrospectively appeared a natural laboratory. Lambert had researched from Westfield College, London, her first Yare publications on *Glyceria Maxima*, 'reed sweet-grass', in *TNNNS* and the *Journal of Ecology* (Lambert 1946a; 1946b; 1947). Pallis and Ellis were her key references, Pallis's Danube plav research offering comparison as 'the only other floating reed-swamp described in detail in literature' (Lambert 1946a: 257; 1946b: 237; Pallis 1911a; 1916; Ellis 1934b). Ellis had taken 'constant interest in the investigation from its inception', with 'much of the preliminary work' at Wheatfen (Lambert 1946b: 230–1). Lambert mapped *Glyceria* between Rockland and Surlingham, solid rafts rising and falling on the daily tide, *Glyceria* 'the main agent in the overgrowth of the drainage dykes consequent upon neglect' (Lambert 1946b: 266). With marsh hay no longer cut, there was 'a condition of vegetational instability', reed replacing *Glyceria* prior to carr growth (265). Lambert also surveyed general ecology on the 'Rockland-Claxton Level' and Bargate nature reserve, the latter acquired by the NNT (Lambert 1948; 1950). Rockland-Claxton was still grazed, with drainage improvements proposed, but at Bargate grazing had lapsed, probably after the 1912 river floods, with embankments, sluices and drains in 'serious disrepair': 'No recognisable elements of the previous pasture cover can be found persisting in any quantity, and the Reserve is covered by a tangle of sweet-grass (*Glyceria maxima*), reed (*Phragmites communis*), and sedge (*Carex* spp.)' (Lambert 1950: 131).

Lambert's April 1952 Presidential Address addressed 'The Past, Present and Future of the Norfolk Broads', beginning:

> I think we may safely say, with legitimate local pride, that the Norfolk Broads, as they exist to-day, are unique ... From the standpoint of plant and animal ecology and natural history, the great areas of natural and semi-natural fenland round the broads are only imperfectly matched by certain small isolated fens, often existing as relicts and much modified by drainage, of which Wicken Fen in Cambridgeshire is probably the best known.
>
> It is all the more surprising, therefore, that so little intensive scientific investigation has yet been carried out within our Broadland heritage. (Lambert 1953a: 223)

Lambert set herself as extending the 'excellent start' made by 'workers at the old Sutton Broad laboratory', and the 'classic pioneer work of Miss Marietta Pallis': 'now, after a lapse of several years while work in Norfolk was concentrated in other districts – particularly the North Norfolk coast – the Wheatfen area is rapidly becoming the centre of a revived interest in Broadland studies, thanks to the active encouragement of Mr. E. A. Ellis' (224). Human influence was emphasised, Jennings and Lambert elsewhere commenting that 'the anthropogenic effect must be given far more prominence than was attributed to it by Pallis herself' (Jennings and Lambert 1951: 118; Lambert 1946b), yet also human neglect. If historically plants had 'their own specific local uses in the life ... of Broadland men' (Lambert 1953a: 240), with 'a considerable bulk ... sent as far afield as London for the use of cab-horses' (238), Broadland was in transition from lapsed exploitation: 'the occasional finding of unexpected overgrown tracks and causeways in deserted stretches of fen and carr is now often our only indication of the former busy activity of the marshman' (239). Succession concealed, making recent human traces quickly archaeological. Lambert concluded:

> we are at a critical transitional stage, with major changes in the utilisation and exploitation of the area even within living memory. It is unlikely that the undrained fens will ever again regain their economic value of the past. Unless a deliberate attempt is made to keep the vegetation open by periodic cutting, as has been done, for instance, in parts of the National Trust property at Wicken Fen, we must resign ourselves to a rapid dominance of trees and bushes. (252)

Lambert, Ellis and others would turn ecological analysis to policy ends through institutional action.

Plant Life in Reserve

Institutional botanics

Nature reserves offered management potential. If ecological analysis showed valued qualities to be a product of lapsed human action, scientific care might restore regional ecological value.

Lambert's *TNNNS* Bargate paper noted the NNT reserve as 'an interesting complement to the Trust's other Broadland properties associated with the Bure and its tributaries', given the 'very marked' vegetational differences between the two valleys (Lambert 1950: 123). The difference was also that, until the RSPB's Strumpshaw reserve was established in 1975, Yare reserves saw plant rather than bird life dominating conservation narrative, whether the NNT's Bargate and Surlingham, or Ellis's Wheatfen. Bargate, 40 acres of marsh within a bend of the Yare, was offered by Judge Daynes in October 1948, its interest 'mainly botanical', accessible only by boat.[18] For Lambert Bargate's ecological value was 'proportionately very great ... The very fact of its instability renders it worthy of further investigation, and the area should provide most valuable material for future biological studies from many points of view' (Lambert 1950: 134). Adjoining Surlingham was purchased from Geoffrey Watling in September 1952, 201 acres of marsh and open water, Lambert advising the NNT on its 'ecological importance'.[19] Ellis was made Warden, with authority to employ local labour: 'from the Trust's point of view the Surlingham Broad property will be valuable mainly as a botanical and entomological nature reserve' (EDP 1952b; Ellis 1952b). As elsewhere, Yare reserves prompted access debate, in this case over parish rights, Ellis addressing the 1953 Surlingham parish meeting: 'No one would be stopped from fishing or amusing himself there quietly, but shooting would be stopped.' Science and shooting did not mix: 'Students would be visiting the Broad during their training and the Trust could not have people "wandering about" with a gun' (EDP 1953h).

Earlier smaller NNT Bure and Thurne properties anticipated the Yare reserves' remit, with landlocked Alderfen Broad a space under botanical watch from 1930. NNNS 1939 excursions included: '12th August. To Alderfen Broad. The fen flora was examined systematically and members took boats on the broad, where rare plants such as greater spearwort and cowbane were seen' (TNNNS 1939a: 3). The NC similarly acquired small research properties, including Calthorpe Broad in 1953, after Robert Gurney's death, where the botanical track record included Cambridge ecologists Godwin and Turner, hosted by Gurney in the early 1930s (Godwin and Turner 1933). Broadland becomes dotted with patches of scientific value for the mind to play on, beyond river influence, public access and immediate public interest. A 1963 Ellis column conveyed a similar sensibility, driving past Mautby Broad in the lower Bure, 'A Forgotten Broad', an unpolluted landlocked 'refuge': 'Perhaps Mautby Broad is still inviolate' (Ellis 1963).

The NC designated the Bure Marshes (including Ranworth, Cockshoot and Hoveton Great Broad) and Hickling Broad as National Nature Reserves in 1958 (George 1992: 456–60). Eight SSSIs were designated in 1954, six more in 1958. Regional Officer Duffey suggested: 'The management policy must restore the habitat diversity, and much can be achieved in this way by redeveloping the reed and sedge beds and mowing or grazing on the mixed fen which produced litter' (BAAS 1961a: 80; Nature Conservancy 1965: 16–17). NNR designation prompted the *EDP* to comment: 'the State, which has as many arms as a Hindu image, has one protective limb in the shape of the Nature Conservancy, to cast around some special portions of the diminishing wilderness, and save them from commercial-minded citizens, from covetous local authorities, or from itself' (EDP 1958c). The NC's scientific remit was personified in its regional staff, Duffey awarded an Oxford doctorate for research on spiders, George with a London PhD on wingless insects. Duffey was Regional Officer for East Anglia 1953–62, based in Norwich, publishing in *TNNNS* (Duffey 1958), was NNNS President 1957–8, and NRC President in 1962–3. Duffey recalled a more cautious relationship with the NNT, its local landowning ethos suspicious of state intervention.[20] Bruce Foreman took over as Regional Officer, succeeded in turn by George (1966–90), who joined the NC in Norwich in 1960 from the Field Studies Council's Pembrokeshire Dale Fort centre. George would become significant in Broadland debate, as NNNS President 1976–7, NRC President 1971–2 and 1978, Broads Society Chair from 1998, and authoring a definitive work on regional conservation in 1992.

From the 1960s NC science shaped reserve presentation, with a Nature Trail at Hoveton Great Broad in 1968, then a novel approach to public engagement (Matless, Watkins and Merchant 2010). George prepared the guide, the trail accessible from a Bure landing stage, visitors walking through vegetational communities. The Guide cover showed a rail sleeper path through carr overlaying several feet of liquid mud: 'Do not step off the sleepers on any account' (Nature Conservancy 1968: 1). Management was explained: 'To show you what a sedge bed looks like we have removed the taller scrub from the area to the left of the path, but retained the remaining area for comparison' (3).[21] Scientific ecology also informed the NNT's Broadland Conservation Centre at Ranworth, opened on 25 November 1976, during the Trust's Golden Jubilee. A 450m 'Nature Walk' boardwalk moved through succession in reverse, oak wood to alder carr, swamp carr, open fen and Ranworth Broad. A floating thatched exhibition building symbolised a Trust at home in its unique environment. The Centre was opened by the Queen, with the Duke of Edinburgh, amid the regional and national nature establishment (NNT 1976b). The royal party flew into RAF Coltishall, boarded a motor launch at Horning, toured the Centre, walked the boardwalk, lunched with the Cators, met the Trust Council, saw Ranworth church, and departed to fly home. Guests included David Attenborough for the Society for the Promotion of Nature Conservation, Timothy Colman as NNT President, Martin George as NC Regional Officer. The Queen was photographed walking the Nature Walk, conversing with Ellis, perhaps on succession.

How Hill: gardened ornament

If Broadland reserves displayed regional plant ecologies, a different plant landscape emerged at How Hill. Norwich architect Edward Boardman, prominent city Liberal, saw How Hill on a 1901 wherry trip, bought the 872 acre estate, and by 1905 built a large thatched house overlooking the Ant, two storeyed with dormer windows and Jacobean detailing within an overall vernacular style. A sun room was added to the south front in 1910 (Boardman 1939; Holmes 1988; 2011; Pevsner and Wilson 1997). How Hill was the most dramatic early twentieth century Broadland riverside development, rivalled only by Cantley's beet factory. Initially a holiday house, car ownership enabled permanent residence after World War I, the Norwich establishment set in Broadland. Shooting parties were held, son Stuart ran a fruit farm from

Figure 29   How Hill, May 2011. Source: photographs by the author.

1926, his twin brother Humphrey, training as an architect, pursued naturalist and photographic interests (noted in Chapter Four). Humphrey recalled seeing his father's car returning from Norwich in the evening, the only car in the area, its lights the only lights across the marshes.[22]

Edward Boardman had been an NNNS member since 1885, and How Hill became an NNNS excursion site, as in 1907: 'On the 18th July, the members were invited to How Hill, Ludham, by Mr. and Mrs. Boardman, and hospitably entertained. An interesting ramble through the grounds and marshes was taken; also a row on the picturesque Crome's Broad. Some rather uncommon plants were found, which provided material for discussion' (Long 1908: 501).[23] Boardman's shaping of How Hill went far beyond marsh and broad maintenance, with 70,000 trees planted, formal topiary fronting the house, and ornamental and exotic planting in 'water gardens' to the north. Boardman's 1939 NNNS Presidential Address, 'The Development of a Broadland Estate at How Hill, Ludham, Norfolk', admitted he was neither naturalist nor scientist, his botanical remarks coming from notes by Alice Geldart: 'I have little knowledge of this subject' (Boardman 1939: 17). Noting that 'I have had a most interesting time developing this place' (21), Boardman commented: 'On the marsh level I have developed a bog garden by digging out the peat and making up beds 18 inches above the ordinary high water level ... The Broadland marshes provide a grand opportunity for those who wish to encourage the plants that hate the lime and like the peat, but require water not far off their roots' (8). How Hill's water gardens included bamboos, rhododendrons, azaleas, lilies. If Wheatfen was at this time itemised for regional botany, How Hill showed a cultivated garden ornamentalism, international water gardens, with one plant story echoing Hope Allen's migration tales:

> A hickory I grew from seed sent me by an American friend has a sentimental value. A labourer … worked in the garden when I first laid it out … As soon as his family all left school he took them off to America, 30 years ago, and we have corresponded regularly ever since. He sends me little packets of seeds labelled 'tree seeds' or 'flower seeds' and I have to grow them to find out what they are. (12)[24]

The Boardman firm dissolved in 1966, with How Hill sold to Norfolk County Council as an education centre, opened in September 1968, with accommodation, camping ground and nature trail. Ellis planted a commemorative *Sequoiadendron giganteum*. The Broads Authority purchased the estate in 1982, with the house and gardens bought by Norwich Union, leased back to a charitable How Hill Trust in 1984 (Holmes 1988; 2011). The Trust runs educational courses, with woodland and marsh field studies, emphasising the human shaping of landscape.[25] The Trust also opens the water gardens for a few early summer peak blooming days. If surrounding estate reed beds are enrolled by the BA into a showpiece story of a sustainably managed future, with electric boat tours, restored wind pumps and a marshman's museum, the gardens stand as historic ornament, set aside from the botanical present for occasional display, though the odd rhododendron escapee blooms by Crome's Broad.

Marsh Harvest

Plant life on the escalator

Plants of marsh and swamp, notably reed, gather cultural stories. Photography, oral history, topographic writing and rural survey have rendered plant life as locus of tradition and livelihood. As reedbed and marsh are cut for thatch and hay, what cultural harvest proceeds?

Raymond Williams' 1973 *The Country and the City* discusses 'A Problem of Perspective', country writing presenting long tradition as always about to pass away. Williams finds the same stance in the 1960s, the 1930s, the 1900s, and 'what seemed like an escalator began to move', on through the nineteenth and eighteenth centuries: 'Where indeed shall we go, before the escalator stops? / One answer, of course is Eden' (Williams 1975: 21). Williams recommends 'the sharpest scepticism' against 'sentimental and intellectualized accounts of an unlocalized "Old England"', though such regularity of narrative pattern indicates for him less 'historical error' than 'historical perspective', itself giving clues to a 'real history' (20). The 'single escalator' turns out to be a 'complicated movement: Old England, settlement, the rural virtues – all these, in fact, mean different things at different times, and quite different values are being brought to question'. Williams' question, 'where indeed shall we go?' is temporally framed, the answer entailing 'precise analysis of each kind of retrospect' (21–2), but can be supplemented with geographic questions of precise location. Where does the escalator lead? What are its Broadland destinations?

It leads to a Ranworth off licence in 1995, selling 'Reedcutter' cider, produced by Ranworth Farms (est 1992), a strong 6.0 per cent, the plastic bottle's label showing a reedcutter leaning against reed bundles, scythe on floor, bottle on floor; a cultural harvest to lubricate holidays afloat.[26] It leads also to How Hill, and reedcutter Eric Edwards (1940–2012), employed from 1967 and BA estate marshman until retirement. Edwards stars on postcards, leaflets, paintings, national park guides, adept at performing traditional labour, with his personal collection of historic marsh tools. On a How Hill Trust postcard, 'Norfolk Marshman, Eric Edwards MBE, surveys the How Hill Reed Beds', scythe over one shoulder, bundle under other arm, ripples indicating just slight movement while in pose. The BA's national park guide has Edwards in sparkling How Hill sunlight silhouette, one leg on a boat, holding reed upright (Tully 2002: 9). Edwards becomes regional icon, choice performer of traditional labour in conservation territory at a time of largely mechanised reedcutting elsewhere. When in November 2008 I helped film an item for BBC East's *Inside Out* on Marietta Pallis, Edwards had a non-speaking role, demonstrating peat digging. I spoke of broads origins, Edwards digging behind up to his knees, and later working the camera close-up, turning peat for viewers.[27] After filming, media notes were compared, Edwards having appeared on Anglia TV's *Bygones*, the BBC's *Generation Game*, and antiques vehicle *Flog It*, working Broadland tradition in accomplished performance.

Figure 30   'Norfolk Marshman, Eric Edwards MBE, surveys the How Hill Reed Beds'. Source: How Hill Trust postcard, purchased 2009. Reproduced by permission of the How Hill Trust.

The escalator leads also to Emerson, indeed pictures of Edwards reprise photographs a century old, Emerson's regional visual grammar still deployed. *Life and Landscape on the Norfolk Broads* showed harvests of hay, gladdon, schoof-stuff and reed, and Emerson's imagery continues to travel, the 2009 British Library exhibition *Points of View: Capturing the 19th Century in Photographs* placing a cropped version of *LLNB*'s 'Coming Home from the Marshes' on its exhibition guide cover.[28] The image also provided the front cover for the UEA's 1986 Emerson exhibition catalogue (McWilliam and Sekules 1986). Emerson's text for 'Coming Home' described a bleak winter 'sublime beauty', with 'typical specimens these of the Norfolk peasant, – wiry in body, pleasant in manner, intelligent in mind … They have just returned from cutting the reed.' The stooping figure is 86, lasting a long hard life. Birds watch their movement: 'So homeward go these Norfolk peasants; a naturalist in his way each one of them' (Emerson and Goodall 1886: 10). The same figures appear in four photographs of 'picturesque labour' showing 'The Reed Harvest', the foreground man pulling a laden boat solo in 'Towing the Reed' and receiving reed on a stack from the second standing man in 'Ricking the Reed', the latter also pictured scything in 'A Reed-Cutter at Work'. 'During the Reed-Harvest', echoing harvest iconography elsewhere in Emerson's work, shows all in labouring ensemble by a dyke, with reedbed, reed boats and reed stacks, the octogenarian stacking, the woman tying the latest bundle (Emerson and Goodall 1886: 59–62). Emerson comes home from the marshes with photographic harvest, presented as generic regional images, without specified location, standing for Broadland.

Figure 31   PH Emerson, 'Coming Home from the Marshes'. Source: Emerson and Goodall 1886, image courtesy of Norfolk County Council Library and Information Service.

Figure 32   PH Emerson, 'During the Reed-Harvest'. Source: Emerson and Goodall 1886, image courtesy of Norfolk County Council Library and Information Service.

Reed industry

Reedcutting between January and April was part of the out-of-seasonal round observed by topographers claiming intimate regional authority. Dutt's *Norfolk Broads* included Southgate watercolours of 'Towing the Reed Harvest' and 'Reed-Stacking', effectively reprising Emerson photographs (Dutt 1903: 84/142). Reedcutting provided labour for underemployed 'eel-catchers, marshmen, millmen, and the men who sail the cruising yachts' (84), but was waning, and soon 'the reed and rush cutters will find their occupations gone' (160). Dutt could still however catch 'a characteristic Broadland scene, … the wide flat reed rafts being quanted up a dyke, the stacks with their light and dark layers of reeds, and the marshmen clad in clothes of a russet hue which harmonised well with the yellow reeds' (141).

If reed harvest was passing, one response was delight. In 1923 HJ Massingham, styling himself as nature writer rather than (as later) an agrarian and craft commentator, revelled in a coming Thurne desolation, 'a land still wild and stealing more wildness year by year', where 'there are no men' (Massingham 1923: 124). Massingham regressed: 'How many centuries ago had I gone hunting through these watery lanes and watched with suspended breath what might appear round the next bend, whether hippopotamus throwing a river off his back as he rose, urus strayed from the herd and sinking in the swamp, Neolithic refugee or Grendel himself … ?' (126). The recent 'decay of the reed-cutting industry' was a boon to re-wilding, reinstating 'the barbaric heart of the reeds' (127–8).

Others proposed industrial revival. Throughout the twentieth century, surveys, plans and institutions sought to revive reedcutting, for monetary harvest, social employment, land conservation, bird preservation. The diagnosis of a fading occupation, warranting revival, suggests an urge to run the wrong way up Williams' escalator. The escalator never stops, and never ends; the effort is worthy, and continues. Thus Volume Two of the Oxford Agricultural Economics Research Institute's *The Rural Industries of England and Wales* considered *Osier-Growing and Basketry and Some Rural Factories* (FitzRandolph and Hay 1926; Brassley 2004; 2006). Helen FitzRandolph (from 1924 Secretary of Gloucestershire Rural Community Council) had surveyed 'East Anglia and Essex' in 1923, including 'Rush, Sedge and Reed Industries' (FitzRandolph and Hay 1926: 88–102). Paralleling the Rural Industries Bureau, set up in 1922, FitzRandolph lauded tradition finding modern function in an arts and crafts inflected vision of unalienated labour, bringing 'work into direct relation not only with his own life, but with that of the community of which he is a member' (v). For FitzRandolph reed had good prospects, its specialist ornamental use travelling far: 'A firm at Salhouse ... sends thatchers, carrying with them their thatching of Norfolk reeds, as far afield as the Thames Valley and Cheshire' (89). FitzRandolph also noted: 'On the Norfolk Broads ... many thatchers carry on basket-making as an alternative industry for wet weather, when thatching cannot be done' (20). Humphrey Boardman, active in the Norfolk CPRE, echoed reed optimism in promoting 'Reed Thatching in Norfolk' in *The Architects' Journal* in 1933, advising technique, and claiming 'benefits to be derived from living or working under thatch and the happier harmony it provides with rural England' (Boardman 1933: 567; Mottram 1935).

For FitzRandolph, if tradition could not adapt, it should die. Rush had a doubtful future, with 'Varied nomenclature' of sedge, reed mace and gladdon confusing, and just one small Norwich rush-plaiting factory and five Norfolk rush-workers surviving: 'two at Stalham, two at Neatishead, and one at Hoveton St. John' (FitzRandolph and Hay 1926: 91). Local practice did not help: 'All Norfolk cutters agree that rushes take about six weeks to dry. The fact that they are left out in all weathers may help to account for the inferior quality of Norfolk rushes' (95). While Women's Institute rush-work, 'a handicraft practised by women who do not seek to earn their entire living by it', might persist, the 'old men' were 'too conservative to learn the new types of decorative and varied rush work'. FitzRandolph wrote a craft obituary: 'The passing of the rush and sedge industries in the march of industrialism will be regretted by all those who value traditionary rural handicrafts and appreciate the picturesque aspects of country life. The district of the Broads and Fens is becoming more and more civilised, and the old marsh men fewer and harder to find. The rush-plaiters belong essentially to this fast-disappearing type' (102).

Rush revival plans nevertheless recurred, with Waveney Apple Growers at Aldeby adopting rushwork from 1946 as out-of-season employment for female apple graders, both tasks involving 'quick, nimble fingers' (EDP 1954b). Reed revival optimism too persisted. The NNT promoted cutting for profit and ecology, Surlingham reclaimed 'from the morass it has tended to become in recent years', reed 'the only harvest which can help to meet the costs of preparing the Broad as a nature reserve' (EDP 1953a). The Norfolk Association of Reed Thatchers, Reed Cutters and Allied Traders was formed in 1950, the Cator family providing Chair and Secretary, membership including 13 master thatchers (EDP 1950). The Woodbastwick estate presented itself through May's 1952 *Norfolk Broads Holiday Book*: '"We've got four skilled thatchers and three learners ... Two are working at Dunstable, Bedfordshire, at this moment, and I'm sending two more to Portsmouth tomorrow." The stack beside the ferry road was due to be loaded on to the lorry to go with them' (May 1952: 140). May found 'ancient craft' succeeding 'in a modern world' (142).

A recent NNNS study of the Catfield estate noted mechanised reedcutting registering different land management philosophy to the 'nature industry' (NNNS 2008: 6); the BA's How Hill imagery is indeed far from most Broadland reed production. Mid twentieth century commentary presented Broadland reed stepping off any memorial escalator, achieving modern functionality, machines entering the harvest from the 1950s, including at the Hickling reserve (George 1992: 212–19). In 1967 the *EDP* featured a new Danish Seiga machine at Ranworth, three balloon tyres allowing rapid amphibious operation: '"Well, that's a sight for you," sighed a middle-aged reed cutter who had come along to witness the walk on the water. "If some of the old hand cutters were to come back today, they'd wonder what planet they were on"' (EDP 1967a). George outlines contrary reedcutting views, ecologists valuing old maintained reedbeds, ornithologists happy with new reedbeds as long as areas remain uncut, cutters wary of poorer quality and preferring new beds on embanked floods, as developed at Ranworth from the early 1920s and Horsey and Hickling in the 1960s (George 1992: 212/469–70). The reedbed's emotional and financial harvest works in variation.

Marsh labour testimony

Marsh harvest has included the testimony of workers, labour taped for regional insight. Thus in 1966 John Stannard interviewed 85 year old Dick Foster of Ranworth for the *EDP*, recalling haymaking before 1914. A 'mystery tour' coach arrived at the staithe: 'I couldn't help wondering what effect the sudden appearance of a fully-loaded hay-lighter would have had on the holidaymakers. There's one thing you can bet on – there would have been a rush for cameras so that a souvenir snap could be taken home of "haymaking on the Broads"' (Stannard 1966).

Memory and historic image also enfolded in the contemporaneous East Anglian oral history of George Ewart Evans, seeking 'the old culture' in rural Britain (Evans 1970: 17; Howkins 2004). Evans investigated 'The Marshmen' on Haddiscoe Island, the 'marsh way of life' not yet 'greatly affected by the recent revolution in agriculture' (Evans 1970: 146).[29] Evans met the Pettingills of Seven Mile House, grazing cattle for different owners and cutting hay and reed. Ninety-four year old Mrs Pettingill and son Leslie (born 1913) were taped for memories. Evans reproduced an unattributed Emerson photograph, 'Quanting Hay on the Norfolk Broads in 1886', plate 16 from *LLNB*, one man quanting an empty boat, another quanting atop a pile of hay; Evans' caption compared to the Pettingills' 1930s methods, Emerson (in 1970 less well known) illustrating uncredited across the years.[30] Leslie also recalled reed and manure taken by wherry, though 'Nearly all the wherries are off the river now' (151). Evans registers his own presence towards the end: 'Just before I left that afternoon Leslie Pettingill told me: "When we've gone it will be the end of this way of things out here.

They won't get any young 'uns to stay out here on the marshes. The cattle will have to look after themselves, I reckon"' (156). Evans leaves an anticipated end of a line.

Williams' 'A Problem of Perspective' cites an unnamed author deploying a 'formula' that: 'A way of life that has come down to us from the days of Virgil has suddenly ended' (Williams 1975: 18). Williams later names his source: 'One of the best recorders, George Ewart Evans, is the man from whose book I took the remark about the continuity from Virgil, and the irony of that is for me, in the end, deeply saddening.'[31] If Evans approaches 'real history', he also boards the escalator: 'Writers I share so much with, in experience and memory, are in an instant of allusion, of a different way of seeing history, the strangers they ought not to be' (Williams 1975: 313). Evans lets Williams down, his work not matching 'real record' and 'true voice', 'direct and unmediated' (314–15). Williams' sadness proceeds from faith that a bedrock of 'real history' might be accessed, but one might also suggest that 'real history' is itself a convention of understanding, of 'perspective', to set alongside pastoral and counterpastoral, these and others making for a rich and inescapable narrative predicament. Evans, FitzRandolph, Emerson, and indeed Williams work through this predicament to different cultural and political effect.

Two recent examples indicate further complexities of marsh cultural harvest, and can conclude this section. In April 2009 The Broads Reed and Sedge Cutters Association (formed in March 2003) issued *Marsh Workers of the Broads*, booklet and DVDs produced by former reedcutter Paul Mace, supported by the Heritage Lottery Fund, giving testimony on lives 'hard' yet 'content': 'some may afford them pity but they would not want or accept it'. Landscape speaks through people: 'The aura of serenity among these people is amazing and cannot be coincidence. The marshes and waterways breathe through them all' (Broads Reed and Sedge Cutters Association 2009: 47). Suffling's 1880s *Land of the Broads* description of 'A Typical Marshman' is taken to accurately render a timeless 'aura of serenity' (2), Mace riding Williams' escalator through Georgic convention. Mace's grandfather Robert, Haddiscoe Island marshman from 1956, is one of 21 respondents, grouped under occupation, including reed and sedge cutter, thatcher, gamekeeper, eel fisher, farmer, millwright, reserve worker, coypu trapper. Two women are interviewed, Pettitts worker Irene Willimott noting Pettitts buying 'anything that moved' in wartime: 'starlings, water hens, coots, shelducks, herons, swans, the lot' (DVD 3).

Workers speak face on to a static camera, with no interviewer voice, most filmed in home armchairs, name and year of birth given. Exceptions include self-taught millwright Vincent Pargeter (also restorer of the wherry Maud (Pargeter and Pargeter 1990)), and the Norfolk Wherry Trust's Mike Sparkes, who stand out as younger traditionalists. Others testify with memory, in Norfolk accents of varying breadth, in neat, sparsely furnished dwellings, the static camera allowing the viewer to take in surroundings. Ecological management is criticised, preoccupation with the bittern adversely affecting other species, Haddiscoe Island likely to be abandoned for 'a duck pond' (Broads Reed and Sedge Cutters Association 2009: 34). Marsh work has another nature; if birds take flight at birdwatchers, they tolerate the working cutter: 'the working man has his own agenda, they can see, sense and live with this' (9).

Marsh workers also featured in Justin Partyka's *The East Anglians*, a 2009 photographic exhibition at the UEA's Sainsbury Centre for the Visual Arts, a gallery harvest of land work. If there were no explicitly Broadland pictures, Partyka, citing Evans on farming as 'the mother and the nurse of the East Anglian culture', followed 'in the footsteps of P.H. Emerson who over 100 years ago was compelled to photograph the country ways of life in East Anglia which he believed were disappearing'. 'Suffolk, 2004' showed a blue overalled man, walking to camera, eyes down, across cut reed, another man carrying bundles behind, Partyka's caption noting the sound of the cutting machine: 'Eventually I emerged into a clearing where the reed cutters were at work. They moved light-footed and quickly across the boggy marsh, never standing still. There was a rhythm to their labour, and as I watched them work it looked as choreographed as a ballet.'[32] A machine operative walks between tasks, labour's routine movement, an image far from Edwards's How Hill handiwork.

The mechanised 'Suffolk, 2004' however belies Partyka's predominant emphasis on a 'ramshackle world': 'Small-time farmers, reed cutters and rabbit catchers: these are the East Anglians, the people of the flatlands who continue to work the land because the need to is in their blood.'[33] These East Anglians are all male and mostly elderly. In 2008 Partyka collaborated with Cambridge-based Robert Macfarlane for *Granta*'s 'The New Nature Writing', Macfarlane turning from his prevailing pursuit of 'wild places' to ride Williams' country escalator, citing Ewart Evans on a culture unbroken from the Middle Ages, until now (Macfarlane 2008: 122–3; Macfarlane 2007).[34] 'Suffolk, 2004' was chosen for *Granta*'s cover, slightly cropped and retitled: 'Reed cutter, Suffolk'. Macfarlane's essay presented 'Ghost Species', mechanised reedwork sitting awkwardly amid an ecological language of loss: 'human ghosts: types of place-faithful people who had been out-evolved by their environments – and whose future disappearance was almost assured' (Macfarlane 2008: 126). Partyka photographs 'rabbit catchers, reed cutters, eel fishermen', various 'different types of ghost', their 'rural ways of life ... brought to the brink of extinction' (112–13).

Reedcutters are cast from their machines into an eastern twilight, perhaps to meet Lubbock's Broadsman, Patterson's fowlers, and the last of other discontinued cultural lines, riding escalators up and down, comparing notes on those who discovered them dying, talking as the sun sets, before it re-rises.

Succession: Experiment in Landscape

Marietta Pallis had led the IPE to the Broads in 1911, making her scientific reputation in the region. After publishing her Danube delta plav study in 1913, Pallis abandoned a research career, her father refusing permission and funds for Amazon travels. Pallis turned to painting, but continued private botany and ecology in Broadland, dividing time between Chelsea and the Hickling marshland property at Long Gores. Pallis summer-rented the cottage from around 1918, buying cottage and marsh in 1935, coming into inheritance on her father's death; Pallis wrote to her friend Joan Wake: 'what can one not do with 76 acres, remote and surrounded by water'.[35] Pallis's Long Gores life, including the 1953 digging of the Double-Headed

Eagle Pool, can conclude this chapter, allowing discussion of the science, politics and proprieties of Broadland plant succession. Pallis judged her own and others' landed abilities, enfolded science and aesthetics, and made her 76 acres an experiment in landscape (Matless and Cameron 2006; 2007a; 2007b).

On buying Long Gores, Pallis wrote to Wake, assessing Broadland property. Long Gores was 'injured so far only by horses', but elsewhere – Horsey, Calthorpe – owners had inflicted 'suburban mindedness': 'Hickling. Lord Desborough. Suburban mind-edness has desperately injured this.'[36] Pallis sought another model, Long Gores less a carefully managed sanctuary than a space of nature released. Beyond converting bullock lodges into a cottage extension, building a substantial thatched studio, and establishing a garden with plants collected on travels, Pallis would not interfere, her eco-philosophy recognising yet lamenting anthropogenic effects: 'In 1935, at Michaelmas, the horses were turned off Long Gores, and henceforth July mowing ceased. By the summer of 1937 several oak and birch seedlings appeared on the level, and the reed and other edible species began to spread anew' (Pallis 1939: 24). Robert Kohler has discussed the idea of 'Nature's experiments in succession' (Kohler 2002: 214), observed by ecologists where 'nature keeps resetting the clock of succession back to start' through catastrophic process (218). Pallis achieved the equivalent through grazing cessation, Long Gores released from active management. In 1955 Pallis would urge Catherine Gurney to act in possible parallel at Calthorpe Broad: 'If the wretched Nature Conservancy do not take Calthorpe you must and let it go wild like Long Gores – just a path for you to get through.'[37]

Figure 33   'Calthorpe Domain', by Marietta Pallis, painting, probably early 1950s. Source: © The Pallis/Vlasto Archive.

Pallis took aesthetic delight in her ecological experiment, her paintings conveying post-impressionist plant landscape. Formally sedate paintings of marsh and grazing horses gave way to energetic images of wood returning, streaks of blue and purple signalling vital vegetational renewal. 'Calthorpe Domain', given by Pallis to Catherine Gurney, showed wood growing out of shallow water, fen turning to tree, with on the reverse a quotation from Rimbaud's 'Les Amis':

> J'aime autant, mieux, même,
> Pourrir dans l'étang,
> Sous l'affreuse crème,
> Près des bois flottants.
> [I had as soon – rather, even –
> Rot in the pond,
> Beneath the horrible scum,
> Near the floating driftwood.][38]

If Pallis had been central to early twentieth century Broadland science, by the 1930s she was marginal, a Bohemian on the edge of the region, cutting Cambridge links and intellectually unorthodox. Vitalist elements increased through her *General Aspects of the Vegetation of Europe* (Pallis 1939), largely discussing Greece and the Broads, and from 1956 a series of pamphlets articulated Pallis's complex eco-theology, the final word of the final pamphlet being 'Transubstantiation' (Pallis 1963: 11; Matless and Cameron 2007b). Pallis had no involvement with Norfolk nature institutions, though maintained individual friendships with Catherine and Robert Gurney, William Percy and Ted Ellis. Pallis admired Ellis's journalism and museum work, suggesting he collect his *EDP* pieces for publication: 'it would serve as a Norfolk case book'.[39] Pallis also appreciated the non-interfering ethos at Wheatfen, visiting the Ellises, Phyllis recalling 'a quite extraordinary woman' in 'a tailored jacket and britches ... very sensible too – she lived in the middle of a reed marsh – what else would you wear?'[40] In 1956 Pallis sent Ellis a draft, later to become *The Impermeability of Peat and the Origin of the Norfolk Broads*, with a protracted correspondence ensuing over whether *TNNNS* might publish it (Matless and Cameron 2006). Pallis withdrew after Editor Ellis suggested excising the more philosophical sections.

If Pallis kept distance from nature institutions, her eco-philosophy rubbed against neighbouring landowners (Matless and Cameron 2007a). In 1941 the Smallburgh Internal Drainage Board, members including Jim Vincent and CA Newman of the adjacent Priory Farm, acquired a Smith Dragline Excavator as part of the 'Hickling Scheme', extending agricultural production in wartime, countering 'the bad state of certain private ditches and drains at Hickling'.[41] IDB minutes record:

> Mr Vincent reported that the work of the dredger had been held up at Hickling owing to the action of a Miss Pallis, who objected to a short length of dyke adjoining her property being cleaned out. It was pointed out that the spoil could be deposited on Mr Newman's side, so that Miss Pallis' willow trees should not be damaged, and eventually Mr Newman agreed.[42]

Vincent objected to Pallis's obstruction being rewarded, but a legal agreement was drawn up, the dyke adjoining Pallis's cottage to be cleaned and widened by hand and not dredger, the IDB not to 'interfere with, trim or damage the willow trees in any way whatsoever'.[43] Landed values clash, tensions sharpened in wartime.

Pallis held woodland to be the 'Global Dominant', trees to succeed marsh in Broadland. Pallis appreciated fen botany, its 'rich jumble of species' (Pallis 1939: 23), but desired it to pass. Grazed Long Gores was an ecology 'in two minds', Pallis persuading it in one direction: 'where interference is in abeyance on fen, trees and shrubs begin to reappear in the attempt (fen-carr) to re-establish the climatic dominant of deciduous forest' (Pallis 1939: 3). Excavating the Double-Headed Eagle Pool, tree trunks were found in peat, showing an oak, willow, birch, alder and sallow past: 'From woodland we come to woodland we go: we have reached the stage in the Broads where woodland is re-asserting itself' (Pallis 1956: 7). Long Gores becomes pioneer territory, with the pool's dramatic interference serving as experimental regional microcosm of Broadland history. For Pallis reed was the 'global sub-dominant' in shallow water, peat digging having given it space, but without interference, reed was lost (Pallis 1958: 14–16). The final stages of Pallis's ten stage 'potential sequence' of regional landscape change signalled the destiny of Long Gores and the double-headed eagle pool:

(7)   Rain-water accumulated as free water in the dug Broads.
(8)   Slud formation
(9)   Aquatic, swamp and fen peat formations.
(10) Woodland gives signs of returning. (Pallis 1961: 13–14)

The pool becomes a regional microcosm, dug to succeed, the symbolic stamping of 'MP' on the landscape healing through successive restoration, its former owner dissolving in the natural order enabled by her benign neglect. From woodland we come, to woodland we go.

After Pallis's 1963 death Long Gores grew into scrub and part wood, the pool hard to make out. From the 1980s however Pallis's great-nephew Dominic Vlasto took residence, enacting different landscape values. Grazing was reintroduced, woodland managed, scrub cleared, the pool connected to the dyke system, the island tended, clear from the air, viewable and filmable from the ground. Nine specimens of *Fraxinus pallisae*, the Balkan ash collected by Pallis and named for her by AL Wilmott for the Linnean Society in 1913 (Pallis 1916: 283–5), were planted around the pool for the nine Greek muses, and a gravestone part carved by Pallis before her death replaced the former metal cross. Familial succession renews 'interference', custodial care putting woodland in abeyance, briars under control.

Icon II
Windmill

Mill Time

As Broadland icons the windmill and wherry concentrate questions of time, technology and air; the wherry on the river, the windmill on the marsh. Most Broadland 'mills' were concerned not with milling but pumping for drainage, though some also ground corn, seed or cement (Apling 1984; Williamson 1997). The progress of drainage, the melancholy of ruin, the harness of nature power, the stump left standing by adjacent steam or electric pumps; the windmill and its remains prompt narratives of a region working or fallen, in health or neglect, for contemplation or restoration. Windmills take the Broadland air to varying cultural effect, with a recurrent scenic movement between close studies of interior machinery, and perspective views of distant sails turning over drained marsh.

The late twentieth century 'Wherry' line rail passenger, journeying between Norwich, Yarmouth and Lowestoft, commonly travelled in a Class 101 DMU (Diesel Multiple Unit), passengers facing forward near the front able to look through the cab glass past the driver to the lines ahead, running seaward, via Acle to Yarmouth, via Reedham and Berney Arms to Yarmouth, or via Reedham to Lowestoft. Past Acle and Reedham marshland opens, the train on pure flat, across the Halvergate level or beside the New Cut and Haddiscoe Island, the only verticals the pylons over the Cut, distant Yarmouth towers and chimneys, and mills, in sail or stump. A recounted August 1992 DMU journey opens WG Sebald's *The Rings of Saturn*, taking the Lowestoft line to Somerleyton (Weston 2011). Sebald's works intersperse photographs into prose to

In the Nature of Landscape: Cultural Geography on the Norfolk Broads, First Edition. David Matless.
© 2014 David Matless. Published 2014 by John Wiley & Sons, Ltd.

spark memory, and his choice here is a picture of a mill stump by a dyke, drainage channel with ex-drainage technology. After Reedham:

> Save for the odd solitary cottage there is nothing to be seen but the grass and the rippling reeds, one or two sunken willows, and some ruined conical buildings, like relics of an extinct civilization. These are all that remains of the countless wind pumps and windmills whose white sails revolved over the marshes of Halvergate and all along the coast until in the decades following the First World War, one after the other, they were all shut down. It's hard to imagine now, I was once told by someone who could remember the turning sails in his childhood, that the white flecks of the windmills lit up the landscape just as a tiny highlight brings life to a painted eye. And when those bright little points faded away, the whole region, so to speak, faded with them. (Sebald 1998: 30)

Tropes of light recur through *The Rings of Saturn*, often flashing surprising former glory; the glasshouses of Somerleyton Hall at night, Lowestoft pier illuminated in its heyday, the gleam of caught herring in a 1950s film of Wilhelmshaven trawlers: 'From the earliest times, human civilization has been no more than a strange luminescence growing more intense by the hour, of which no one can say when it will begin to wane and when it will fade away' (170). Sebald finds a region faded from technology passed, conical stumps missing their sail finish, marshland diminished, unlit.

Wind Technology

From mills in ruin to a windmill in a ruin. St Benet's Abbey, with a working windmill built into the former abbey gatehouse, was a common subject for 'Norwich School' painters. Today the gatehouse is joined to the redundant mill stump, ruined regional emblems propping one another.

The St Benet's mill was a brick tower structure, likely early eighteenth century, its cap and sail blown off in 1863, with some dispute as to whether it was initially used to crush seed for oil, or always a drainage mill (Apling 1984; Williamson 1997). Early nineteenth century paintings depicted current mill technology amid picturesque ruin, Hemingway noting the juxtaposition of medieval decay and modern development as one of the 'staple tropes' of topographic imagery (Hemingway 1992: 270). Emerson's later *LLNB* photographic contrast of derelict windpump and new steam pumphouse, 'The Old Order and the New' (Emerson and Goodall 1886: 35), is anticipated in different form, St Benet's sails contemporary against abbey medieval. John Sell Cotman's 1831 'St Benet's Abbey, Norfolk', in Norwich Castle Museum, shows the mill built into the ruin, a wooden platform over the walls, eel trappers by an inlet in the foreground. In Henry Bright's 'Remains of St Benedict's Abbey on the Norfolk Marshes – Thunderstorm Clearing Off' (c.1847, also in the Castle), sunlight casts the abbey ruin as scenic focus, the mill working in shade, another mill standing tall across the river, a ruin spectacle in a working landscape (Allthorpe-Guyton 1986: 34). Stark's *Scenery of the Rivers of Norfolk* shows a sunset St Benet's image, moon rising, Robberds meditating on ruin, the gatehouse 'at once an emblem and a monument of fading glory',

Figure 34   'St Benedict's Abbey, Norfolk', by James Stark. Source: Stark and Robberds 1834, image courtesy of Norfolk County Council Library and Information Service.

the mill 'an unsightly and intrusive superstructure' (Stark and Robberds 1834: plate 6). If, notwithstanding Robberds' 'unsightly' comment, the mill might carry some picturesque quality in pictorial composition and ramshackle wooden platform, this was nevertheless a working object, functioning for decades, its tower solid, its sails still turning.

Windmills accompanied wherries in 'Norwich School' river paintings, turning the elements to commerce. Alfred Stannard's 1846 'On the River Yare' (Norwich Castle Museum), likely downstream of Reedham, works through striking verticals of moored wherry sails, a wooden drainage mill on a brick base, and the smoking chimney of a brickworks behind, different orders of technology and machinery in concert. Broadland windmills could also evoke art and technology beyond the region, Norwich painters taking a cue from Dutch landscape art, and their subjects matching scenes across the North Sea, then commonly termed the German Ocean. The comparison held into the twentieth century. In 1962 Pevsner would comment: 'WINDMILLS. There are to the S and SE of Runham along the river Bure five drainage mills, all brick tower-mills, with and without sails. From certain points the view is much like in Dutch C17 landscape paintings' (Pevsner 1962: 306).

Drainage mills are prominent in the major recent work of Broadland landscape history, Tom Williamson's 1997 *The Norfolk Broads*, casting the region as functional industrial landscape; its lakes originally dug for peat, with further large scale eighteenth and nineteenth century shallow peat extraction, its marshes drained, its rivers channels of trade (Williamson 1997; Matless 1999b). If the twentieth century also saw mills claimed for tradition against the present, as discussed below, Williamson, working from the UEA, finds historic Broadland modernity, wary of formulations of heritage presented by the BA and others. The drainage mill is a primarily Victorian industrial archaeological artefact, the most visible relic of an industrial landscape including extractive industries of chalk and lime, brickyards and cement works. Williamson's double-page spread of 59 mill towers in profile, 'Comparative elevations of Broadland tower drainage mills', runs from Oby in 1753 to Horsey in 1912, electric pumps and reorganised drainage under IDBs from 1930 leading to final wind abandonment. For Williamson the 72 surviving drainage mills signify former modernity: 'the mills are not some timeless adornment of the scenery. They are practical industrial features with a *history*: of technical ingenuity, mediated by economic demands' (135–6).

Williamson's mill starting point was Rex Wailes, windmill historian whose 1954 *The English Windmill* devoted a chapter to Broads drainage mills (Wailes 1954: 72–81), work extended in his 1956 *Newcomen Society Transactions* complete Norfolk windmill survey, first presented at the Science Museum (Wailes 1956).[1] Wailes registered the windmill within an emerging industrial archaeological sensibility. If windmill study was a field 'aesthetic, topographical, social, antiquarian and technical' (Wailes 1954: xxiii), Wailes emphasised technology, the mill as regional fulcrum: 'When you see a windmill, don't think of her as just a pleasant part of the landscape. Think of the thought and devoted craftsmanship that has gone into the designing and building of her, and in the upkeep since then. And to appreciate their value to the full you must experience the fascination of windmills at work.' Wailes urged appreciation not only from outside, seeing sails turning, 'but from the inside too', evoking sounds of working machinery, its 'swish', 'rumble', 'clatter', 'rattle', 'clap' and 'tapping', 'a fine blend of cross rhythms' (xxii). Technical drawings conveyed tower structures, caps and curbs, winding gear, sails, shafts and wheels. Wailes found millwrights' individual skills readable in specific structures: 'The millwrights were all fine, ingenious and resourceful men and it is to the eternal disgrace of the Norfolk County Council and the Catchment Board that they have not thought fit to preserve some of their work, much of which has been wantonly destroyed since the war' (Wailes 1956: 168–9). Norfolk and Suffolk drainage mills had now 'almost all ceased work' (Wailes 1954: 76); the last, Ash Tree Mill on the lower Bure, lost its sails in the gale of 31 January 1953 (Grimes 1954). Like Sebald, Wailes could only evoke prospects lost: 'Everyone who visited the Norfolk Broads before 1939 must have their own memories of the windmills that at one time dominated the landscape. My own especial impression is of the scene from the Acle "New Road" to Yarmouth from which, in the mid-twenties, nearly a score of mills could be seen in working order at all points of the compass' (Wailes 1954: 72).

Restorative

Wailes presented the windmill as (formerly) working object, whose technological appreciation could inform possible restoration. As with the wherry, a 'restoration movement' emerged post-war, public and private bodies salvaging regional exemplars (Scott 1977: 15–26). Wailes advised public bodies on restoration, as at Berney Arms, the remote Yareside marsh community, accessible only by boat or rail, its mill built in 1865 for grinding cement as well as drainage, the cement works closing in 1880.

Figure 35   Berney Arms Mill and electric pumping station, August 2011. Source: photograph by the author.

The Ministry of Public Buildings and Works took over Berney Arms Mill from the local IDB in 1951, opening it to the public in 1956; Wailes produced the souvenir booklet on 'the finest example of the old marsh mills', over 70 feet high, an isometric drawing cutting away the exterior to show the workings (Wailes 1957: 5; Wailes 1954: 77). In the early 1960s public funds from Norfolk County Council helped restore Stracey Arms, while privately owned mills at Thurne Dyke and Horsey were also restored, sometimes with commercial boat hire firm support (Scott 1977).

As with the wherry, restoration activism could present past technology as a refuge from modernity, and mill loss as indictment of the present. A week before the *EDP* letter announcing the intention to form a wherry trust, Roy Clark had responded to press correspondence on marsh mills by noting a possible role for Society for the Protection of Ancient Buildings' windmill section, and asking why, given there was a forest conservation body called the Men of the Trees, there should not be a 'Men of the Mills' (Clark 1949). Kenneth Grimes called for 'a vigorous policy of restoration', lamenting the 'wanton destruction of one of England's loveliest landscapes', in what used to be 'a land of windmills – a "Little Holland"' (Grimes 1954: 60–1). For Grimes: 'A Broads National Park without marsh mills at work is a fantastic absurdity ... The first restoration could begin in this historic year of Coronation. What a splendid inauguration of the Second Elizabethan Age this would be for Broadland' (Grimes 1953a). *EDP* editorials noted enthusiasts' 'agitation', Norfolk behind other counties where it ought to lead: 'it also possesses the Broads where, well within living memory, the landscape was diversified in every direction by the revolving sails of the mills which were so beloved by the Norwich School of painters' (EDP 1953o). The paper also noted other mill suggestions: 'Mr. H. T. Percival, of Horning, has suggested as a boat-letter that the bases of the mills might be turned into water closets and bathrooms for holidaymakers on the Broads: the tops of the towers being then decorated with light open-work facsimiles of the original sails. That is a highly practical suggestion, but it is hardly what the enthusiasts want' (EDP 1953l). Percival's inventive reuse, facsimiles for public convenience, would not count as restoration.

Wailes spoke in Norfolk in April 1954, urging County Council leadership. Immediate attention, with campaigning by CPRE, focussed on Billingford Mill, a corn mill near Diss, whose case prompted the County Planning Officer to list mills for potential restoration, with 18 selected for grants from 1960. Pevsner noted of Broadland: 'Norfolk is the county of windmills more than any other, and the recent decision of the County Council to preserve permanently at least eighteen out of the hundreds whose tapering brick towers penetrate the skyline is a telling token' (Pevsner 1962: 14). The term 'telling token' may indicate both heritage significance and an ultimately lame gesture, just 18 saved for what will no longer be a county of working mills. In 1963 the Council established the Norfolk Windmills Trust, with Broads windpumps a key focus; the Friends of the Norfolk Windmills Trust was launched in 1976 to further enlist public enthusiasm, windmills fostering an environmental activism of work-party labour alongside expert millwrighting (Scott 1977). Arthur Smith's *Drainage Windmills of the Norfolk Marshes*, based on 1971–8 fieldwork, catalogued the scene, photographing all 72 sites; some crumbling for lack of attention, others proudly restored, others in functionless suspension, clear shapes evident in flat landscape, attracting the eye if nothing else, Smith spotting all from car and bicycle (Smith 1978).

The Norfolk Windmills Trust co-operated with the new BA, managing properties including 12 Broads drainage mills.[2] In 1995, three years after Sebald's DMU trip to Somerleyton found melancholy in stumps (and in the same year as *The Rings of Saturn*'s first German publication), the Norfolk Windmills Trust and the BA proposed a counter-melancholy move, a restorative £1.1m 'Project for the Millennium', creating a 'Land of the Windmills' from the restoration of 13 derelict mills on the Halvergate marshes, 'rejuvenating the historic Broads landscape and creating a strik-ing visual feature to mark the millennium in the Broads' (Norfolk Windmills Trust and Broads Authority 1995: 1). The project brochure gave line drawings of each mill in its current state, local artist Mel Harris visualising the outcome through a fold-out cover painting, with a wherry in sail, and distances foreshortened to show 13 mills, several looming close, others not far distant, as if mills were on the march to restore an older Broadland, crowding to make a millennial point. Restoration aggressively polishes a faded region, scrubbing ruin away.

'Land of the Windmills' was turned down by English Heritage and the National Lottery in January 1996 (EEN 1996). In 1997 Williamson reflected on BA judge-ments, noting how things labelled ugly when new were now deemed traditionally attractive, and commenting on lost late nineteenth century structures:

If the sprawling cement works at Berney Arms in the heart of Halvergate had survived longer, would it too have become part of the 'traditional' landscape, perhaps utilised as a heritage centre? Many of the drainage mills in Broadland are little older: some are much more recent, and all are pieces of practical industrial plant. Current proposals to restore, almost to working order, vast numbers of these redundant structures are more than faintly bizarre. Mills like Martham or Horsey are younger than my grandmother.

Williamson ended his landscape history with other possible futures: 'After all, if global warming proceeds, much of Broadland will become once more a network of estuaries and saltings. Perhaps within the lifetime of some alive today, the tides may flow freely once more across Halvergate, and waves lap the walls of the drainage mills' (Williamson 1997: 165).

New Sails

From the early twenty-first century, if the land of windmills remained mostly stumps, walkers along Breydon's north shore had other sails visible beyond Yarmouth, the turbines of the offshore Scroby Sands wind farm notable over the urban spit. New sails had made the scene after all. From the early 1990s, Thurne boaters could not help but spot new sails above Somerton, wind farmed on Blood Hills above the village, ten 30m high three-blade turbines constructed in 1992. The latest variant of Broadland wind power prompted contrasting aesthetic and political judgement. In 1995 the Lion Inn at Somerton sold cards showing the recently built turbines, drawn by a Winterton artist, signed 'N Ward, '93'. One of Nicholas Ward's pencil drawings showed all ten before a landscape stretching to Winterton, a pile of tyres on a foreground fieldside verge. Another pictured a single turbine by Somerton church, Horsey Mill tiny in the distance, while 'Wind Turbines – West Somerton'

Figure 36   'Wind Turbines – West Somerton', by Nicholas Ward, 1993. Reproduced by permission of Nicholas Ward RE.

had all ten on the skyline from a gateless field entrance; an ivy-covered gatepost, a harvested beet pile, farm behind farm. Nicholas Ward's drawings, in their composition and line, integrate turbines into Norfolk landscape, a new presence given appreciation, an addition in workaday keeping, their lines drawn in the manner of field and hedge, farm and farm in grain.[3]

The BA had opposed the Blood Hills wind farm, which lay just outside their planning authority boundary, but was visible within, their 1993 Broads Plan including a photograph within the 'Landscape' section of two turbines, with Somerton church small behind: 'Winds of change? This wind farm was opposed by the Broads Authority on landscape grounds. But some people believe the wind turbines have a certain elegance.' While welcoming 'alternative energy sources', the BA, with its 'first duty … to protect the landscape value of the Broads', stated: 'The development of wind turbines will not normally be permitted in the Broads area' (Broads Authority 1993: 111/151). In 1994 the BA rejected a plan for 14 turbines at Bure Loop, near Yarmouth, a site near to Ashtree Farm Mill, one of the proposed 13 millennial windmill restorations (EEN Eastern Evening News, 1994). Mill heritage and turbine futures receive contrasting landscape valuation. In 2000, a 65m 1.5MW turbine was built half a mile west of the Somerton group towards Martham, marking the ridge overlooking the Thurne, the Blood Hills group appearing in contrast a minor construction from an earlier phase, almost traditional after 20 years. Operators Ecotricity note that 'old style windmills' were 'alien' when first introduced: 'What could be more natural than a windmill on the Norfolk Broads.'[4]

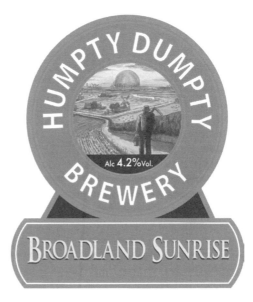

Figure 37   'Broadland Sunrise', beer pump clip, Humpty Dumpty Brewery, Reedham, designed by Cherry Ann Burns-Salmond, c.2008. Reproduced by permission of Humpty Dumpty Brewery.

If 'Wherry' has gained a drinking reputation, mills and turbines too feature in microbrewed imagery, new sails and old harnessed for brand recognition, as familiar and popular signs. Thus Norfolk Square brewery of Great Yarmouth offers 'Scroby', three turbines on the label, while the Humpty Dumpty Brewery of Reedham sells 'Broadland Sunrise' (alongside 'Swallowtail' and 'Reedcutter') through wind-powered landscape. Cherry Ann Burns-Salmond's 'Broadland Sunrise' label pictures a version of Halvergate, a worker walking east onto the marsh at dawn with scythe and beer. Two windmills in full sail stand by rivers and dykes, neat farmsteads tend marshes, sheep graze in meadows, and sun rises orange over sea, silhouetting offshore wind turbines. Broadland powers conjoin under the sun, cutter's strength and beer, sails and blades, air turning technology old and new.

Chapter Six
The Ends of Landscape

The Balance of Waters

Broadland is shaped by the circulation of waters. Fresh and saline mark the region via diurnal tidal movement upriver, or unusual sea incursions, the region defined as delicately balanced, vulnerable to disturbance. Alongside silting and succession, considered in earlier chapters, pollution, drought, drainage and flood have signalled possible circulatory imbalance, ends of Broadland, a region no longer itself. This chapter gives greater attention to recent decades, as a sense of environmental disaster has overtaken the region, notably with the eutrophication of Broadland waters. Narratives of climate change and sea level rise, termed by the BA 'the greatest challenge' for a region 'in the front line', have further heightened the sense of Broadland at risk (Broads Authority 2011: 13/17; cf Daniels and Endfield 2009), although evaluation of past accounts of sea flood indicates other possible regional messages, whether of resilience against threat or secret relish of inundation. Whether styled as danger or possibility, the ends of landscape shadow the present.

Two mid twentieth century descriptions of regional circulatory balance can serve as stepping off points here. Both call to mind David Nash's 'Wooden Boulder' sculpture, a carved form placed in a North Wales river in 1978, filmed in its 26 year journey to eventual estuary, washed by flood, stuck under bridges, bobbing into sea, stuck on sand (Nash 2010: 114–19). Thus in February 1959 Ellis described 'Bottle's Voyage in Broadland':

> We know comparatively little about the circulation of the waters ebbing and flowing in the rivers of Broadland, so when on January 26th I found a Smallburgh address in a

In the Nature of Landscape: Cultural Geography on the Norfolk Broads, First Edition. David Matless.
© 2014 David Matless. Published 2014 by John Wiley & Sons, Ltd.

bottle washed up in a reed bed at the point where our Fen Channel enters Rockland Broad, I wrote to the sender of the message. He informs me that he placed the bottle in the River Ant at Wayford Bridge on September 18th, 1958. On the face of it, one assumes that the bottle travelled down the Ant and Bure to Yarmouth, then up the Yare to Rockland Broad on a strong flood tide. If this journey was not 'assisted,' it reveals an interesting state of affairs so far as tidal influence in the Bure and Yare is concerned. (Ellis 1959)

Likewise May's 1952 Broads guide delighted in circulation, tracking a floating 'hover', probably of the *Glyceria* studied by Lambert, 50 yards round with resident sedge warblers. Detached by high tides in Womack Water, the island cruised 'erratically up and down the Bure and the Thurne for two months', blocking Thurne Dyke, Womack Water and Potter Heigham bridge, before it 'wandered off disconsolately downstream, a magic island that nobody loved, to drown itself in the sea. Another little shift in the process of Broadland change was ended' (May 1952: 159–60). For Ellis and May, objects moving of their own accord through Broadland mark the normal peculiarities of regional circulation.

The Death of Water

Crisis diagnosis

The only Broadland waters remaining clear through the past 100 years are at the head of the Thurne, in Martham Broad and beyond. Water plants remain abundant, visible, at Somerton staithe.

By the 1970s, the term 'eutrophication' captured policy and popular imagination, high nutrient levels changing Broadland waters, clouding becoming symbolic of ecological crisis (George 1992: 99–162; Moss 2001: 186–230). Eutrophication offered a historical narrative, the BA's 1982 *What Future for Broadland?* summarising:

Phase 1. This is the situation which existed in most of the broads prior to 1900. The vegetation was fairly sparse and of low growing species. The only broads retaining this flora are those which do not receive any sewage or livestock effluent and are separated from arable land by an area of unreclaimed fen. The fen filters off the nutrients from the waters flowing into the broad.

Phase 2. As nutrient levels increased the broads and rivers moved into this phase. The higher nutrient level encouraged a more luxuriant growth of taller plant species. The extent of this growth was often so great that the Port and Haven Commissioners had to cut channels through the weed for navigation …

Phase 3. The high nutrient level of the water encourages abundant growth of algae which have shaded out the submerged aquatic plants (macrophytes) …

   With the loss of the aquatic plant life has come a decline in the whole range of aquatic animals which rely directly or indirectly on them. (Broads Authority 1982: 14)

Most broads were classified Phase 3, with local factors occasionally adding to enrichment, as with gull colony excreta at Hickling. Somerton's plant life would signal Phase 2, not some pristine state. Early accounts of floral and faunal abundance appeared in retrospect less touristic promotional exaggeration than confirmation of Phase 2 'high biological productivity' (George 1992: 112).

What Future for Broadland? included a telling errata slip for its eutrophication illustration: 'Diagram opposite Phase 1 should be opposite Phase 2 and vice versa'. The figure as printed suggested decline, from Phase 1 abundance to sparse Phase 2 and a plantless Phase 3, with a dead duck floating (Broads Authority 1982: 16). The first two had however been swapped, sparseness where luxuriant growth should have been, the error perhaps indicating an assumption that death would follow decline rather than abundance. Eutrophication instead suggested Broadland destroyed by excess, matching for some an equivalently excessive post-war human conduct, abundance all round destroying the region.

Initial 1960s concern for water pollution had focused on sewage, boating interests commissioning Roy Hoather of the Counties Public Health Laboratories, London, to analyse pollution from river craft. Little was found, James Hoseason concluding that 'anti-pollution by-laws for the Broads are totally unnecessary' (EDP 1965c). Hoseason's interpretation of 'Norfolk Broads purity confirmed' was however contested, critics urging 'Cleansing the Broads' of direct sewage discharge (EDP 1965e; 1967b), and by 1971 by-laws prohibited direct discharge, floating faeces and used contraceptives (as described in Day's 1967 tripper critique of 'the thin grey scum of contraceptives on the water') disappearing from the regional scene (Day 1967: 36; George 1992: 122). Sewage disposal was not though simply an amenity question. The NC's 1965 *Report* noted 'Personal communication' from Ellis that 'some naturalists' regarded raw sewage as 'relatively harmless to wild life as compared to chemicals contained in the effluent from large sewage works' (Nature Conservancy 1965: 56). For Ellis new household detergents in sewage waste were tipping regional balance, making broads 'sterile'. Whitlingham sewage works, upstream of Wheatfen, processed Norwich's sewage into the Yare, Ellis noting 'the disappearance of most of the submerged and floating water weeds from the Yare Broads and tidal fen waterways ... in the past eight years or so' (Ellis 1958a). Ellis restated the point in columns and broadcasts in the early 1960s, finding 'demonstrable change in the microscopic planktons' (Ellis 1960b): 'we don't know where it's all going to end'.[1]

Ellis's Yare observations echoed Hickling concerns, Piggin reporting clouding and plant death in May 1962, with the NC asked to investigate 'the effects of (a) detergents and (b) diesel oil on aquatic vegetation'.[2] By 1974 Hickling water plants had disappeared, with coot numbers falling from thousands to a hundred (Moss 1977a). Hickling had capricious ecological history, AJ Rudd's 1943 NNNS Presidential Address having noted: 'This is a strange water and its weeds are constantly changing. Fifty years ago *Chara* predominated, then came a *Zygnema* which killed it all. This was followed by the so-called "Hickling weed," a wretched alga, *Cladophora Sauteri*' (Rudd 1943: 381). That current change might however be from human action rather than natural fluctuation was suggested by parallel animal deaths, avian botulism killing waterfowl in 1975 and 1976, and major fish kills by *Prymnesium* algae in Hickling and

Horsey in August–September 1969 (Nature Conservancy Council 1977). The Broads appeared as diseased nature, with Phase 3 eutrophication a narrative to tie local episodes into general decay. Fish kills remain a reference point for angling interests, as with protests when 2010 and 2011 BA dredging plans were seen to threaten new algal blooms, 1969 a traumatic folk memory (EDP 2010; 2011c).

Ecological enquiry found academic co-ordination via the UEA's Schools of Biological Sciences and Environmental Sciences. UEA research funders included the BA, Nature Conservancy Council (NCC), Anglian Water Authority (AWA), Department of Environment, Natural Environment Research Council, and the Soap and Detergents Industry Association; by 1984 the BA could issue a Broadland Research Register, marking Broadland's emergence as research region (George 1992: 489). In 1970 the UEA awarded Ellis an honorary Doctor of Science, registering connection between university and regional naturalism, and the Broads would help make the university's regional and international reputation, notably through the work of Brian Moss in Biological Sciences and Tim O'Riordan in Environmental Sciences, both interweaving ecological and social analysis. Moss would author the second New Naturalist volume on *The Broads*, gathering specialist ecological research into regional overview, and his work is discussed further below (Moss 2001). O'Riordan's 1967 Cambridge doctorate had considered 'Multipurpose Water Resource Management on the Northern Rivers of Broadland', and social scientific geographical expertise would inform his work as a Countryside Commission BA nominee. O'Riordan presented Broadland as a case study in the fledgling discipline of environmental management, echoing the NC's 1965 *Report* on zoning and multiple use (O'Riordan 1969; 1971). Conduct was mapped onto types of Broads user, from first timers to 'dedicated Broadland lovers' possessing heightened 'Broadland consciousness' (O'Riordan 1969: 45), the latter 'the most articulate and reliable guide in the formulation of a future environmental policy for Broadland'. O'Riordan came however to melancholy conclusion: 'As Broadland becomes increasingly popular, these solitude-seeking groups will be forced more and more toward the outer fringes of the season, and eventually may be lost from the Broadland scene for ever' (49). The discerning visitor becomes a species under threat, rare as a bittern in a popular future.

The first published UEA Broads ecological research was Mason and Bryant's 1975 *Freshwater Biology* paper on 'Changes in the Ecology of the Norfolk Broads', from a 1972–3 survey of diminished flora in 28 broads, with modern farming an identified culprit: 'the run-off of agricultural fertilizers may be largely responsible for the loss of macrophytes' (Mason and Bryant 1975: 268–9). In May 1975 the NNT joined public debate, a statement on 'The decline of wildlife in the Norfolk Broads' citing UEA research, with nutrient enrichment suspected: 'Something is clearly wrong and, in the only way she can, nature is sending distress signals' (EDP 1975). The NCC funded UEA work, including historical survey, AM O'Riordan compiling *A Broadland Bibliography* (1976), abstracts of Broadland literature grouped under 'Physical Science', 'Botany and General Ecology', 'Zoology', and 'Other Stuff', an audit for a newly environmentalist age to succeed GA Stephen's 1921 bibliography of regional discovery (Stephen 1921; O'Riordan 1976; George 1976). On 2 July 1977 Moss organised a public forum at the UEA on *The Future of Broadland*, with speakers from local authorities, conservation, tourism and farming bodies, the AWA and GYPHC, George speaking for the NCC, Ellis for the NNT (Simmonds 1978). The NCC's *The Future of Broadland* summarised 'ecological degradation' as 'symptomatic of a general malaise', with eutrophication 'by far the most important single threat to the Broadland environment'. A complex diagram, reprinted in the summary of the UEA proceedings, and in George's 1978 Presidential Address to the NRC, showed '"Cause and Effect" Inter-Reactions in Broadland'. Thirty-seven interconnected boxes formed a maze without exit, with independent input from 'natural processes', 'increased agricultural productivity', 'sewage effluent' and 'seagull roosts (Hickling Broad)' (Nature Conservancy Council 1977: 7; George 1977; George 1978; Simmonds 1978; also Moore 1978; Moss 1979; O'Riordan 1979a; Moss 1983). Bombarded, Broadland becomes trapped in cause and effect.

In 1983 Moss noted 'an unexpected aspect', with 'the activities of the scientists ... under scrutiny as subject matter for the research of social scientists' (Moss 1983: 553). O'Riordan studied Broadland science in action, Moss, George and others shaping policy, with scientific doubt and ambiguity allowing 'vested interests' (AWA, MAFF, boating groups) to challenge university ecology (itself by implication beyond vested interest), with 'the serious ecological researcher ... caught in the cross fire' (O'Riordan 1979b: 809). More recent climate science controversies come to mind, O'Riordan concluding that ecologists needed to 'become politically literate' (811; O'Riordan 1978). Political and commercial implications had indeed emerged from Moss's earliest papers, a 1973 turbidity survey suggesting boats were a minor factor, the environmental blame lying elsewhere: 'It is concluded that increase in turbidity is a function of increased nutrient loading from human activities in the catchment area' (Moss 1977a: 95). Science here shifts blame from leisure. Moss also took science into radical environmentalist fora, writing in *The Ecologist* on 'The State of the Norfolk Broads': 'Man and the Broadland have been happily married for centuries and it is only in the last few decades that the union has foundered.' Six labelled slides, some dated, illustrated a dying waterland; rich plant life in an isolated broad (August 1969), turbid water (May 77), *Prymnesium* cells through a microscope, a floating pike dead from *Prymnesium* (July 77), bank erosion (May 74), 'ugly' bank piling (March 77) (Moss 1977b: 324–5). Scientific, political and aesthetic commentary combine.

The public profile of UEA science was evident in Anglia TV's 1979 film *No Lullaby for Broadland*, written and directed by Geoffrey Weaver, and worthy of detailed discussion for its public crisis diagnosis, with regional history deployed in present critique, and leisure and pollution enfolded in dystopic fashion. Anglia viewers on Tuesday 20 November could begin their evening with the *UK Disco Dancing Championships* (7.00–8.30), sit through the comedy of *George and Mildred* (8.30–9.00), and end in a boat with Brian Moss.[3] Narrator Eric Thompson (also the voice of children's animation *The Magic Roundabout*) soothingly described crisis and chaos, the 'Chemical Cocktail' (EDP 1979) of eutrophication gathered into a critique of regional destruction by consumer excess.

No Lullaby opens with a montage of imagery and activity, tripper critique framing chemical argument. A Hoseason advert adapting the song 'Messing about on the River' is intercut with vox pop and holiday shots to imply idyllic dreams debased.

A Jarrold 'Birds of the Norfolk Broads' cover appears over two swimsuited women. Midland hire coaches descend, boats crush to a soundtrack of 'Roll out those lazy, hazy, crazy days of summer'. Commentary states: 'Beyond causing physical damage the boating industry, along with poor planning, has led to aesthetic corruption. Potter Heigham announces an image, but what of the nastier reality?' At this moment Boney M's 1978 hit 'Rivers of Babylon' kicks in over images of litter, commentary describing 'bungaloid growth and ugly boatyards', 1930s critique renewed. Hoseason walks into a villain's role, his advertisement opening Part 1, and Part 2 prefacing his contribution by stating: 'Intellectual ovaltine tranquilises ecological nightmares'. Filmed before red phones in his booking office, Hoseason hymns freedom over 'doomwatch journalism', narration noting 'Mr Average' as concerned more with cruiser facilities than surroundings. A grandmother reads the 1973 reprint of Sampson's 1931 *Ghosts of the Broads* in a cruiser cabin: 'The ghosts of the broads hold more fascination than its geography' (Sampson 1931; 1973). A Midlands-accented young man at Broadshaven, beer can in hand, comments: 'Landscape? You can't see a bloody thing for the reeds.'

No Lullaby sets leisure against conservation, speaking for Broadland as nature region through voices from the UEA, NC (Martin George in the field) and NNT, with Ellis filmed at Wheatfen lamenting 'steep decline'. John Buxton acted as wildlife cameraman. Halfway through Part 1 Moss appears in a rowing boat on Upton Broad, holding a jar of water, intercut with microscopic views of algae, explaining water clouding, fish kills and avian botulism, a calm reasoned voice. The problem is not the poisoning of plants but that they have grown all too well. Moss also appears at Stalham sewage works; supermarket scenes show consumers buying detergents. UEA science complements naturalist enthusiasm, with a party filmed carrying telescopes: 'Specialists can be flushed out of the reedbeds.' Here things become curious though, the film turning to excess in its own method. None of the naturalist party are named, and may be actors, and the film then cuts to figures in Edwardian naturalist costume, tweedy binoculars leering over 1970s sunbathers. Human–nature harmony is re-enacted as costumed actors cut and transport reeds, effectively acting out Emerson's photography, then subject to re-evaluation (Middleton 1978). An Edwardian couple punt in hazy focus.

Decline is dated to 1918, evoked by films of interwar female leisure, the region gendered frivolous, Hullabaloos arriving. The present gives little hope, with local authorities deemed self-interested and amateur, the GYPHC an 'antique' authority absurdly concerned for navigation as a 'democratic right of every Englishman'. The new Broads Authority is predicted to be insular, Moss describing 'a bunch of amateurs'. The film leaps from parochialism to end with naturalist Frank Fraser Darling on a Scottish hillside, his visionary words on land as community not commodity spoken over closing shots of NNT educational work: 'There's a therapy in the green leaf.'

No Lullaby's move across nature, science, history and leisure characterised much Broadland debate, including in radical conservationist argument. On 12 October 1979 Breydon Friends of the Earth (FoE) produced a four page 'special', the *Eastern Mourning News* (parodying the Norwich *Eastern Evening News*) on 'Broads Disaster'. The 'process known as Eutrophication' was turning waters to 'a turbid brown or green mess'. The Anglian Water Authority had installed only one sewage phosphate stripping plant, the GYPHC allowed overcrowding to 'Bursting Point', the BA were a compromise body representing vested interests: 'SOD the formalities, it's time for POSITIVE ACTION!' Anglia's ironic 'Broadland lullaby!' was noted.[4] Ellis contributed a 'Natural perspective': 'while to the modern visitor the general scene is in its simpler aspects still quite delightful, a rot has set in so far as much of the special wildlife is concerned' (Breydon Friends of the Earth 1979). Broadland's future clouded, redemption unclear.

Restoration Authority

The Broads Authority first met in September 1978, with members from local government, AWA, River Commissioners and the Countryside Commission (O'Riordan 1979a; 1979c). Aitken Clark was appointed as Broads Officer in 1979, after two appointees failed to take up the post. Previously Professor of Regional Planning at Clemson University in South Carolina, Clark was without prior local connection; he retired in 2001, having decisively shaped the BA's role, and died in 2010 (Adeney 2010). In April 1989 the BA gained statutory authority equivalent to National Park status, responsible for navigation, conservation and recreation, the Broads Bill prompted by Countryside Commission proposals in 1984. Transfer of navigation powers was opposed by GYPHC (reduced to a port authority), IWA and boat hire groups, while the NNT withheld support as the Bill (unlike in other national parks) did not make conservation an overriding priority. The Broads Bill passed in March 1988, the *EDP* hailing: 'The Broads: Dawn of a New Era' (EDP 1988).

Established during crisis diagnosis, from the start the BA promoted experimental 'restoration of aquatic plant life', accompanying the cultural restoration work noted in Chapter Three (Broads Authority 1982: 18; Moss 1983; George 1992: 488–520; Moss 2001). Images of suction dredging, water jetting, settlement lagoons, fowl exclosures, electro-fishing, became restoration currency, reprising in high mechanics the hefty hand labours which dug the broads. In 1983 Moss reviewed 'experiments in the restoration of a complex wetland' for *Biological Reviews*, setting the UEA as enquiring regional successors to Pallis, Gurney, Ellis and Lambert, with added field science and water chemistry. Animal and plant studies combined, as when from 1975 a UEA project studied a silted broad at Brundall, where a flight-crippled cormorant's fish predation allowed an increase in Daphnia zooplankton grazing on phytoplankton, with resultant plant growth (Moss 1983). Moss's later research emphasised zooplankton, with eutrophication Phases 2 and 3 presented as alternative rather than sequential stable states, switches triggered by changes affecting phytoplankton or zooplankton, whether toxicity, salinity or predation: 'In the loss of plants from the Broads, it was now clear that animals as much as nutrients were important' (Moss 2001: 211).

For the BA, then styling the Broads as the 'Last Enchanted Land', Cockshoot Broad, owned by NNT, provided showpiece restorative re-enchantment (Moss 2001: 199/296–302; George 1992: 502–8). Suggested by George as a possible site, restoration was guided by Moss, with phosphatic sediment removed from 1979 and the broad isolated from the Bure. The water cleared, and when reclouding indicated a lack of grazing zooplankton, bio-manipulation reinstated clarity. A visitor boardwalk showed the 'restoration of a dead broad': 'how all the broads would have been like, 100 years ago' (Broads Authority 1987). Visitors passed a dam blocking river water: 'Notice how murky and lifeless the River Bure is, compared to Cockshoot Dyke' (Broads Authority c.1990). How though might clarity extend to the main stream?

Restoration turned to Barton Broad, with suction dredging between 1995–2001, supported by £200,000 from the Soap and Detergent Industry Environmental Trust, with the southern end isolated from fish to encourage zooplankton. Barton would be restored 'From Darkness to Light', with 'rapid re-establishment' of diverse plants, including *Najas marina*: 'It is possible that ancient seed, exposed after dredging, germinated' (Broads Authority 2005: 15). The work became a millennial 'Clearwater 2000', with £1.15m from the Millennium Commission, a boardwalk constructed to the broad's edge (though clear water is not always evident), and a Freshwater Ecology Norfolk Broads Study Centre opened at How Hill (Broads Authority 1997). The BA style themselves a visionary environmental authority.

For Moss restoration was possible on broads but very difficult in rivers, their upper catchment beyond BA control. The Waveney was 'least damaged', the Yare 'the lowest priority of all' (Moss 2001: 324), and the Bure 'a nightmare for ecological restoration' (317), effectively 'a single convoluted lake' (320), Cockshoot signifying only 'small victories in an unwinnable war' (324). Restoration would require 'a major change in land practices' (323), and broad closures: 'Isolation might prove illegal if challenged in the courts' (321). On the Thurne, Martham needed 'merely protection of its present favours' (317), and then there was the strange 'self-restoration of Hickling Broad' (309), water clearing in 1998, possibly from gull excrement nutrients declining when the nearby Martham waste tip was covered. Moss foresaw future 'isolated nuggets of high quality in a bedrock of mediocrity', though extreme weather events accompanying climate change might intervene: 'They might change entire landscapes to unpredictable extents and obviate all manner of previous restoration work or achieve more in a week than has been possible over a generation' (327). Revolution might overtake restoration. Moss's visions of Broadland future, and the anxieties and excitements of climate change, are discussed below.

Broadland Drained

In January 1948 Ellis described a new moon 'Low Tide on the Broad', with a southerly gale holding back the flood tide: 'we experienced a continuous ebb in the Yare waters for twelve hours on end. Mudflats became exposed in the Broads round about mid-day, and by 8 p.m. the water had reached the lowest level we have ever known here. Rockland Broad came to look like Breydon, with high shelf-like mud banks intersected by branching channels' (Ellis 1948; 1982: 19). Weather and tide turn circulation to strange effect. Broadland was also noted dry in summer 1921, Bird and Gurney reporting on drought for *TNNS*, and Patterson recording Breydon's marshes like 'cokernut matting', with salt tides pushing barnacles to St Olaves and a dogfish to Thurne Mouth, eels and freshwater fish dead and gulls 'gorged with carrion' (Patterson 1929: 109–11; Bird 1922; Gurney 1922a). Natural circulation tilts to imbalance, drawing mordant curiosity.

If tide or drought might bring strange ecology, Broadland as humanly drained landscape could spark conflict. Drainage has long shaped Broadland, and if reedbeds and open waters are deemed regionally characteristic, so too is land drained to grazing marsh, with distinctive scenery and ecology (Williamson 1997). The Halvergate marshes are not a visitor showpiece, seen only from the Yarmouth train or the A47 Acle Straight, or touched by boat at Berney Arms, but in the 1980s they saw national political battle. In 1980 the Lower Bure, Acle Marshes and Halvergate Fleet Internal Drainage Board proposed the deep drainage of three marsh levels, enabling conversion to arable. Private and state agricultural interests (MAFF, IDB, National Farmers' Union, Country Landowners' Association) lined up against private and state conservationist concerns (Department of Environment, BA, Countryside Commission, NCC, CPRE, NNT, FoE), with local and national bodies on both sides, Halvergate debated from field to parliament. MAFF's Agricultural and Development Advisory Service, noting different viewpoints, suggested farmers were 'not in business to subsidise other peoples' scientific interests and recreational pursuits or to act as curators of an unchanging landscape' (ADAS 1984: 76). The 1981 Wildlife and Countryside Act focused public debate, the perception of the farmer shifting from rural guardian to country destroyer (Mabey 1980; Shoard 1980), and Halvergate became a national environmental cause célèbre, also shaping the BA's conservation identity (O'Riordan 1980a; 1986; 1993; Broads Authority 1982: 42–50; Purseglove 1988: 266–78; George 1992: 277–312). Years of argument produced pioneering compromise, MAFF and the Countryside Commission paying farmers to maintain landscape via the 1985 Broads Grazing Marsh Conservation Scheme, incorporated into the Broads Environmentally Sensitive Area (covering all valleys) in March 1987 (George 1992: 481–8). For O'Riordan this saved 'the last remaining stretch of open grazing marsh in eastern England', 'one of the most precious landscapes in the nation', whose 'value lies in its total experience' (O'Riordan 1986: 267–8).

Halvergate brought together ecology, aesthetics and direct action, with landscape a key diagnostic term. In 1980 the BA had defined landscape character as 'typicality and heritage' (Broads Authority 1982: 36; O'Riordan 1986: 278–9), O'Riordan's UEA inaugural lecture outlining four Broads landscape character types, with the Halvergate marshes Type 1, 'most characteristic', of 'extensive open' character. O'Riordan tabulated 'Conditions Characteristic of a Broads Landscape', the 'Sense of a Landscape Unit' and 'Indications of Pastoralism', including 'reed fringed rivers' and 'a "big" sky'. Humanistic geography was effectively deployed against marsh drainage, 'landscape character' providing 'local resident and visitor alike with a sense of continuity and perspective and ... sense of belonging ... This is the feeling that differentiates "place" from mere "space", for attachment to place leads to a sense of belonging and caring' (O'Riordan 1980b: 18–19). O'Riordan suggested drainage interests misunderstood landscape argument, thinking in terms of scenic amenity rather than landscape character, though his own arguments struggled to keep beauty and character apart: 'The important distinguishing feature is character, not beauty, even though it is extraordinary [sic] difficult not to attach some sort of qualitative aesthetic judgement' (20). The BA's 1982 *What Future for Broadland?* included sketches contrasting 'traditional grazing marsh landscape' with the 'effects of agricultural improvement'; diminished bird life, power lines to pumping stations, dykes deepened, crops in order (Broads Authority 1982: 44–5). Open, rickety charm disappears. Social scientific 'character' judgement cannot help but enact aesthetic value, beauty seeping back.

Halvergate's ecological value was traced to dyke ecology, noted by specialist eyes, highlighted by the NCC in scheduling SSSIs, and threatened by deep drainage (George 1992: 265–76). The BA's 1980 'Botanical Survey of Marsh Dykes in Broadland' identified 108 aquatic plant species, 'virtually all the plants and most of the related animals which have been lost from the broads', the dykes plant 'reservoirs' to 'restore the main broads and rivers when their nutrient enrichment problems have been overcome' (Broads Authority 1982: 38/46). Nature Conservancy natural historian Rob Driscoll, later of the Castle Museum, had surveyed northern Broads dykes between 1972–5, publishing extensively in *TNNNS*, with historical records indicating floral continuity under traditional management (Driscoll 1982; 1983; 1985). Driscoll's photographs of Scaregap on the lower Bure, north of the Acle–Yarmouth railway, showed a marsh dyke in October 1979, filled with earth in September 1981, by August 1982 an arable field (George 1992: 270–1).

Halvergate also saw direct action. Driscoll's 1970s survey assistant was Cardiff graduate Andrew Lees, born in Yarmouth in 1949, and by the early 1980s Chair of Broadland Friends of the Earth. Halvergate helped make Lees' and FoE's campaigning role, Lees taking up a national position in 1985. Lees died working in Madagascar against Rio Tinto Zinc in 1994. A memorial plaque on the ground just outside Wickhampton churchyard, by a lane leading to Halvergate marshes, commemorates:

ANDREW JOHN LEES

1949–1994

GLOBAL ENVIRONMENTALIST

HE SAVED THE HALVERGATE MARSHES

A TRUE FRIEND OF THE EARTH

"Who speaks for the butterflies."

Direct marsh action was prompted by owners threatening to plough in June 1984 unless the BA bought their land in compensation. FoE obstructed contracting equipment and staged protest meetings, the story making the *Observer* and prompting parliamentary questions. Agreement was reached following CPRE lobbying, Lees hailing 'a momentous day for conservation', FoE holding a celebratory public marsh ramble (EDP 1984a). The *EDP* suggested protest 'had a crucial and probably deciding influence' on the 'Marsh Deal', peaceful direct action contrasting to the conflict around the 1984 miners' strike, marking out FoE from 'the unacceptable face of the conservation lobby':

'A nation, sickened by political cant and public violence, warms to them and their methods' (EDP 1984b).

Visions of Broadland landscape did not only inform anti-drainage argument. Drainage for arable was itself informed by an aesthetic of improvement; land made good, put to profitable use and to be admired for productive qualities. EEC financial support under the 1976 Land Drainage Act extended a post-war productivist ethos of landscape reshaped through technology. If Halvergate put a wedge between self-consciously progressive farming and Broadland landscape value, agricultural interests have recently sought accommodation. The NFU's 2010 *Why Farming Matters to the Broads* gathered nature into an agricultural 'human future', with naturalist Chris Packham, presenter of BBC TV's popular *Springwatch*, emphasising Broadland's 'manscape' shaped by 'farmers and land managers' (National Farmers Union 2010: i). Packham's presence helped banish Halvergate ghosts, the report presenting agriculture as evolutionary stewardship, balancing arable and pastoral in a landscape of 'high quality foods' and 'high level environmental benefits' (8). BA Chief Executive John Packman and NWT President Sir Nicholas Bacon praised farming, Henry Cator (third generation farmer, Chair of the Broads IDB) stood before his rare cattle. Threats from climate change could be met by drainage expertise, contrary to those suggesting that 'if the Broads environment cannot be protected in situ, it must be recreated elsewhere' (6). For the NFU, Broadland should remain in Broadland, as farmed land.

River Flood, Broadland Balance

Water has frequently topped Broadland riverbanks, whether from rainfall, high tide or tidal surge upstream; George lists flood events since 1287, tabulating 17 between 1954 and 1989, mostly over the Yare and Waveney marshes (George 1992: 318–25; Storey 2012). Sea incursion through coastal defence is considered below; this section discusses river flood, proposals for a tidal surge barrier, and studies of Broadland's everyday balance of fresh and salt water.

The NNNS noted flood as well as drought. Arthur Preston, contributor of annual 'Meteorological Notes', described 'the Great Norfolk Rainstorm of 25th and 26th August, 1912', Norwich flooded to 17 feet, 'a level never before reached'. After a wet month, four inches fell over 15 hours across upstream catchments, with Brundall recording 8.09 inches, Moulton 7.13, Coltishall 6.88, Ormesby 6.59, Great Yarmouth 5.12, Geldeston 4.48: 'the gauges at Dilham and Acle overflowed' (Preston 1913: 552). Drainage was damaged such that some marshes remained in disrepair, reverting to fen. Chadwin noted Cantley construction disrupted:

> Those who visited Norfolk immediately afterwards witnessed a scene never to be effaced from their memory. The factory seemed to be in the very centre of the storm, and suffered severely; the river overflowed its banks and flooded the whole site; the rains washed down mud and debris which filled all the excavations, and at one time it was feared that the factory foundations would be undermined. Traffic had to be suspended on the railway; bridges and culverts were destroyed; roads were rendered impassable; and for the moment chaos reigned supreme.

Boilers and engines survived, though, and the first 'factory campaign' commenced on 12 November (Chadwin 1913: 577). Mottram recalled 1912 as the reassertion of estuarine 'ancient conditions': 'a small boat could be sailed from Norwich to near Brundall across what are usually dry enough meadows' (Mottram 1952: 34).

The major 1953 east coast floods of 31 January/1 February did not break Broadland's coastal dunes, but flooded valleys through upstream tidal surge. Ellis recorded: 'we heard that a porpoise had been seen leaping and plunging in the river at Surlingham, right in the vanguard of the sea's assault' (Ellis 1953a). Breydon's walls were breached in ten places, flooding 6000 acres and cutting road and rail, with 10,000 evacuated from Yarmouth's Cobholm district, and nine killed (EDP 1953b; 1953c; George 1992: 321–3). Robin Harrison described Breydon's 'awe-inspiring spectacle', high waves smashing, houseboats broken, belly-deep cattle 'mystified' (Harrison 1960: 20). Patterson had recorded earlier Breydon floods for *TNNNS*, the 13 January 1916 event tearing timber from Yarmouth quays and breaching Breydon walls: 'This, under the circumstances, was fortunate, for it eased the pressure upon the town banks, or Southtown had been worse placed than ever.' Freshwater fish perished, sea creatures went inland, water lay on marshes for days: 'When going to St. Olaves by train, on the 16th, ... the flood water shimmering in the sunlight, and rippling right up to the railway metals on one side, gave one a curious impression as of discovering a newly-formed Broad' (Patterson 1916: 166–7).

Patterson recorded flood and drought as natural variations in landscape curiosity. Others sought flood protection via the separation of fresh and salt, proposing surge containment by a tidal barrier, installed either below Yarmouth's Haven Bridge to protect all Broadland, or at the Bure mouth to protect the northern rivers, and either for emergency use or as part of a permanent Yarmouth lock. Implications for Yarmouth varied, the town potentially vulnerable to water backed behind any Haven barrier. Barrier argument accompanied the Halvergate controversy, risk reduction proposed to support drainage investment. The NFU's 2010 study indeed revived the possibility, suggesting a 'scoping study into the construction of a barrier across the River Yare' (National Farmers Union 2010: 7).

Lambert's 1952 NNNS Presidential Address noted the 'most controversial' suggestion of 'an artificial lock at Yarmouth': 'If this were done, we must accept the fact that the whole balance of the region would be changed. The unique character of the broads largely depends upon their water circulation; elimination of tidal rise and fall must have far-reaching physical and biological effects.' Lambert's argument that a barrier might 'destroy rather than improve the amenities of the region' (Lambert 1953a: 253) alluded to contemporary boating arguments, Arrow envisaging a 'revolutionary' extension of sailing behind a 'master lock', countering 'the well organized and ably briefed naturalists who regard Broadland as their special province' (Arrow 1949: 99). May also registered the 'ultimate "dream"' of a Yarmouth lock, making Breydon 'one of the finest freshwater lakes in Britain' (May 1952: 188). A master lock appealed as a key to regional management, Broadland a controlled estate, Edward Boardman drawing a How Hill parallel: 'If lock gates could be put at Great Yarmouth they would act as beneficially as do my sluice doors in their small way' (Boardman 1939: 10). National park debate prompted barrage and lock proposals, for some the Broads' salvation, for others regional ruin, Mosby and Sainty foreseeing the 'delicate balance' of fresh and salt 'utterly broken and irrevocable' (Mosby and Sainty 1948; EDP 1948c). Sainty predicted: 'the end of Broadland as we know it. This alone is sufficient reason to condemn utterly the proposal for the lock' (Sainty 1948b).

From 1955 consulting engineers Rendel, Palmer and Tritton produced feasibility reports, suggesting an emergency Yare barrier, though GYPHC hostility countered possible implementation (EDP 1955f; George 1992: 325–40). The barrier resurfaced alongside 1970s drainage debate, an AWA report emphasising psychological value for arable conversion. A January 1976 tidal surge prompted an RPT survey for the AWA, a proposed Haven Bridge barrier again foundering on GYPHC opposition, the scheme dropped in 1984. Debate set the NFU, Country Landowners' Association and IDBs against the GYPHC, NCC, Countryside Commission and UEA scientists, with the BA reserving judgement. The FoE's *Eastern Mourning News* opposed the 'Breydon Barrier' as 'A flood prevention scheme that could DESTROY THE BROADS', a 'guillotine' turning unique marshland to 'a typical slice of East Anglian corn prairie' (Breydon Friends of the Earth 1979).

O'Riordan's entitled his November 1980 inaugural UEA lecture *Lessons from the Yare Barrier Controversy*. The barrier spotlit the BA's 'political credibility': 'At present, most of the likely cherished areas of Broadland are extremely vulnerable to individual land drainage schemes' (O'Riordan 1980b: 1/27). O'Riordan reviewed schemes and surges over 100 years, sea level now 15cm higher, and arable conversion increasing potential financial loss: 'the spectre of tidal surge still haunts the Yare basin farming community' (6). The barrier made the 1981 New Year cover of *New Scientist*, carrying Catherine Caulfield's article, 'What Can we Save of the Broadlands?'.[5] Ken Cox's cover showed a bowler-hatted artist-official by a stream, colourful flowers, butterflies and birds around, his easel scene a matching segment in lifeless black and white, the phrase 'official concept' stamped bottom left, and a sluice barrier cutting the stream, the Yare barrier a Magritte-like 'giant guillotine' (Caulfield 1981: 28). Caulfield asked: 'Does it make sense to spend millions of pounds of public money on a project designed to increase production of wheat? We already have a wheat mountain in the EEC' (30). Rejected in the 1980s, the barrier has resurfaced through new futures of climate change. In 1992 George speculated that with new projections of sea level rise, and reduced incentives for arable conversion, the BA and NCC might even support a surge barrier (George 1992: 339). A 2011 suggestion presented a flood barrier incorporating a tidal energy harvester, navigation not impeded, regional circulation maintained, the Yare barrier greened (EDP 2011a).

If for some a master lock promised a Broadland secure from the sea, naturalists valued Broadland *for* its salinity, producing distinctive flora and fauna through variable tidal movement. At the SBL Gurney researched salinity and tides with AG Innes, documenting 'The Crustacea of the East Norfolk Rivers' (Gurney 1907; 1911b; 1949a; Innes 1911; Ellis 1965: 151–60). Gurney traced the 'Chemical Tide', distinct from the 'tidal wave', with the Bure up to Six-Mile House 'purely estuarine' (Gurney 1907: 411–12, Gurney and Gurney 1908: 6–8), and up to Acle Bridge 'a region of transition from salt to fresh water' (Gurney 1907: 415). Occasional maritime specimens appeared above South Walsham (and Oulton Broad and Cantley on the southern rivers), 'comparable to the migratory birds which occasionally straggle to our islands outside their ordinary range' (422). On 3–4 September at Stokesby, with a full moon and high tide, Gurney saw fauna switch from freshwater to maritime, with species lists mutually exclusive (Gurney 1911b: 238–42).

Parts of Broadland could appear essentially saline. At the RGS in December 1910 Gurney outlined 'Some Observations on the Waters of the River Bure and its Tributaries', noting 'unstable conditions', with 'at certain times the old estuarine conditions' re-established (Gurney 1911a: 295). Gurney was followed by Pallis discussing salinity in the Norfolk Broads, their papers published together in the *Geographical Journal*. Pallis considered 'a system of broads, situated between the sea and the village of Hickling, whose waters have a distinctly saline taste' (Pallis 1911b: 284), digging holes to trace the cause, and challenging Eustace and Robert Gurney's suggestion of salt springs at Horsey: '(1) there is a salt-water table in the coastal region ...; (2) this salt-water table appears to stand in actual relation with the permanent level of the sea ...; and (3) the broads and dykes of the district owe their salinity to the presence of this layer of salt water'. Pallis concluded: 'the district is fundamentally saline', with fresh water merely 'a modification' (288), and salinity derived from 'direct underground communication ... between the sea and the waters of the Thurne district' (284). Four decades later, digging the Double-Headed Eagle Pool, Pallis wrote to Lambert: 'in the deep holes the water is very salt – much saltier than in Horsey or Hickling, impossible to drink. A pleasant confirmation of my Geogr. Paper.'[6] Pallis found coastal Broadland shaped by a direct land–sea relation. Permanent seepage also however alluded to a more violent sea–land connection, by flood.

Death by Water: Sea Flood

Estuary transgression

Origin stories highlighted marine transgression at Broadland's base, a stratigraphy of marine clay beneath silt and peat, later diggings flooded from medieval sea level rise. Transgression also shaped the imagined regional future, as when Ellis remarked on stagnant hollows: 'Were a new marine transgression to take place on the same scale as that which led to the formation of the Broads, these rotten fens would become broads; and they would be the first areas to recover their freshwater character' (Ellis 1949: 32). In 1992 George would suggest: 'if ... sea levels rise by a metre during the next 100 years (instead of *c*.10 to 15 cm per century during historic times), and by 5 m during the ensuing 200 years, it would seem likely that Broadland will once again be subject to a marine transgression' (George 1992: 340). Trangression might repeat, a regional sea cycle.

Marine history haunts Broadland at two points, the dunes between Winterton and Sea Palling (discussed below), and Breydon Water as estuarine relic. Dutt began his *Norfolk Broads*: 'Even if the general aspect of the district were not enough to indicate it, there is abundant geological evidence that the valleys of East Norfolk were formerly the bed of a large estuary or tidal waste of waters connected with the North Sea' (Dutt 1903: 1). Dredging could reveal marine archaeology, Gurney exhibiting Roman

'Estuarine Shells from Ludham' to the NNNS: 'The excellent preservation of these specimens, the two valves being often not separated, and the minute size of some of the individuals show that they were not the refuse from St. Benet's Abbey, but must have lived where they were found' (Gurney 1909: 855). The Roman name 'Garienis Ostium' entered common topographic parlance, Patterson showing two maps, one of 'Garienis Ostium reduced to its present area (Breydon)', the other of 'Norfolk Waterways in Roman Times', with an estuary to Norwich, and Potter Heigham, Caister and Burgh Castle on offshore islands: 'I have often pictured to myself what a magnificent estuary this water-covered plain must have been in the ages long ago, when the Roman galleons sailed up *Garienis Ostium* to their camp at Burgh Castle, signalling "All's well!" to the camp at Caister, as their vessels ploughed the sea that then rolled over the site of Yarmouth, as the North Sea now sweeps over Scroby Sands' (Patterson 1907: 1/3/9).

The 1924 Ward Lock Guide to *The Broads* reproduced 'an interesting eleventh century map', 'Yarmouth on an Island' at an estuary mouth, Norwich inland, sea creatures in the water, Yarmouth an empty sandbank: 'From this it appears that there was a time when the site of Yarmouth had no existence' (Ward Lock 1924: 17/34). Made in Elizabethan times to depict the eleventh century region, the 'Yarmouth Hutch Map' was drawn on a sheep hide, with south at the top, land coloured green, sea blue, and sandbank golden. The map marked civic prehistory for AW and JL Ecclestone in their 1959 *The Rise of Great Yarmouth: The Story of a Sandbank*, their redrawn version adding Haddiscoe, Reedham, Halvergate, Acle and Ludham as

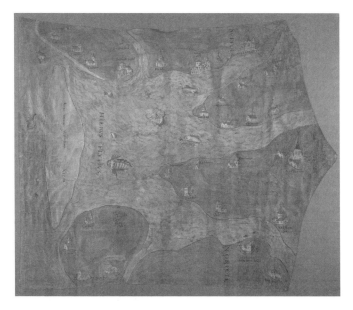

Figure 38   The Yarmouth Hutch Map, Norfolk Record Office, ref. no. Y/C 37/1. Reproduced by permission of Norfolk Record Office.

coastal villages, with St Benet's an island abbey (Ecclestone and Ecclestone 1959: 20–1). The Ecclestones mapped the 'Seven Havens of Yarmouth', former outlets north and south of the present town, and different entrances cut through the spit formed across the estuary mouth between 1347 and 1567. Yarmouth was made 'From out the Azure Main', its origin 'rather like Scroby at the present time' (17), the Hutch Map taken to denote a precarious history suggesting vigilance was needed over accretion and erosion.

Dune defence

Five miles north of Horsey is Eccles, the village covered by sandhills in the seventeenth century, the church tower left standing on the beach, finally falling in a gale on 23 January 1895.[7] Storms in 1947 revealed the layout of ruins for the first time in years, the former burial ground holding human remains (EDP 1947a). Pevsner noted in 1962: 'The village has long disappeared in the sea, and of the church whose tower old people still remember there is now no more than two heaps of flint on the beach the size of two beginners' sandcastles' (Pevsner 1962: 125). Eccles joined other ex-settlements in the sea; Dunwich in Suffolk, Shipden off Cromer. Happisburgh, on soft cliffs north of Eccles, has been the focus of recent erosion debate, houses falling with defensive neglect.

The dunes between Winterton and Sea Palling have a history of sea breach, George giving a rollcall of major flood events; 1250, 1287, 1608, 1617, 1622, 1717, 1718, 1720, 1791, 1897, 1907, 1938, 1953. One hundred and eight died at Hickling in 1287, seven at Sea Palling in 1953. From 1609 Sea Breach Commissions oversaw defence, with a permanent Commission, prompted by the protection of newly enclosed land at Potter Heigham and Hickling, formed in 1802, succeeded by the East Norfolk Rivers Catchment Board in 1931 (Cornford 1979; George 1992: 314–16). After the 1622 flood a proposal had envisaged raising the causeway to Potter Heigham bridge as a flood bank, allowing permanent flooding of marshland to the east. Seventeenth century visions were reprised in 2008, when Natural England's projections for a future of sea level rise ranged from defending the present coastal line to allowing 'embayment of the Upper Thurne'. The new bay would cover Horsey and Hickling, and public debate ensued, with emotional and financial attachments to property and habitat at stake. Village meetings attracted hundreds, and lobbying from local officials, the Broads Society and MPs secured government assurance that embayment would not be countenanced, the local press hailing eventual 'Victory for the People' against 'proposals to surrender 25 square miles of the Broads to the North Sea' (EDP 2008a; 2008b; 2009). The BA notes current policy as 'to "hold the line" along the coast for the next 50 years' (Broads Authority 2011: 21).

A specific past river–sea connection haunts the upper Thurne. The former outlet of the Hundred Stream, at the Winterton–Horsey parish boundary, saw breaches in 1601, 1608, 1625, 1651, 1715, 1791, 1897 and 1938. A coastal 'Floodgate' is indicated on topographic guide maps, even lingering, though long gone, on Bartholomew's 1973 popular 'Map of the Norfolk Broads' (Dodd 1896; Emerson 1898b; Dutt 1923). A possible derivation is Faden's 1797 county map, the word 'Flatgates' written south of the Hundred Stream, though Cornford also notes an Elizabethan map showing coastal sluices (Cornford 1979). Flood history and flood anticipation shapes landscape narration, Burrell noting in the 1914 *Flora of Norfolk*: 'It is estimated that forty thousand acres of land would be jeopardised by the failure of the sand barrier at this weak spot, and one hundred and three parishes are liable to contribute to its maintenance' (Nicholson 1914: 21). Dutt discussed the 1897 flood, recording a marshman critiquing Commissioners' neglect: 'it's their bisness to luke arter th' merrimills, an' if they'd a-done it as they should ha' done, th' sea 'ud never ha' got tru'. Dutt comments: 'Sea Breach Commissioners make feeble attempts to withstand this siege, but the enemy is far too strong for them' (Dutt 1903: 182–4). Conversely, after severe March 1947 river floods, engineer SW Mobbs suggested the restitution of the Hundred Stream outlet might offer flood solution, with 'a gravity sluice at the old Hundred drain at Somerton or opposite Horsey Mere' for emergency use, taking excess to sea (EDP 1947b).

The proximity of coast and broads allows a popular walk from Horsey Mere to dunes and beach. John Betjeman's poem 'East Anglian Bathe', published in 1945, thus celebrated a holiday stroll to a cold sea bathe and return to Horsey's 'warm freshwater ripples' (Betjeman 1974: 136–7). The adjacent sea is a common topographic and literary subject, whether as calm enigma or fearful force:

So clear and so light was the summer-air, that the one cloud in the eastern quarter of the heaven was the smoke cloud left by a passing steamer three miles distant or more on the invisible sea. (Wilkie Collins)[8]

On the lapsing land that recedes as the growth of
    the strong sea strengthens
  Shoreward, thrusting further and further its out-
    works in ...
  (Algernon Swinburne)[9]

And only the desolate, wind-sculptured hollows of the dunes keep the blue waters from mixing, and gaining the mastery of the sopping land, where the flocks feed, and the watery tribe of birds cradle their young. (PH Emerson)[10]

The white sand-hills on the coast were plainly visible, and the thunder of the surf was audible. (GC Davies)[11]

The sea like some raging hungry wild beast is constantly fretting and striving to break down the frail sand barriers which confine it. (Emma Turner)[12]

For some time I had been half conscious of a sound coming from the direction of the eastern horizon – a sort of steady rushing sound, scarcely distinguishable from the silence itself, rather like the moaning of a mighty hurricane, heard from a great way off. And then all at once, I knew what it was, and I called out to Hugh in the cabin – 'Hi!' I said, 'I can hear the sea!' (EHT Jukes)[13]

On Long Gores marsh at night, a continuous drone sounds, as if a motorway ran nearby, belying a B-road region. Pallis lies with her companion on the eagle island, to the sound of the sea.

Horsey 1938

The only major twentieth century breach at Horsey came in 1938. The meeting of natural history, geography, history, documentary film, tourism and land management over flood and its aftermath warrants detailed discussion, the flood concentrating minds on landscape's past, present and future.

On 13 February 1939 the RGS discussed 'Recent Coastal Changes in South-Eastern England', speakers including Stamp and Steers. JEG Mosby (Norfolk director of Stamp's Land Utilisation Survey) discussed 'The Horsey Flood, 1938: An Example of Storm Effect on a Low Coast': 'On the night of 12 February 1938, during a north-westerly gale, high waves were driven against the sand-hills of the east Norfolk coast with such force that they demolished 517 yards of the dunes at Horsey. Salt water poured through this breach and flooded 7500 acres of low-lying hinterland' (Mosby 1939: 413; Storey 2012). Mosby mapped the flooded area: 'the distance along the outside margin 43 miles, while the combined length of the shore-line of the islands amounted to 9 miles' (418). Horsey formed one of the islands. The gap was filled, breached in early March, and again on 3 April, washing away 'piles, sandbags, and the light railway that had run along the top of the barrier' (417). Sea water lay over some areas for four months. There were no human fatalities.

The NNNS took flood interest, with Anthony Buxton, *TNNNS* Editor, resident in the flood at Horsey Hall. The Junior Branch visited on 28 April 1938 (TNNNS 1939b), and adult members on 2 June 1939 (TNNNS 1939a), while the March 1939 *TNNNS* gave a substantial review. Sainty described 'Past History of Flooding and the Cause of the 1938 Flood' (Sainty 1938), Mosby 'Mapping the Flooded Area' ('I am particularly grateful to Mr. Jim Vincent who walked round with me' (Mosby 1938b: 347)), Buxton 'General Effects of the Flood', and Ellis 'Detailed Observations' (Ellis 1938).[14] Sainty identified the Hundred Stream as historic weak point, its seaward exit artificially blocked two centuries ago, while at the RGS Mosby noted: 'Some of the old inhabitants affirm that sluice gates were placed at the mouth to prevent the sea water from entering the stream at high tide, and Mr. S. E. Thain stated that he saw the old posts of the sluice exposed soon after the recent flooding' (Mosby 1939: 415).

Buxton's 'General Effects of the Flood' was the first of six annual *TNNNS* reports, beginning:

An inland sea that turned into a desert is my general impression of the flood, and its worst feature the resulting lack of life. The process of natural recovery will prove interesting, but the devastation and the general air of a great flat rubbish heap on which nothing can thrive, produce a feeling of intense depression. The limit between flooded and unflooded

Figure 39   'Flooded area, Horsey and district, 1938'. Source: Mosby 1939. © Royal Geographical Society (with the Institute of British Geographers).

area was during the spring and summer blatantly abrupt – from bright green life to red brown death; the contrast between red and green was reminiscent of an impressionist picture, and a bad one at that. There may have been, and indeed there was, some beauty about the floods, particularly at sunrise and sunset and by moonlight; there was none in the aftermath.

Inland waterway levels had been relatively low, enabling much flood water to drain away, 'which in fact entered Norfolk at the breach and left it at Yarmouth unassisted by man' (Buxton 1938a: 349; 1939b; 1940; 1941; 1942; 1943; 1946: 104–16). Flood occurrence however indicated defensive neglect: 'The Catchment Board ought to have taken the clear warning given them by the storm of 1936, which swept the beach to such an extent that it disclosed the foundations of a farm, 50 yards on the seaward side of the present sandhills, and domestic animals' tracks, that must have been hidden for at least 200 years' (Buxton 1941: 267).

The flood threw marine creatures inland: 'One of the millmen secured a 2 lb. cod at Horsey mill and a grey mullet was caught at Hickling. A crab was seen to walk across a ploughed field into a dyke' (Buxton 1938a: 360). Birds were confused, the bittern off-kilter: 'The effect on bitterns was peculiar … for a long time they were

Figure 40   Horsey, views out to sea and inland from the site of the 1938 sea breach, Bramble Hill. Source: photographs by the author, August 2011.

quite incapable of producing a proper boom ... it was not until June that they regained their voices' (358). From late May millions of barnacles appeared in Horsey Mere, attracting an October article in *Nature* (Holmes and Pryor 1938). Ellis identified *Balanus imporvisus*, a brackish-water barnacle: 'This species has long been present in the lower reaches of east Norfolk rivers; Gurney recorded it from the River Bure at Muckfleet more than thirty years ago ... it was present at South Walsham Broad in 1935, when the water was almost fresh' (Ellis 1938: 386–7). *Balanus imporvisus* indicates Broadland's land–sea complexity, Ellis and Buxton monitoring the restitution or otherwise of a sub-maritime flora and fauna. Given the area's relatively high normal salinity, Ellis noted that 'the truly aquatic life of this district was better prepared for trial than that of the higher ground reached by the flood' (Ellis 1938: 374). A general story of eventual recovery included however loss on the way, Ellis recalling a freshwater sponge which had lived in a brackish dyke opposite Horsey Mill since 1929: 'it has now disappeared' (385).

Buxton reported as both naturalist and agriculturalist:

> Speaking as the Chairman of an Internal Drainage Board I advocate much deeper cleaner dykes, and plant sufficiently powerful to pump them dry, not merely to skim the top water off, according to Norfolk standards. Speaking as a Naturalist I must confess that the result of good drainage is hideous and uninteresting. Trying to speak in both capacities, the best plan seems to me to drain the better portions of the land really well and not to spend money on the rest, but to leave it as a good hearty bog. (Buxton 1941: 260)

Two aesthetics, of productivity and benign neglect, cohabit; Buxton hoped for a post-war co-existence of progressive agriculture and nature reserves (Buxton 1943: 418–19). For the former Buxton took Dutch inspiration, visiting Holland in November 1938 for advice on sea defence, salt removal, replanting and deep drainage: 'it is impossible to persuade people in Norfolk, that the Dutch know the business of recovery from salt from A to Z and that we do not' (Buxton 1940: 156; Buxton 1938a; Buxton, Hornor and Mann 1938). The Dutch advised not to plough until drainage was improved and salt reduced, and in 1942 Buxton contrasted two 'experiments', applauding Mr WA Alston at Martham Holmes, draining with a view to grazing, but less convinced by Mr HP Neave, buying 'derelict' marshes between Waxham Cut and the Hickling Wall, next to Long Gores, installing concrete access roads and burning and ploughing with Ministry approval: 'Mr. Neave's experiment is a direct flaunting of Dutch advice' (Buxton 1942: 336–7). A year later however Buxton came round, his pessimism misplaced: 'I ... am glad to say that I think I was wrong' (Buxton 1943: 413).

The 1938 flood was also a story for local, regional and national media, Buxton appealing in *The Times* for those with 'happy memories' from Broads holidays to contribute to the Norfolk Flood Relief Fund (Buxton 1938b; Daley 1938). Journalists got through, wiring reports from Horsey post office, kept open by Mrs Goose and daughter Freda (Post Office Magazine 1938). The *EDP* reported Rev Hart's post-flood Horsey service: 'The congregation consisted of Mrs Hart, Mrs J Brown, 15-year-old Ernest Bentley, and some reporters' (EDP 1938c). Horsey residents achieved national and local celebrity, millman Arthur Dove appearing on BBC radio's *In Town Tonight* (EDP 1938a), others venturing out to aid flood relief, 12 villagers, including Mrs Goose and Freda, and keeper Crees, attended a giant fundraising whist drive at Hemsby Holiday Camp, 2100 players raising £400, the villagers home at 4 am. (Yarmouth Mercury 1938d). The flood itself was a day trip spectacle, with long traffic queues through Somerton on Sunday 20 February for the ridge view, the AA estimating over 10,000 cars (EDP 1938b). Sunday 6 March brought 'sightseers from all over England', walking the sandhills from Winterton to view the breach, 'Showplace No. 1 in the Eastern Counties' (Yarmouth Mercury 1938c).

Film crews also arrived, Pathe Gazette issuing a 'Sea Invades Norfolk' newsreel on 17 February (Cleveland 2009: 124), and East Anglia Films producing *The Sea Breaks Through*, praised in *Sight and Sound* as 'the most dramatic documentary of the past quarter' (Rice 1938). The General Post Office Film Unit, key in the documentary film movement, also visited, director Pat Jackson's *The Horsey Mail* exploring flood through efforts to deliver the post (Matless 2011; 2012).[15] The film titles credited 'Postmen' Bob O'Brian and Claude Simmonds, referred to by first names throughout, O'Brian driving from Yarmouth to help local postman Simmonds, the flood extending his round from two to 15 miles, including six miles of rowing.[16] *The Horsey Mail* is nine minutes long, the titles rolling over a shot of Somerton Mill from the ridge, with opening and closing newsreel style sections giving aerial and ground views, including close-ups of labourers filling the breach. The middle two-thirds presents Simmonds and O'Brian delivering the mail, Victor Yates's musical accompaniment setting variously jaunty and sombre moods, driving past flood refugees, rowing the line of the main road past protruding vehicles to Horsey's dry street, Simmonds walking to outlying dwellings, the two taking a boat to marooned Ford's Farm.

The Horsey Mail is unusual in GPO documentary in including O'Brian's locally accented voice within commentary. The Received Pronunciation narrator mentions his name at the Yarmouth depot, and he interjects: 'Yes, that's me there. They've put off my winter leave so I could take Claude by van to Waxham. I got my thigh boots from the boy Herbert at the office.' If the newsreel-style sections are narrated in monovocal R-P convention, mail delivery warrants shared commentary, accents of landscape enacting geographical scales of meaning. Unlike some other GPO films, such as *The Saving of Bill Blewitt* (1936), a story of Cornish fishing, O'Brian as local voice never speaks on camera, but shares commentary, achieving a form of narrative equality, different in effect to being interviewed as a testifying local. Norfolk voice occasionally provides comic effect, as when Simmonds trudges off through the water and O'Brian begins to sing 'I Do Like to be Beside the Seaside', the R-P narrator reluctantly joining in, but there is also serious comment on salt, farming and flood: 'They used to tell me as a youngster that the sea always found its own level. Well she's found it this time – about ten mile inland.' R-P narrator and Bob are equally if differently articulate voices, the audience hearing a carefully enunciated Norfolk-accented analysis. One local voice conspicuously absent however is Buxton. If 'Arthur Dove the millman' is shown preparing to join other refugees at Horsey Hall, Buxton is never mentioned, the Hall never shown, Jackson shooting the ordinary person, in keeping with a GPO ethos of democratic service, social hierarchy not registered.

After a first showing in the GPO's pavilion at the summer 1938 Empire Exhibition in Glasgow, *The Horsey Mail* received local circulation, part of a GPO display in Norwich's October 'Civic Week', celebrating the new City Hall (EDP 1938d), and shown in Yarmouth's Little Theatre from 7–11 November, the display opened by the Mayor: 'Post Office workers of the Yarmouth district saw themselves on the screen' (EDP 1938e; Yarmouth Independent 1938; Yarmouth Mercury 1938e). Simmonds and O'Brian had also received instituional recognition:

On Monday, October 31, a pleasing ceremony took place in the Great Yarmouth Sorting Office when Messrs. Claude Simmons and R. J. O'Brien, Postmen, were each presented with cheques for £5 ... These postmen also figured prominently in the Post Office film displays which opened at Great Yarmouth on November 7, for they were the principal actors in the film 'Horsey Mail'. (Post Office Magazine 1939)

Simmonds had already featured in the *Yarmouth Mercury*'s post-flood coverage, the March flood bringing 'Postman's Dash Across Breach', with Simmonds the 'last man to cross the gap' (Yarmouth Mercury 1938b; 1938a). In May 1938 the *Post Office Magazine* featured 'Postman of the Flood', Simmonds pictured with bicycle, flood water half way up the wheels. Neither *The Horsey Mail* nor O'Brian were mentioned, Simmonds the lone hero: 'The little boat was caught half-way along the road by a terrific current, and came within an ace of being capsized. The wind dashed foam-laden water over its sides, and, with Horsey in sight, the postman was compelled to turn back. Wet through, he was still undefeated.' Simmonds kept going, changed clothes, cycled to Winterton, three miles along the sandhills, across the sea breach at low tide, inland to the flood, to deliver the Horsey mail (Post Office Magazine 1938).

Simmonds maintains his daily round, Buxton monitors flood effect, Ellis itemises species, the GPO Film Unit surveys, emergency motorists view. The 1938 flood concentrates landscape, Broadland rendered at once as leisure spectacle, rural community, productive land, valued natural history, region under sea threat.

New defence

After 1938 2km of sea wall were built at Horsey, and 2.4km at Eccles. Under military control in wartime, coastal defence neglect was highlighted in post-war debate, with Horsey 1938 a reference point for consequences of inaction (Sainty 1952). River engineer Mobbs noted 150 yards of firm ground between sea and marshes at Caister, predicting a breach would flood Broadland up to Norwich (EDP 1947d). Mardle saw the sea 'intent on reopening the old Grubb's Haven, and converting Yarmouth into an island. If a breach were made there, the River Bure would have a new outlet to the sea, and most of the Broads might be flooded by salt water.' A 'tremendous agitation' to local and national authorities, 'during which the women of Caister petitioned the Queen', had however secured a concrete wall and groynes (Mardle 1950). Mardle commented: 'The study of coast erosion is one or our local hobbies in Norfolk and Suffolk – sometimes to the annoyance of the engineers of the defence works, who find themselves criticised by amateurs' (Mardle 1952).

The 31 January/1 February 1953 east coast floods killed 307, flooding 24,000 houses and damaging over 1000 miles of coastline, with 1200 breaches (Steers 1953a; Summers 1978). The Horsey wall held, but seven died from the dune breach at Sea Palling, with the Brograve level flooded. The 1938 flood remained a reference point, the *EDP* reprinting Buxton's 1938 Dutch report, and Ellis and Buxton giving 'Lessons from Horsey' (EDP 1953d; Ellis 1953b; Buxton 1953). The emergency iconography of 1938 was reprised along the east coast, and indeed in Holland. An English edition of the Dutch *De Ramp/The Battle of the Floods*, issued to support the Netherlands Flood Relief Fund, showed community rallying, stranded puzzled animals, trudging refugees, breaches seen from the air, orderly relief (*The Battle of the Floods* 1953). Wranglings over cost after 1953 allowed other voices to enter debate, R Baden Gladden of Sutton Hall (Lambert's origins opponent) arguing for a defensive earth bank between Happisburgh and Winterton, and noting the old Hundred Stream as a warning: 'History may well repeat itself and the sea once again flow over its ancient valley bed' (Gladden c.1956; EDP 1954d). Buxton urged supplementary Horsey pine woods, as at Holkham on the north coast, proposing to offer the Forestry Commission 'full rights of planting in perpetuity'; Ellis and the Nature Conservancy objected on botanical grounds (Buxton 1955a; 1955b; EDP 1955e; Ellis 1955a). By November 1958 however the dunes from Happisburgh to Winterton were protected with reinforced concrete walling; the 1938 Horsey section would be replaced in 1988. The *EDP*'s February 1959 supplement on 'Sea Defence in East Anglia' captured a defiant outlook of walls completed, construction firms advertising skills of 'Attack! and Defence in Depth' (May and Gurney) and 'Tackle to tackle anything!' (Maclean of Cromer). Mardle reflected on 'The Invading Waters':

> Norfolk is a county that can be defended from flood only by artifice. If there were no coast defence and drainage works, we should become a peninsula ... To the east, the rivers Yare, Bure and Waveney would empty themselves separately into a shallow bay, and Yarmouth, if it existed at all, would be a precarious settlement on a sandbank. (EDP 1959b: I; Mardle 1953)

Eccles remained a reminder, River Board Chief Engineer Cotton showing an engraving of 'Eccles Church, about 1840', to illustrate formerly 'Soft Shores'. The coast was now happily defended by 'interlocking sheet steel toe piling with parabolic section stepped apron with wave return parapet', softness removed (EDP 1959b: IV).

Community and vulnerability on soft shores was dramatised in Hester Burton's 1960 children's flood story, *The Great Gale*, set around a fictional east Norfolk 'Reedsmere'. Events from 1953 are relocated, Burton's heroic American airman Rod Cooper based on Ries Leming's rescue of 28 people from Hunstanton bungalow roofs (Burton 1960: 207). Mark and Mary Vaughan, children of the Reedsmere doctor, lead the narrative, their cross-class friendships showing a co-operative social order, rallying in the flood, and noted (without irony) as they return to boarding school, leaving homeless working class friends behind: 'It made them realize that they belonged to each other – that not one of them was alone or set apart' (199). Horsey 1938 is reprised, with dunes breached, and Reedsmere Hall a rescue centre. From a 'ruined countryside', birdless, with livestock dead and river weed indoors, comes restoration. The Queen visits the hall, Harold Macmillan the dunes, the broad reappears: 'the landscape that they knew so well was taking shape again' (164). Mark's personal 'museum', including his stuffed bittern, survives the flood. A dead coypu lies under the sofa; Mark will have the 'wonderful specimen' stuffed (108).

Something beyond fiction, on which I can find nothing more, appears in the minute books of the Happisburgh to Winterton Internal Drainage Board, chaired by Buxton at Horsey Hall on 4 July 1955. 'Matters arising' included 'SEA DEFENCES':

'Mr AUTONOMOUS'
The Chairman said that he was constantly receiving affectionate enquiries about the gentleman known throughout the district as 'Mr. Autonomous'. This gentleman had shortly after the sea breach at Palling, arrived on the beach in a car driven at furious speed: the car had got stuck in the sand and was extricated by various heavy vehicles, whereupon the occupant had made a speech to the multitude at Palling, starting with 'I am autonomous', which had been greeted with loud applause.

Mr. Alsop replied that members of the River Board staff had spent a good deal of their time 'rescuing' Mr. Autonomous from various beaches and passing him back to safety inland. The gentleman had now unfortunately ceased to protect the coasts of these islands and had opened an office in London to deal with other business, the nature of which was somewhat obscure.

The Board expressed universal regret at the news that they would be unlikely to see him any more disporting himself on the beaches or to hear his invigorating speeches.[17]

Tilting England

Flood events were shadowed by diagnosis of longer term change. In 1926 Anthony Collett reflected:

If the whole coast is gradually subsiding, as it undoubtedly has subsided in times past, it is difficult to see how the breach at Horsey can fail to be overrun, and a great part of the Broads country again become a tract of tidal marsh, with emerging islands ... when subsidence is persistent, at last comes the high tide which does not retreat when the storm blows over. (Collett 1926: 84)

Sainty's *TNNNS* Horsey 1938 account noted the 'fundamental' question of relative land and sea level, Ordnance Survey geodetic levelling of Britain from 1912–21 showing 'a definite tilting movement, resulting in the elevation of the north and the depression of the south' (Sainty 1938: 340; Jolly 1939). Horsey had in geologically recent time been 40 feet above sea level, the North Sea once land, yielding undersea archaeological finds such as the Leman and Ower harpoon, trawled in 1931, held in Norwich Castle, exhibited there to the Prehistoric Society of East Anglia in 1932, and upheld by Godwin, Grahame Clark and others as evidence for Mesolithic culture over a North Sea land (Fagan 2001; Stringer 2006). Sainty noted post-glacial isostatic uplift in northern Britain balanced by a 'slow and long-drawn' East Anglian depression, such that 'equilibrium may easily be disturbed. An abnormally high tide will tower above the marshlands and only the stability of the sea bank can defend the district from inundation' (Sainty 1938: 340–1). The tilting of England ghosts engineering effort, undercutting coastal defence. If twenty-first century coastal debate is inevitably entangled in climate narratives, tilting provides the equivalent mid twentieth century frame of inexorable change. If however human-tainted sea level rise tends to prompt guilt, and an ameliorative joining hands with nature in cautious retreat (or a contrarian urge to be defiant in denial), the absolutely non-human spectre of tilting generates a mixture of acceptance and defiant defence, modern technology seeking to hold a familiar coastal line against inevitable dip.

Tilting also informed Broads origin accounts, Sainty presenting Horsey 1938, and the 1928 Thames floods, as 'reminders that the post glacial transgression of the sea is still continuing' (Sainty 1948a: 374). In 1959 origins researcher Green contributed to the *EDP* sea defence supplement, describing 'Norfolk Coast Changes', the 'Remarkable Records of Land's Ups and Downs', with tilting a threat 'perhaps the graver in that it was scarcely recognised' (EDP 1959b: II). Mardle noted models in the Castle Museum's Geological Gallery, showing Britain's 'geological seesaw', tilting Norfolk downward (EDP 1959b: I). The publication of Steers' New Naturalist book on *The Sea Coast* in February 1953 prompted an *EDP* review by Ellis, headlined 'Is Britain Tilting?': 'it discusses a matter which is very much in our minds at the moment – whether or not we are sinking gradually beneath the sea' (Ellis 1953c; Steers 1953b).[18]

Tilting also shadowed coastal policy, Steers sitting on the government Waverley Committee, responding to the 1953 floods, its report noting Cambridge-based Hartmut Valentin's study of vertical land–sea movements (Departmental Committee on Coastal Flooding 1954: 40; Valentin 1953). Britain tilted around a hinge line, with East Anglia on the wrong side, Sainty noting Norfolk 'near the axis of turning', sinking by 'an inch or two per century' (Sainty 1953a). Sainty also looked to research on Arctic ice retreat: 'Spitzbergen, which in 1900 had only four months ice-free, has this year enjoyed seven' (Sainty 1948a: 374). The Greenland Scientific Expedition's diagnosis of 'The Melting Ice Cap' put the 1953 floods into perspective: 'If this is so, we may expect smaller ice fields and increasingly high sea levels, necessitating defences of a magnitude we dare not contemplate. Behind it all lies the vital question. "Are we now witnessing the passing of the Ice Age, or are we living in a minor Interglacial period?" The future alone can answer this' (Sainty 1954). The Horsey dunes, and new sea walls, achieve possible insignificance.

Broadland Sealand

In 1962 Nikolaus Pevsner commented: 'With the Wash and the Fens on one side, and the North Sea on the other two, Norfolk seems only just able to keep above water, and there can be very little difference between the exposed surface of the county and the invisible sea-floor beyond its coasts' (Pevsner 1962: 15). Tilting raised the sea, waiting behind dunes to renew transgression and make a new sea-floor. For most, this was a future to be fought, yet others found prospects of delight, the sea a slate cleaner or restorer of an older Broadland, wetter and wilder, the modern human wiped aside. Particular sites and spatio-temporal scales achieve regional resonance here; the dry North Sea around the ancient Rhine, the offshore sandbanks of Scroby.

Restyled from a threat, climate change can open new futures. Moss ends *The Broads* with an imaginative 15 page 'Postscript', the term chosen to give his publisher a reassuring sense of literary distance from the preceding science (Moss 2001: 346–60).[19] Moss presents two futures, an 'epoch of belief' and 'spring of hope' (phrases from Dickens' *A Tale of Two Cities*), both regional recoveries after catastrophe, with artist Alan Batley commissioned to picture alternative scenarios. A cyclone devastates Britain in 2025, with London drowned, and government shifted to Manchester, but Broadland survives, Halvergate triangle taking the brunt of flooding. Moss describes two 2050 Broadlands, the first a heritage fantasy, the second a new vernacular working landscape, his sympathies evidently with the latter. The 'epoch of belief' sees concrete embankments lining rivers and coast, and a Yarmouth barrier guarding a defensive and exclusive region, a 'Broads Heritage Area' enjoyed by the rich in costumed style. Ellis's 'breathing space for the cure of souls' becomes a brand slogan for 'Broadland Heritage', Emerson photographs signify the recreated past, Hickling is a theme village, a new broad is dug historically by hand near the Muck Fleet, marshes are restored through 'Windmill Heritage' (346–53). Moss projects late twentieth century aspects of BA work into a negative future. Batley's painting shows a church and restored staithe, visitors escorted by Emersonian guides, reeds cut by hand.

The 'spring of hope' finds the cyclone transforming ideas and power structures, in a manner akin to the revolution in William Morris's *News from Nowhere*. Moss implicitly appears, an 'old man in his nineties', noting alternative ecological states, the cyclone accomplishing an equivalent social switch to a Broadland of environmental taxation, good design, new economy, and reduced managerialism under a Broadland Catchment Authority. The lower valleys are a tidal estuary, as is the Thurne. Broadland becomes a working area rather than 'a museum of the past', with less intensive farming, local and offshore renewable energy, and broads allowed to grow up, though the Trinity Broads are preserved as an open air museum. 'Fortress' nature reserves are abandoned, reed and carr cut by machine, and tourism thrives in a new working landscape, 'interrupted in the doldrum years between 1960 and 2025' (353–60). The cyclone jolts Broadland to health. Batley's painting shows hard-hatted operatives taking logs from carr, a yacht sailing distant, plant life in the waters, water buffalo content in stream and meadow.

A final 'Epilogue' leaps further. Ice retreats in AD 34,300, and Spanish archaeologists work on eastern English moraines. Square worked flint blocks from 33,000 years ago match a picture (Moss's own plate 9) still held in Europe's 'southern libraries', 'in an area that had been called 'The Broads' in a twenty-first-century book of that name' (360). Landscape shuttles over long duration, such that projects to, say, restore 'Clearwater 2000' to a corner of Barton Broad appear as small, even trivial, millenial presents. Cyclone, sea flood, even ice, will leave another landscape for future fascination.

Moss's futures echo earlier Broadland accounts, scientific and popular, looking to the ends of landscape. Miller's *Seen From a Windmill*, for example, presented a Broadland doomed to pass, one way or another: 'These rivers and lakes are all that remain of what was a vast estuary of the North Sea. A gradual silting up is taking place, and each year they grow smaller. Eventually – in a thousand years or so – they will disappear, unless the sea breaks through the sand hills along the coast and once more claims Broadland as its own' (Miller 1935: 19). Distant pasts and possible futures met also when Robert Gurney identified 'A New British Prawn Found in Norfolk', *Leander longirostris*: 'Perhaps it is a relic of the time when our rivers were tributaries of the Rhine, and the larvae have the instinct to return here just as the migrating bird is able to find its way to its winter quarters' (Gurney 1922b: 312). In Dutt's *Norfolk Broads*, FW Harmer's 'A Sketch of the Geological History of the Broadland' enacts a dioramic scheme, with proto-Norfolk Room visions of pre-glacial Broadland: 'Our first panoramic glimpse of the geological history of the Broadland area is that of a deep rolling sea, an arm of the Atlantic, with no land in sight. / The scene changes. East Anglia has emerged, and forms part of the European continent' (Dutt 1903: 268). Broadland takes shape on the Rhine's estuarine west side, Dutt elsewhere tracing 'the scenery and flora of the ancient Rhine banks of East Anglia' (Dutt 1906: 342–3). Broadland becomes former Rhineland. The prospect of the sea's return conversely offered Dutt potential floral and faunal restoration: 'I know that unless the sea breaks through the marram-banks and transforms the lowlands of East Anglia in wild wastes of morass and reedy salt marsh, it is hopeless to expect to see again such abundant bird life in Broadland' (Dutt 1901: 155–6) / 'such swamps as these flowers grow in in rank luxuriance are rare, and in a few years, unless the sea breaks in again upon the low-lying levels of Broadland, there will be none of them left' (Dutt 1903: 160).

Pallis also envisaged the sea resetting succession, with flooding a 'catastrophic interlude' restoring natural order: 'So it is in the Broads, and so it is everywhere. In the Broads we have irruption from the sea, and Interference, and now the return of the woodland if Interference is in abeyance' (Pallis 1956: 9). After 1953 Pallis mused: 'The line of weakness across England is in East Anglia, at the Wash and the Yarmouth ingress; joined they would cut Norfolk into an island ... The Norfolk floods may thus be of more than local significance.' Pallis reflected: 'Thus even great tracts of land come into the position of danger (such as Atlantis)' (Pallis 1961: 20). For Pallis: 'The question here arises, how are irruptions distinguished from transgressions? It is, I suppose, a question of time' (18):

> There are comings and goings, sinking, emergings, submergings, at various depths, and for various lengths of time, shiftings, sweepings away, disseverings, siltings, shoalings, dryings up, drownings, disjoinings, severings, excavations, extensions, eddyings, tackings, tailings, capturings, all of these involving the passages, channels, gateways, gaps, roads, tides and currents, constant compensations and movements in one direction or another; spreadings outward to sea, and close to the land shallowing or deepening, displacements because of depositions, bringing the water level up, rises or falls in the water level more or less far flung, even if not always measurable, and in Norfolk the trend is, I think, up not down, the sea coming further in and shallowing. (5)

Broadland and undersea topography made sense together: 'The archipelago structure shown by the submerged sandbanks round Yarmouth, is repeated in the glacial upland round Hickling, though here the islands have emerged' (16). Pallis assumed peat crossing the coastal line, underlying the sea bed as it underlay Long Gores. In 1943 Godwin had mapped undersea stratigraphy, with coastal peat beds exposed in submerged forest, and peat 'moorlog' dredged from the North Sea bed. Godwin imagined what 1938 might have produced: 'Only repair of the banks and sustained pumping got rid of the sea water, and had human interference not prevailed the fresh-water fen peats growing round the Broads in 1937 might already in 1943 be exposed as 'submerged peats' upon the Norfolk shore well within tidal range, and all this with no necessary change in the relation of land- and sea-level' (Godwin 1943: 204–5). In the absence of sea walls, Horsey peat becomes prospective moorlog.

Pallis's on-offshore archipelago includes Scroby, the offshore sands visible from Yarmouth, as Broadland's enigmatic sealandmark. Robin Harrison's 1959 'pictorial souvenir' of *Scroby Sands* presented Yarmouth's 'natural breakwater', with a natural history of terns and seals, wrecks embedded in the sand, a delight for Yarmouth boat trippers, 'like some South Sea island minus its vegetation' (Harrison 1959: 1). For Harrison tern sound 'always reminds me of the signature tune to a popular radio programme, *A Sleepy Lagoon*, by Eric Coates', Scroby evoking desert island light music (10). Scroby's comings and goings also attracted the 1924 Ward Lock *Broads* guide, sixteenth century Scroby supposedly a 'grass-covered' excursion spot, recently re-risen: 'In 1922, Scroby Sands again rose above water, and became clearly visible at low tide from the beach. Motor-launches run from Britannia Pier and the beach to the sands, which are the resort of large numbers of gulls. It is a weird experience to walk dry-footed over a place that for centuries has been under the sea. The cause of this sudden upheaval is not known' (Ward Lock 1924: 17).

In 1929 the *Mariner's Mirror*, journal of the Society for Nautical Research, carried H Muir Evans' 'The Sandbanks of Yarmouth and Lowestoft'. Sandbank names included various 'holmes', echoing inland former island 'holm' names, at Cobholm and St Benet's Holm. Sea and land names make topographic echoes. For Evans Scroby had begun off Scratby and shifted south to Yarmouth, and from 1578 was dry land, with entertainments and cultivation, shipwrecked goods making it disputed legal territory, until a storm swept all away: 'to the east of the Scroby the sea receives the surprising name of Barley Pightel or Pickle – this is an old term for an enclosure – and when we read that grass and other vegetables grow thereon we must conclude that a field of barley became drowned and gave its name to the water' (Evans 1929: 270). Pallis delighted in Evans' catalogue of sandbank names: 'There is bank, bar, knole, knock, head, barber, rigg, ridge, warp, ower and many others' (Pallis 1961: 4). Pallis recalled the Midland and Great Northern Railway line into Yarmouth from Catfield, closed in 1959, running from Broadland by the sea:

> Thus the bluffs at Scratby and California ... furnished the soil to be spread by the sea for the Barley Pightel ... There the railway of the past used to emerge, close to the crumbling brink, to disclose suddenly – coming out of darkness as it were into light – the great light-filled gulch of the sea below, the Yarmouth Roads, road and ridge parallel to the shore alternating, placid blue passage and yellow crested and spumed, the point of vantage of the whole coast. (6; Adderson and Kenworthy 2007)

Exit Broadland, south of California. Today's road past California's beachware shops and chalets stops at the cliff, a few hundred yards north of the old railway brink, 30 offshore turbines now clearly marking Scroby, defences keeping the beach, holding the land line. Postcards for sale include 'Scroby Sands Wind Farm', pleasantly turning behind surf, and a cartoon satire against wind energy. A dolphin leaps from blue water, with an added sticker: 'Souvenir of California'. Older cards achieve a poignancy, or an irony, unanticipated at their making, with southern views over unbuilt sea, the coast as was. The sand is golden, the sea is blue, horizons uninterrrupted, marine roads clear.

Broadland Signs

Figure 41 Broadland Signs I (Clockwise from top left: How Hill field studies sign, May 2011 / Wildlife Sightings Box, Heron's Carr, Barton Broad, May 2011 / Patterson Close, Great Yarmouth, August 2010 / Andrew Lees memorial plaque, outside Wickhampton church, May 2011/ Discarded RSPB tag, Strumpshaw Fen, May 2011 / Display board, The Thatch, Wheatfen nature reserve, August 2010). Source: photographs by the author.

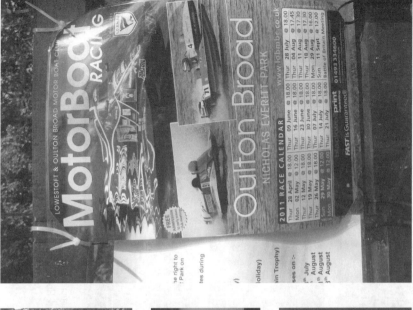

Figure 42 Broadland Signs II (Clockwise from top left: Water ski sign, River Yare, Strumpshaw, May 2011 / Ranworth village sign, August 2011 / Motor Boat Racing poster, Oulton Broad, August 2011 / Potter Heigham riverside bungalow, August 2011 / Swallowtail butterflies notice, by Martin George, Strumpshaw, May 2011 / Everitt Road, Oulton Broad, August 2011). Source: photographs by the author.

Figure 43 Broadland Signs III (Clockwise from top left: Berney Arms railway station information board / Oulton Broad, August 2011 / Berney Arms Mill ancient monument direction sign (detail) / Wherryman's Way marker, Berney Arms, August 2011 / Bridge over Muck Fleet, Stokesby New Road, August 2011 / Reedham Ferry, May 2011 / Wayford Bridge, August 2011). Source: photographs by the author.

Chapter Seven
Concluding

Bibliography

Origin, Ghosts, In pictures;
Adventure, Scribblings, Birds;
Handbook, Marsh leaves, On the;
Portrait, Charm of, Plan.

Books on, Gang on, Pilot;
Idyls, What to do;
Summer, Sport, In colour;
What future?, Making, Smiles.[1]

Definition: Regional Cultural Landscape

Pallis paints carr, Day downs a duck; Ellis noses fungi, Terry Scott squints. Transects and records show Lambert's new origin; Cox sings tradition for Moeran to arrange; Woods and Hoseason make waterland business. And in 1959, Rev ED Everard, vicar of Ranworth, shows geographers to the top of his tower to take in the view, the IBG seeking regional prospects.

The 1959 IBG excursionists, noted at the outset of Chapter Two, took a Broads Tours launch into a winter landscape, viewing, in the company of Lambert and Smith, and in advance of the 1960 RGS origins publication, the field sites of stratigraphic,

In the Nature of Landscape: Cultural Geography on the Norfolk Broads, First Edition. David Matless.
© 2014 David Matless. Published 2014 by John Wiley & Sons, Ltd.

botanic and historical enquiry. This book has made a geographic return, approaching Broadland as regional cultural landscape, with landscape, culture and region all terms significantly re-thought, in geography and beyond, in recent decades. Regional cultural landscape continues to direct attention to those studies in stratigraphy, botany and history which preoccupied the 1959 IBG tour, but these take their place as cultural practices alongside forms of human conduct which might have passed the earlier discipline by. An expanded sense of culture and landscape, and a contested sense of region, transforms regional cultural landscape, which grows to encompass natural history, painting, planning, wildfowling, science, song, boat hire, history, comedy; all elements making up cultures of landscape. The orbit of cultural geography takes in scientists and singers, botanists and cruisers, gunners and planners, commerce and paint, with individual biographies sometimes marking connections across fields; Pallis the scientist and artist, Ellis the botanist and broadcaster. Institutional structures shape the region; boundaries for planning, reserves for nature, companies trading pleasure, trusts salvaging the relic. Leisure structures jostle regulatory authority, tensions of landscaped pleasure surfacing in boatyards and bungalows. Architecture and material culture register contrasting and conflicting senses of landscape value, with icons of landscape such as wherry and windmill rewarding cultural scrutiny as emblematic regional shorthand. Buildings from before the planning era stand out, wonders or sore thumbs which might never lately have been permitted; a shack with columned veranda by the Ant, a mill stump in St Benet's abbey ruin, Cantley's sugar factory.

Regional cultural landscape demands movement across fields of human conduct often kept apart; a ranging, sometimes trespassing, form of enquiry. Confinement to one region makes study rove. Variety is embraced not simply for its pluralistic sake but for the charge of juxtaposition, the drawing out of contrast, the understanding of each in its setting of others. Regional cultural landscape also entails movement across human and non-human geographies, animal and plant landscapes configuring Broadland, as birds inhabit reed, gunners kill birds, weeds sing in water, coypus thrive and pass. Twentieth century Broadland sees Hickling's nature showpiece, Wheatfen's botanic treasure, Long Gores' overgrowth; human properties nurturing non-human qualities in nature's name. Elsewhere, land deemed marginal is reclaimed for workable profit. Halvergate's ditches, once seldom viewed or studied beyond the marsh worker and curious botanist, become a late twentieth century political frontline, gathered as valued landscape through drainage possibility, political protest and planning formulation. Regional cultural landscape also enacts trajectories of past, present and future. Narratives of origin and projections forward shape present argument, whether in peat digging discovery, marine transgression detection, plans for digging new broads, scoping studies for a Yare tidal barrier, or twenty-first century scenarios for upper Thurne embayment. The ends of landscape haunt today, actual and potential destruction informing the present. Eutrophication prompts ecological enquiry, historical analysis and social comment on modern excess. The 1938 Horsey flood triggers defensive review, agricultural experiment, historic survey and tilting speculation. Events of flood and pollution flush out contested visions of the region. The sea's return concentrates landscape minds.

Examining Broadland as regional cultural landscape also draws attention to the cultural work done by regional definition, or indeed by the definition of an area as county, village, world city, etc. Qualities of regional landscape are identified for development or defence, territories staked for control, names debated with consequence. If indeed you never moved beyond the regional label, if you took only the name, prolific analysis could follow, of regional extents denoted, of landscapes connoted, of spelling convention; Broads, Broadland, broads, Norfolk Broads, Norfolk and Suffolk Broads. Broadland here parallels the heath and forest region of Breckland, also crossing the Suffolk border in south-west Norfolk. While 'breck', meaning 'a tract of heathland broken up for cultivation from time to time and then allowed to revert to waste' (Clarke 1937: 1), was a commonly remarked nineteenth century feature of the district, the regional name 'Breckland' was only coined in 1894 by antiquarian WG Clarke, in a *Naturalists' Journal* article entitled 'A Breckland Ramble' (Clarke 1894). The term stuck, signifying a region remote and primitive, overlooked save by those with specialist natural history or anti-quarian interest (Matless 2008b). Broadland likewise moved from being a nineteenth century area whose unusual lakes, locally called broads, drew occasional comment – Lubbock's 'region of the broads' (Lubbock 1879: 175) – to being the 'Norfolk Broads', although here the name denoted a region newly visited and popular, filling with narrative (*On the Broads, Broadland Scribblings*), quickly 'beguided' (Nature 1887), as other tourist regions before. From holiday spot, Broadland carries into natural history and field sports (*Broadland Birds, Broadland Sport*), indeed becomes something around which battles can be joined, one interest defending the region against another; leisure against restrictive property, natural history against destructive sightseeing. Regional definition puts cultural landscape at stake, with place names coming in the process to stand differently for the region, cultural pressure points where contrasting versions of Broadland become apparent; Potter Heigham's leisure vernacular, Ranworth's conservation overview, Halvergate's political ditches, Yarmouth's port and varieties.

If considering only the regional name triggers cultural complexity, wider questions of geographic particularity have run throughout this book, and warrant further comment in conclusion. The theme of landscape voice will likewise be revisited, before six vantage points close the book.

Provision: Geographic Particularity

When Broadland was discovered as a leisure region, accounts commonly asserted its uniqueness. Here was a landscape unlike anywhere else, distinctiveness waiting to be seen. The same writers also commonly made comparison with, and noted resemblance to, Holland. Broadland was peculiar, yet similar to a coastal region directly east across the North Sea. This seeming inconsistency, Broadland unlike-yet-like, denotes less a slack reasoning in promotional writing than something characteristic of geographical particularity; sites made sense of through comparisons and connections, and uniqueness thereby affirmed. Consideration of Dutch connections, and other links of Broadland and elsewhere, can further illustrate the geographic particularity of the Norfolk Broads, and the provisional quality of regional identity.

Holland, specifically the regions of Friesland, Texel and the Zuiderzee, served as Broadland's cross-sea equivalent, as when HM Doughty published on genteel late nineteenth century wherry excursions, from *Summer in Broadland* to *On Friesland meres* and *Our Wherry in Wendish Lands* (Doughty 1889a; 1889b; 1891). Regional touristic dialogue was fostered, GC Davies following his Broads handbook bestseller with advice on *Cruising in the Netherlands* ('My advice is, see the Broads first and Friesland afterwards as, coming after the great Meres of Friesland, the Broads appear insignificant' (Davies 1894: 12)), while Norfolk boatyards such as Loynes opened hire agencies in Friesland. Anna Dodd reported a Wroxham boat hirer stating: '"Holland's better known. It's been more writ up; an' there's nothin' like writin' up a country to make it known." (In one man's mind, at least, literature had its solid uses.)' (Dodd 1896: 21). Dodd quotes in part to claim distinction for her own regional writing up, though 1890s Broadland was hardly short of words. Holland shadows Broadland also in Robert Gurney and Emma Turner's comparative Texel ornithology, Norwich School painters' deployment of Dutch landscape convention across Norfolk scenery, Demangeon's geographical visit, Clifford Smith's search for historic peat digging parallels. Overseas landscape lends ways of seeing and comparative experience to make sense of Broadland.

The same sea too shapes both sides. The 1953 North Sea surge flooded eastern and western shores, Holland's 'Low Countries' label all too appropriate, East Anglia battered and Sea Palling residents killed, although Broadland's adjacent coastline held:

Mary saw, in her mind's eye, a huge tidal wave moving slowly and relentlessly down the North Sea, like some gigantic prehistoric monster risen from the depths of the ocean.

'And it'll go crash against the Dutch sand dunes at about breakfast time tomorrow,' she thought.

'Oh, Mother, poor Holland! Poor us! Poor everybody!' She gasped aloud. (Burton 1960: 91)

In 1938, Horsey under water, Buxton headed to Holland for advice on defence and restoration. Early twentieth century Dutch expertise also shaped Broadland's largest built structure, the Anglo-Dutch Sugar Corporation's Cantley, sugar beet planted in international arable exchange. And Holland and Norfolk have historically been drained by windmills, geographic architectural icons balancing waters, managing circulation. Windmills, especially the many mills standing across the Halvergate marshes, trigger regular Dutch comparison, Broadland viewed as something other than typical England; flat, open, big skied, looking east. Yarmouth port schemes continue to hope for North Sea Dutch ferry connections, bringing direct passenger traffic, though the current boat to Holland runs well south from Essex, through Harwich to the Hook.[2]

Contemporary visual culture sustains Dutch comparison. Richard Denyer's 2012 Norwich photographic exhibition, *Neither Land nor Water*, revisited the Broadland documented in his 1989 *Still Waters* for a project on 'water, land and sky in Broadland and the Netherlands'. Photographs of Broadland and Friesland mixed, the regions presented in mutual recognition through lives led by water, landscapes of drainage and threatened flood, boating pleasures and waterside dwelling, wetland ecologies shaped by or defended against human construction. Sharp light characterises Denyer's tripod-shot, long exposure images, structures standing clear while water and sky move into blur. Denyer's photographs tend towards a comparison of Dutch neatness and more ragged-edged England, tidy canalsides against makeshift riverbanks, and show Friesland as accommodating a waterside industrial and residential modernism, with only Cantley structurally so bold in Broadland. The display of Broadland and Friesland together variegates both, Broadland cast into variation; 'Brundall waterfront' and Thurneside bungalows, neat and bespoke Horning and Wroxham riversides, the New Cut straight as a Zuiderzee causeway.[3]

Geography far beyond the cartographic edge of the region has shaped, and will continue to shape, its landscape. County, national and supra-national governmental policy shapes the region from Norwich, London and Brussels, while the Broads Authority makes its plans from the bank of the Wensum in Norwich, just outside its own territorial jurisdiction. Place-name fabric denotes older, pre-Conquest, movement for regional settlement – the Danish suffix -by, the Anglo-Saxon -ham and -ton: Filby, Mautby, Clippesby, Oby; Runham, Waxham, Ludham, Stalham; Hoveton, Sutton, Winterton, Somerton (Sandred 1996). Broadland sites and distant geographies make the region, Wroxham and London joining through the Blakes Agency and the Great Eastern Railway, Horsey and Geneva meeting through Buxton and Crees. Broadland and North America link through Hope Allen's life and stories, and Boardman's How Hill hickory tree, seed sent by a former worker migrating transatlantic. Cambridge and Southampton foster Lambert's discovery of original artifice. Hickling shelters Whitehall ornithologists and royal gunners, Lord Desborough's second country estate gathering elite metropolitan outdoor culture. Next door, Pallis lands Byzantium in Broadland, signs from 500 years ago carved into a private marsh. Meanwhile, blithe to birds and history, Yarmouth brings national summer season stars, passing in Variety; Hudd, Lloyd, Scott.

Movements make the region, furnishing rather than diminishing geographic particularity. Never complete, always refurbished, regional identity cannot help but be provisional; provision denoting not only tentativeness, but sustenance.

Colloquial: A Geography of Voice

Broadland regional cultural landscape is shaped by a colloquium of voices, some carrying scientific expertise, some bearing colloquial knowledge, operating in dialogue or in ignorance of one another, speaking to and across shared terrain, articulating voice through speech, writing, image, sound and noise. Voices of varied accent and authority articulate the region. Broadland's landscape colloquial is styled through a cultural geography of voice, with qualities of affinity, authenticity, distance, engagement and detachment affecting hearings given and authority gained. Social hierarchies shape and are shaped by articulations of the region, in voicings of land ownership and land work, gunning and science, boating joy and planning rule.

Geographic affinity may be voiced through residence, research, ownership, ancestry. Lubbock conveys a landscape of amateur natural history, hobby observations in keeping with his parochial role, care for parish matched by regional observance. Ellis as resident specialist observes that which passing acquaintance would not perceive, noting the seasonally shifting and the microscopic, and able to spot and relish eccentricities by virtue of intimate familiarity with the common. A city university takes a nearby region within its research remit, research time and money permitting systematic field study, UEA results taken beyond the region to foster international scholarly reputation, while embedding the university within its environs; just as Cambridge science did at further remove in earlier decades, Smith or Godwin journeying to Broadland, and back again, for field science. Boardman takes residence at How Hill to make a house and estate commanding the Ant, supplementing professional city status with country dwelling; Miller converts a Bureside mill, sizing down from American banking to make a writing life, setting himself in the scene to mark his narrating role. Turner achieves bird intimacy through basic Hickling living, her scientific status secured by time spent in the field, and the intellectual and financial ability to take a break from it, travelling to a Cambridge home, or a Linnean Society hall. Day takes Broadland as his shooting space, claiming friendship with residents, occasionally renting shooting rights, before retreating to his Essex home to write up regional devotions. Holiday voices may claim authoritative affinity – May with his suggestion for a holidaymakers association, Ransome with his carefully mapped adventures – or may happily declare on the region through brief experience. Keble Howard's fast gentleman, or Anna Bowman Dodd's cosmopolitan reverie, voice another form of geographic affinity, voices confident enough in their class of pronouncement to narrate the region through jaunty impression.

Voices may mutate through regional inflection. Emerson moves into dialect, demonstrating his sense of general capability, landing on regional language with social and intellectual superiority. Hope Allen listens to Hickling residents for dialect phrases to shape modernist short stories, words sent to North America for broadcast. Eric Fowler writes as Jonathan Mardle to gather dialect authority, moving across classes within the county. Patterson shuttles between the self-conscious performance of Yarmouth living and the deployment of a scientific vocabulary, seeking a tone of voice through which he might be heard. The Duchess of Bedford's patronage of Patterson provides an acute case of the tensions and anxieties present in attempts to speak and be heard across class and geographic scale. A voice so evidently Yarmouth depends on county and national patronage for a hearing, whether at the NNNS or the Linnean Society, and the several meanings of patronage come into play. Other voices may mutate through career trajectories, as with Lambert achieving national scientific standing over her home region, having moved away from it through educational opportunity, speaking back on Broadland from Cambridge and Southampton, defining regional origins. Pallis, meanwhile, moves contrariwise, meeting Lambert across generational and geographical trajectories, beginning as Cambridge scientist in the Broadland field, pinning down regional ecology, and ending as devoted Bohemian Broadland owner-resident, letting regional nature take its course, and noting dialect for literary experiment. Pallis's unpublished novelistic tale, noted in Chapter Three, of a language teacher seeking to standardise regional speech, ironing out dialect only to get his comeuppance as children and adults revolt in the name of place, offers a characteristically singular take on the geography of voice, which can conclude this section. Pallis has her 'country lady' affirming the native, with voice mutating in glee:

'You should have heard the grand old Canute-Nudds affirm themselves' my naughty fenny friend gave out in gleeful spite. 'No haf an haf abaut ut' – suddenly she was wholly local – 'Doant yu go a' meddlin wi ma North sea brogue'; suddenly she was pugnacious, and, just as suddenly, reflective …

'And I raced across the levels right in the teeth of that gale of salted English', she ended with a skirrrll of infinite delight.

'So native speech finds life, fanned by the breath of passion. That war voice production wi a wengeance'.[4]

Six Vantage Points

Concluding, six vantage points can serve to look back across the book, and into the region, specific spots to supplement the sites described in Chapter One. Three are on the geographic edge of Broadland, three are well inside. Ordnance Survey grid references are provided; all have free access for the visitor.

Yare Mouth (OS 534037)

See Broadland meet the North Sea at Yarmouth harbour, the Yare's mouth best viewed from the Gorleston side, fine sand beached to the south side of the pier, swirling water to the north. Swimmers trying a crossing are hauled out, if lucky. Rains from Horsey and Coltishall, Norwich and Bungay, all run here. The entrance shown as awkward by Stark in 1834 remains; a new outer harbour built from Yarmouth beach hopes to lure larger trade. No jetty over there, the north side now restricted port territory, harbour strollers accessing Gorleston only. Suffolk sits south over a coastal rise, to Broadland's Lowestoft lock exit. A brick lighthouse overlooks the inbound 90 degree Yare turn, sea view turning to river view. Industrial quays run to Haven Bridge, drifters and herring long gone, vessels servicing gas and wind. River ebb meets sea current, the wash of the region heading for Scroby.

Heron's Carr Boardwalk, Barton Broad (OS 359207)

Take a 610 metre boardwalk through Norfolk Wildlife Trust land, BA-built in 2003 through Heron's Carr, opening space once visited only for science. Wood succeeds over reedbed on deep mud, evident in exposed dykes. Tossed coins sink, barriers and injunctions guard against temptation; just how deep, how liquid? A viewing platform gives north across the mile of Barton Broad, cruisers making for the Ant north or south, sound chugging over water. Waterfowl attend, crust hopeful. Black pen-on-wood lettering announces 'Wildlife Sightings', the lid of a bolted box opened to find a margarine tub, relabelled to hold a wildlife notebook, falling at random: 'Today I saw a teradactal having a bird fight with a Bald Eagle, all over a piece of brown bread. Mr M Mouse' / 'You can tell when it is half term, can't you?' / '<u>March 7th</u>. Kingfisher – 2 Goldeneye – 31' / '<u>March 12th</u>. remains of a ½-eaten bird on one of the red posts (gull?) – presumably the left-overs of a Peregrine meal.' Bird life on public record, the instant archival broadside.

Norwich Yacht Station (OS 238086)

View Broadland's seventh river, scarcely mentioned in the guides after Bure, Thurne, Ant, Waveney, Chet and Yare. The Wensum, coming all the way from Fakenham, meets the Yare above Thorpe, and drops its name. London passengers would alight at Norwich's Thorpe Station, just by the Wensum, to join another train for Wroxham, or Yarmouth; the city river sidestepped, a Broadland dead end. Excursion boats pass the city's river backs, or head downstream for scenery; woods at Bramerton, Yare's finery. No hire craft beyond Bishopbridge, where Kett crossed to take the city. Private school meadows frame cathedral prospects. 1912 rain spilled the Wensum through August streets. Late summer Saturday green and yellow passes the moored, downstream to Carrow Bridge. Distant cheers fill open canopies at the head of navigation.

Muck Fleet Bridge, Stokesby New Road (OS 422117)

Turn off the A1064 towards Stokesby along the 'New Road', not on the 1910 OS map but there by 1932, running straight to the village with one bend, over Bure marshes grazed and cropped. You might miss the Muck Fleet. White painted metal railings mark a bridge, the road barely rising over something once sailed. One hundred years an ex-navigation; to think boats came this way, making for the Trinity Broads, now out of circulation. Above the bridge, a sluice backs a pool, weed green. Below the bridge, plant life overgrows. Railing signs state: 'Canoeing prohibited'; somebody must have tried, injunctions warning intrepid paddlers. A minor road crosses a dyke, road navigating over river.

Bramble Hill (OS 480224)

Walk from Horsey Mill to the dunes, turning right to the remotest Broadland coast, sea held by sandhills held by marram. Pass beach seals and naturists, and just before the Bramble Hill Gap you cross the Horsey–Somerton parish boundary, finding the sea at the old Hundred Stream outflow. No sign now. Groynes hold the beach, plaques in Gaps report parliamentarians unveiling wall reinforcements, the coastal line held. The dune height sees from Happisburgh to Winterton, towers north and south, out over sea and in over marsh, the 1938 flood land and the heights of Martham. Dunes run taut across a possible bay. Marsh waters head strangely inland, the coast a watershed.

A fence at an angle runs the parish line, doglegging to the Sock Drain and the Hundred Stream, into the Thurne. A summer ditch puddle marks a source of sorts, water looking to Yarmouth.

Thurne Squint (OS 405156)

Ascend from Thurne village, to St Edmund King and Martyr, church standing alone, old English patronage. Eye level on the west wall of the fourteenth century tower finds a hole; a squint or hagioscope, antique curiosity, allowing sight through the wall from outside to the altar, and from inside giving onto the marshes of Thurne and Bure, dead straight to St Benet's Abbey ruins. Begin with Ranworth's tower panorama; end with a squint. On grass looking in, or on the chill tower floor peering out, Thurne catches Broadland as region of sightlines. Lepers, or others barred from inside, might partake in worship by sight, if not touch or taste. And the squint might signal to St Benet's, a shining hole light across the marsh. Viewing out, land falls to Thurne's level; solids and river, marsh and walls. A sliding wooden cover blocks the inside against bird entry. Drawn back, twigs hold a pale egg, the squint a nesting shelter. The shell remains another year. Slide the cover shut.

Notes

Chapter One: Cultural Geography on the Norfolk Broads

1   Clout's survey of French regional monographs, whereby geographers made their careers and discipline through field and archival study of regional territory, identifies Demangeon's 1905 *La plaine picarde* as the first in the genre (Clout 2009).

2   George (1992: 46–7) lists 50 broads, the Nature Conservancy (1965: 64–7) counts 42.

3   Appleton's work is nonetheless sensitive to history and geography, including in relation to his own life. Appleton's autobiographical *How I Made the World* (1994) includes discussion of his north Norfolk childhood, and Appleton indeed accompanied his academic geographical career at the University of Hull with the writing and broadcasting of Norfolk dialect stories on the BBC Midland Home Service in the 1950s (letter, Jay Appleton–David Matless, 20 November 2002).

4   Attempts to restore former extensions to navigation, now lost, continue, as when the Chet was reopened to navigation in 1958. A recent project seeks to investigate and potentially restore the Aylsham navigation, badly damaged, like the North Walsham–Dilham canal, after severe floods in 1912. The Bure Navigation Conservation Trust, with the Aylsham Navigation Project, linked to landscape history research at the UEA, organised a centenary sailing of the wherry Albion upstream to Coltishall on August bank holiday 2012, with cargo brought down from Aylsham by canoe (EDP 2011b; Spooner 2012; the Aylsham Navigation website is at: http://aylsham-navigation.norfolkparishes.gov.uk/ (accessed August 2012)).

5   Robberds argued that the valleys had become land by a slow and continuing fall in sea level of 40 feet over 6000 years, a fall which should be accounted for in any port plans. Robberds made his argument from shell beds found 40 feet above sea level on the valley sides, evidence whose value was contested the following year by RC Taylor, who suggested the shell beds were part of the Crag formation and unrelated to any estuary, and that the estuary might have infilled without significant sea level change (Jennings 1952: 9). For further discussion of debates on landscape origins see Chapter Two.

In the Nature of Landscape: Cultural Geography on the Norfolk Broads, First Edition. David Matless.
© 2014 David Matless. Published 2014 by John Wiley & Sons, Ltd.

6   Paasi posits an 'analytical distinction' between 'regional identity' and 'identity of a region', the former denoting regional consciousness, the latter those features used in classificatory 'acts of power' to distinguish a region from others (2003: 478), and notes that much work on regional identity thereby finds a 'major difficulty' in the ways in which 'narratives on regional / "our" identity become constituents of the interpretations of what identity means', with such interpretations 'deeply political categories' (480). This may however be best regarded less as a difficulty than a productive predicament for enquiry, indeed only becomes a difficulty if one seeks to maintain an analytical demarcation of regional identity and identity of a region, rather than treating matters of region and identity as 'performative discourse' (Bourdieu 1991: 223) through which such distinctions might indeed be produced.

7   For Kirkpatrick Sale bio-regionalism denotes: 'a life-territory, a place defined by its life forms, its topography and its biota, rather than by human dictates; a region governed by nature, not legislature' (Sale 1985: 43).

8   *South Riding* was made into a feature film in 1938, and has received various television adaptations, most recently in 2011 by Andrew Davies.

9   Royal Society of Literature website: www.rslit.org (accessed January 2014).

10   Everett similarly notes a contextual post-war 'devolutionary shift' in British culture in film, television, fiction and drama, which Bunting 'responded to and helped to promote' (Everett 2007: 316). Bunting's interest in regional/dialect poetry is evident also in his 1976 edition of nineteenth century north-east collier poet Joseph Skipsey (Skipsey 1976).

11   The approach taken here to regional cultural landscape indeed could also apply to the county scale, should conjunctions of landscape and culture work in county form; the Broadland story includes varying articulations of Norfolk and 'Norfolkness', themselves shaping regional argument.

12   The LCA, detailing the 31 areas, giving key characteristics, history, features of note, and overall 'condition', is available at: www.broads-authority.gov.uk/planning/landscape-character-assessment.html (accessed January 2014).

13   Similar versions of these accounts appeared in Matless 2010.

14   The assumed request may have been prompted by the experience of participating in Simon Pope's WG Sebald-inspired project *The Memorial Walks*, where various commentators scrutinised paintings in Lincoln and Norwich museums, before being driven into the nearby countryside, walked over fields, and asked to describe the painting they had viewed. The Norwich walks included Broadland sites, but I participated in Lincoln, describing 'Landscape with Tree' by Nicholas Van Schoor while standing in a field near Scothern. The event is commemorated in a book (Pope 2008), and at the time of writing the field recording retains an online presence behind painting number four at: http://www.waterlog.fvu.co.uk/simonpope/ (accessed January 2014). The experience of description out of situ, the insistence that a detailed account of the painting, saying what you saw and nothing more, was all that was required, may have informed these subsequent exercises.

Chapter Two: Origins

1   Brady and Robertson had named a newly discovered ostracod shrimp, *Polycheles Stevensoni*, for Stevenson, 'to whose kind interest in the objects of our visit to the district we are indebted for much of our success' (Brady and Robertson 1870: 26).

2   Pallis's analysis appeared within Cambridge ecologist AG Tansley's edited collection *Types of British Vegetation*, discussed further in Chapter Five below (Tansley 1911a). Tansley had earlier led students from University College London on field excursions to the Broads; a copy of his introductory lecture on 'The Physical Features of the Norfolk Broads' for the July 1903 UCL 'summer ecological excursion' shows him following Gregory's theory of origin. Tansley, lecture notes, UCL, 1903; Tansley Archive, University of Cambridge – I am grateful to Laura Cameron for this reference.

3   The table was part compiled with AE Ellis (no relation), on whom more in Chapter Five below.

4   Interviews with Joyce Lambert, 10 October 2000, 3 May 2003.

5   Letter, Joyce Lambert–David Matless, 15 April 2003.

6   Lambert would again serve as NNNS President in 1961–2.

7   Interview with Clifford Smith, 5 February 2002.

8   Letter, Joyce Lambert–David Matless, 3 June 2002.

9   Interview with Clifford Smith, 5 February 2002.

10 Interview with Clifford Smith, 5 February 2002.

11 *Mystery of the Broads*, Midland Home Service, 15 February 1957, 2215–2245; BBC Written Archives, Caversham Park, Reading.

12 Interview with Joyce Lambert, 3 May 2003.

13 Hunter's novels are discussed in Chapter Three below.

14 Letter, Joyce Lambert–David Matless, 3 June 2002.

15 Letter, Joyce Lambert–David Matless, 3 June 2002.

16 Letter, Marietta Pallis–Catherine Gurney, 10 August 1953, Pallis/Vlasto Archive, Hickling.

17 Letter, Marietta Pallis–Joyce Lambert, 27 August 1953, Pallis/Vlasto Archive, Hickling.

18 Letter, Marietta Pallis–Joyce Lambert, 11 October 1953, Pallis/Vlasto Archive, Hickling.

19 Letter, Marietta Pallis–Joyce Lambert, 14 October 1953, Pallis/Vlasto Archive, Hickling.

20 Letter, Marietta Pallis–Joyce Lambert, 1 April 1957, collection of Joyce Lambert.

21 Interview with Joyce Lambert, 10 October 2000.

22 Letter, Marietta Pallis–Joyce Lambert, 2 April 1956, Pallis/Vlasto Archive, Hickling.

23 Letter, Joyce Lambert–Marietta Pallis, 27 April 1956, Pallis/Vlasto Archive, Hickling.

24 Letter, Joyce Lambert–Marietta Pallis, 27 April 1956, Pallis/Vlasto Archive, Hickling.

25 Pallis/Vlasto Archive, Hickling.

26 Translation from the Danish for Joyce Lambert made by Dr Sigurd Olsen; the cartoon is reproduced as figure 2 in Matless 2003a.

27 *The Parliamentary Debates* (Hansard), Fifth Series, vol. 202, House of Lords, Third Volume of Session 1956–57, column 651.

28 A photograph of the University Broad in front of the UEA's ziggurat style buildings, designed by Denys Lasdun, appears in Manning 1980, facing page 97.

Chapter Three: Conduct

1   Davies's *Mountain, Meadow and Mere* had included a story of 'Breydon Jack' (Davies 1873: 28), but the meres described were in Shropshire, with the Broads barely mentioned.

2   Blakes was bought by Thomson Travel in 1998.

3   Suffling remarked on the Horning Ferry Inn that: 'A glance at its visitors' book will show that many notable persons have used this inn as their rendezvous, ... Lord Suffield and Mark Twain among others' (Jennings 1892: 25).

4   Dutt (1870–1939) was a journalist in Lowestoft and London before retiring to Lowestoft due to ill health, and wrote extensively on East Anglia in general, including Breckland (Matless 2008b; 2009b).

5   Emerson has less regional presence in mid twentieth century publications, although the *Shell Guide to Norfolk* (Harrod and Linnell 1957) reproduces three photographs from *Life and Landscape on the Norfolk Broads* (Emerson and Goodall 1886).

6   Goodall's painting 'The Bow Net' (1886) echoed Emerson's *LLNB* photograph 'Setting up the Bow Net', while his painting 'Rockland Broad' (1883) anticipated *LLNB* compositions.

7   Emerson would partially renounce this claim in an 1890 pamphlet, *The Death of Naturalistic Photography*, on the basis that a lack of artistic control of tonal relations undermined its claim to naturalistic art (Emerson 1890c).

8   In an article on Lee's career, Bearman (1999: 630) casts doubt on the plausibility of Graves' account, though there appears no direct testimony from Lee to confirm or deny the story.

9   A further 1927 manuscript source gives 63 pages of 'Songs Collected in Norfolk', with Broadland prominent (Moeran 1927; Karpeles et al 1931; Moeran et al 1931).

10 Tune and words are close cousins of the song elsewhere collected from Hampshire as 'The Banks of Sweet Primeroses' (Vaughan Williams and Lloyd 1959: 17), later recorded by Shirley Collins and others.

11 The recording appears on *Harry Cox: The Bonny Labouring Boy*, 2000, Topic Records TSCD512D. Cox described 'old Snuffers' hiring the broad for private use, but failing to prevent others eel-babbing.

12 Quoted at www.moeran.net/Audio/GoodOrder.html#sleevenotes (accessed July 2012), billed as the 'worldwide Moeran database'.

13 I am grateful to Stokesby local historian David Trowbridge for this reference.

14 KH Tuson, 'Cruises on the Broads', diary extract from 1923; Norfolk Heritage Centre, Millennium Library, Norwich.

15 'Child's Diary', Norfolk Record Office, MC 1249/1.

16 Emerson would also collect and publish Welsh folk tales (Emerson 1894).

17 The draft is included in a letter, Marietta Pallis–Joan Strachey, Box X85, Joan Wake Collection, Northamptonshire Record Office. The quoted phrase appears on the page of the draft numbered 124.

18 Letter, Hope Allen–Dr M Day, 17 March 1949, Bodleian Library, Bod MS. Eng. letters d.217.

19 Letter, Hope Allen–Dr M Day, 17 March 1949, Bodleian Library, Bod MS. Eng. letters d.217.

20 Letter, Joan Wake–Hope Allen, 1 June 1933, Pallis/Vlasto Archive, Hickling.

21 While Sylvia was referred to in reviews as Mrs Miller, the two were not married, simply a couple who shared the same surname. I owe this information to their American relative Peggy Rand, who, having seen my reference to Miller in an academic article, kindly sent a copy of a 16 × 18 inch layout made by Sylvia; at the top the title of *What to Do* 1936, with her own watercolours and sketches (she illustrated Drew's writings), press clipping reviews, and Drew's inscription to Sylvia's daughter: 'To Babbie / With love and many thanks for the "punch" you have contributed to "Seen From a Windmill". / From the author – Drew Miller. / 21st February 1935.'

22 *Gloria!*, broadcast on ITV, 22 August 1993.

23 Thriller writer Mitchell would also set *Wraiths and Changelings* (Mitchell 1978) on the Broads, featuring ghosts and murder along the Bure.

24   The Gently stories have in recent years been revived for BBC television, relocated to north-east England. Hunter produced *The Norwich Poems* (Hunter 1944), including a 75-verse boating poem, 'Swan Song' (38–48), and 'Saturday at Wroxham' (36–7) on holiday arrivals, and wrote dialect stories for broadcast, running a secondhand bookshop in Norwich before his publishing success (EDP 1955d; Earwaker and Becker 1998: 106–8).

25   'Child's Diary', Norfolk Record Office, MC 1249/1, 2 August 1926. Also KH Tuson, 'Cruises on the Broads', Norfolk Heritage Centre, Millennium Library, Norwich.

26   'Child's Diary', Norfolk Record Office, MC 1249/1, 18 August 1926.

27   EHT Jukes, 'The Cruise of the Scamp', Norfolk Heritage Centre, Millennium Library, Norwich.

28   Broads Tours leaflet, 2012; http://www.broadstours.co.uk/ (accessed January 2014).

29   *I'm Alan Partridge*, BBC DVD, 2002. The sequence, not filmed in Broadland, would seem to have been shot in southern England, with a pub umbrella advertising the then Salisbury brewery Gibbs Mew.

30   Films held at the East Anglian Film Archive, Norfolk Record Office, Norwich. *Let's Get Away From it All* and the 1972 Ladbrokes film are available online at www.eafa.org.uk (accessed July 2012). Hudd, resident in Suffolk and a historian of variety and music hall, has donated his collection of sheet music to the UEA, having received an honorary doctorate in 2007; less welcome Norfolk publicity came in 2006, with Hudd reported hurling abuse at Broads Authority officials while struggling to moor at Reedham: 'Witnesses also report seeing him shaking a 4ft boat hook at the waterside attendant' (EDP 2006).

31   1966 and 1967 regional television reports on this invented custom appear on the British Film Institute's 2011 DVD collection *Here's a Health to the Barley Mow: A Century of Folk Customs and Ancient Rural Games*, alongside longer-standing customary practice, including a 1979 film of Norfolk step dancing. Dwile-flonking survives, the 2011 'World Dwile-Flonking Championships' being held at Ludham in Broadland.

32   Daley's writings had promoted the Broads as a landscape for sailing (Daley 1935; 1949).

33   EA Ellis Archive, Castle Museum, Norwich, Box 57: Broads Society.

34   Boardman's frivolous SPBBPS suggestion may have alluded to Herbert Woods' Broadland Protection Society, with its promotion of navigation, discussed below.

35   *The Parliamentary Debates* (Hansard), Fifth Series, vol. 444, 1952–1960; EDP 1947c. Evans also spoke in the debate on the national parks Bill, supporting a Broads national park; EDP 1949b.

36   The plaque, still at the main entrance to the Castle Museum, states that it was 'placed here by the citizens of Norwich in reparation and honour to a notable and courageous leader in the long struggle of the common people of England to escape from a servile life into the freedom of just conditions'.

37   The BPS had also been promoted in Rooke's Blakes-published *Let's be Broad-Minded!* (Rooke 1948: 104).

38   'Life on Mars' appeared on David Bowie's LP *Hunky Dory*, released in 1971.

39   Knights had also participated in the contemporary revisioning of Emerson's photography, exploring the relationship of class and aesthetics in 'Emerson's social order' (Knights 1986: 18).

40   'The Arthur Ransome Trail', leaflet issued by Norfolk County Council Library and Information Services, c.1995, giving a two-part tour of the Broads by car, 'featuring places of interest in "Coot Club" and "The Big Six"'.

41   The brand was devised through the Broads Tourism forum, set up in 2004.

Icon I: Wherry

1 In 1952 the Norfolk Wherry Trust had noted: 'It used to be said that no one had been to Norfolk who hadn't seen a wherry and a windmill; both were veritable hall-marks of the county' (Norfolk Wherry Trust 1952: 3).

2 Letter, Sylvia Townsend Warner–Charles Prentice, 5 September 1931, written from 'Houseboat Memories, P.O. Thurne'; reproduced in Maxwell 1982: 18.

3 Robert Malster, author of a key text on the wherry (Malster 1971), was editor of *The Norfolk Sailor* (1958–67), newsletter of the Norfolk Nautical Research Society, formed 1954, with chiefly maritime interest, but including Malster on 'The Norwich River' (Malster 1963) and 'Norfolk Navigations' (Malster 1966).

4 'Three Ways' here denotes the junction of Wherryman's Way with Weaver's Way and Angles' Way.

5 Doyle (1995) notes however that Russell Colman was the only significant figure in the pre-1914 Norwich Liberal establishment to make the political switch to the Conservative party.

6 Exceptions include Thirtle's 1815–20 pencil and watercolour sketches of 'Hoveton Little Broad' (Hemingway 1979: 40; Blayney Brown, Hemingway and Lyles 2000: 130), and 'Cottages, Cattle and Figure by a Broad' (Allthorpe-Guyton 1977: 54), Cotman's late unfinished works 'Alder Carr, Norwich' and 'The Edge of the Broad', and some chalk drawings by Stark.

7 Examples include Stannard's 'Boats on the Yare, Bramerton, Norfolk', 1828 (Blayney Brown, Hemingway and Lyles 2000: 149), or Thirtle's 'Thorpe Staithe', c.1829 (Allthorpe-Guyton 1977: 65).

8 The painting, c.1813–17, also known as 'Boats: Junction of the Yare and Waveney', is displayed in Manchester Art Gallery. The Yare steam barge was destroyed by an explosion in 1817, killing nine passengers (Edwards 1965).

9 Norfolk Wherry Trust website: www.wherryalbion.com (accessed July 2012).

10 Mayhew would also become first President of the Broads Society in 1956, serving for 12 years, with the Society's Lady Mayhew Trophy later awarded for organisations bringing young people to the Broads (Dicker 1998). Broads Society founder Ramuz was also an early Wherry Trust committee member.

11 Borrow also appeared as Broadland spirit in the work of Dutt, his 1907 *Some Literary Associations of East Anglia* noting Borrow's heath wind (Dutt 1907: 117), but also dwelling on Borrow's Oulton Broad residence, Borrow remembered as an enigmatic figure roaming the lanes, reclusive in late life, his house now gone, but the summer house by the broad remaining, 'still seen by cruisers on the Broad, and overshadowing it are the same dark pines amid which the wind made mournful music while he sat in his lonely retreat' (180; Adams 1914).

12 Borrow's *Lavengro* description of the view from Mousehold over the 'fine old city' has also given Norwich a motto, shown on road signs at the city boundaries (Borrow 1924: 90; Dutt 1907: 122). The cover drawing of a 1913 Norwich 'George Borrow Celebration' souvenir, by Alfred Munnings, showed two figures overlooking the city in a breeze, captioned: 'There's a wind on the heath brother, who would wish to die – ?' (Hooper 1913).

13 In 1953 Green included Albion in a list of wherries 'still remembered' (33–8), but the Norfolk Wherry Trust was mentioned only when the 'Despatch' was noted as 'One of the vessels for which the Wherry Trust made an offer to purchase' (49), their Albion efforts ignored. Green was born in Norwich in 1877, lived in South Africa and Australia, and from 1946 in Birmingham. The Lowestoft Museum holds some of Green's paintings, while a photograph from 1944, held by Norfolk County Council, shows an elderly Green in nautical cap and jacket;

viewed at the 'Picture Norfolk' website hosted by Norfolk County Council; norlink.norfolk.gov.uk (accessed July 2012).

14   The wherry stood in a case representing Norfolk and Yorkshire boats, alongside a 'Norfolk Keel' and 'Humber Keel'. The Science Museum shipping galleries were closed to the public in May 2012, and the displays dismantled for storage. Material here is taken from site visits before their removal, and documentation held in the Science Museum archive. The full story of the 'British Small Craft' displays is told in James Fenner's subsequent University of Nottingham PhD on the topic. On the history of the Museum see Morris 2010. 'Lowestoft Trader' is one of the wherry names engraved on signage along the Wherryman's Way footpath, appearing on a signpost at Berney Arms.

15   Science Museum archive, file 1927/822, 'Norfolk Wherry'.

16   The Museum file on the wherry model (1927/822) also includes a 9 March 2009 email from Norfolk Wherry Trust Membership Secretary Trevor Hipperson, 'amazed' at seeing the maritime displays on a recent visit, and asking if mention of Albion and the Trust could be added to the display information card. This did not happen, and the displays were removed in 2012.

17   Science Museum archive, curator clipping file, 'Small Craft England The East Coast Thames Estuary to Cromer'.

18   Letter, Walker–Clowes, 28 June 1927, Science Museum archive, file 1927/822, 'Norfolk Wherry'. Malster notes William Hall as one of the Reedham wherry-building family, builders of Hathor among others (2003: 117).

19   Letter, Clowes–Walker, 29 June 1927, Science Museum archive, file 1927/822, 'Norfolk Wherry'.

20   Letter, Darby–Clowes, 22 July 1927, Science Museum archive, file 1927/822, 'Norfolk Wherry'.

21   Letter, Darby–Clowes, 31 July 1927, Science Museum archive, file 1927/822, 'Norfolk Wherry'.

22   Letter, Hall–Clowes, 5 August 1927, Science Museum archive, file 1927/822, 'Norfolk Wherry'.

23   Letter, Hall–Clowes, 7 September 1927, Science Museum archive, file 1927/822, 'Norfolk Wherry'.

24   Science Museum archive, file 2616. A draft label noted Hall basing the model on 'measurements taken by him and four others, some sixteen years earlier, of a Norfolk Keel which lay partially buried in the river-bank near Postwick Grove, some 4 miles below Norwich' (Science Museum archive, file 1929/884, 'Norfolk Keel'). Hall's Bridewell keel was also based on the Postwick boat. Hall would also make the Science Museum's 'Beach Yawl', used between Winterton and Southwold, completed in August 1932 and later included in the 'Norfolk and Suffolk' British Small Craft case.

25   Letter, Hall–Clowes, 16 August 1927, Science Museum archive, file 1927/822, 'Norfolk Wherry'.

26   Science Museum archive, curator clipping file, 'Small Craft England The East Coast Thames Estuary to Cromer'.

Chapter Four: Animal Landscapes

1   Broadland, Breydon and the Loke were complete by 1933, Breckland and the coast added by 1935, and the Yare Valley later in the decade (Leney 1933; TNNNS 1935a).

2   Item 1.990.1.058, 'Norfolk Room', notebook, EA Ellis Archive, Castle Museum, Norwich.

3   Item 1.990.1.019, 'Fauna and Flora of Norfolk (Norfolk Room)', notebook, EA Ellis Archive, Castle Museum, Norwich.

4   Item 1.990.1.058, 'Norfolk Room', notebook, EA Ellis Archive, Castle Museum, Norwich.

5   Lubbock was rector of Eccles in south Norfolk from 1837, and earlier curate in Rockland and Bramerton.

6   Dye, Fiszer and Allard (2009) have recently provided a rich archaeology of species' earliest appearance in Norfolk in *Birds New to Norfolk*.

7   *Birds, Beasts and Fishes of the Norfolk Broadland* presented Emerson as a writer. The photographs, mostly by TA Cotton, showing stuffed specimens or blurred images 'from life'.

8   Du Maurier's sister Isabel was the first wife of Clement Scott, late nineteenth century populariser of the north Norfolk resort of Cromer as 'Poppyland', another case of metropolitan dreams coming to rest in/on Norfolk.

9   Interview with Phyllis Ellis, Wheatfen, 28 June 2001.

10   Laski stated in the High Court that at the conclusion of a Newark market place speech, with the crowd beginning to disperse: 'At that point a gentleman wholly unknown to me, clad in what I believe to be plus fours, strode a little forward, being about 40 to 50 yards from the platform on my left, and began asking me, in a very challenging voice, certain questions' (Daily Express 1947: 64). Day accused Laski of supporting revolution by violence, and Laski unsuccessfully sued the *Newark Advertiser, Daily Express* and others over their reporting. Day recounted the incident, and his 1945 Nottinghamshire election work as Conservative 'publicity adviser' (228), in *Harvest Adventure*, castigating the 'condescending' and 'urban-minded' intellectual Laski, with hostile allusions to his Jewishness, the 'British idea of freedom' being 'not in his blood' (Day 1946: 330–4).

11   Day here anticipates the prominence of gambling in a later Right cultural politics pursued by figures such as James Goldsmith and John Aspinall, another nature enthusiast.

12   A rhyming recipe accompanied each fattened bird: 'Take three pounds of beef, beat fine in a mortar, / Put it into the Swan – that is, when you've caught her' (Dutt 1906: 262). The same rhyme appeared in 1890 in Stevenson's account of attending the 1871 upping with Southwell, with attribution to Southwell's relative Rev JC Matchett, 40 years earlier (Stevenson 1890: 88–111). The rhyme also made the 1904 *New York Times*, in Laura Starr's piece of picturesque Englishry on 'Annual Upping in an Old English Cathedral Town' (Starr 1904).

13   The Norfolk Naturalists Trust, founded in 1926, took over the committee from 1933. The NNT and NNNS were closely linked; NNT founder Sydney Long was NNNS Secretary for 24 years, and Honorary Secretary of the Norfolk Wild Birds Protection Committee from its foundation (George 1992; George 2000b).

14   William Grenfell (Lord Desborough), ennobled in 1905, shared a maternal grandfather with Lady Lucas. Desborough's two other sons, including war poet Julian Grenfell, were killed in action in 1915 (Mosley 1976).

15   Form of Recommendation for a Fellow of the Linnean Society of London; Linnean Society archive, London.

16   Letter, Duchess of Bedford–Linnean Society, 11 February 1935; Linnean Society archive, London.

17   'Breydon in 1755' is a chapter in *The Cruise of the 'Walrus'* describing a month in May, 'In Norfolk Bird Haunts in A.D. 1755' is a 100 page collection of *Norfolk Chronicle* articles detailing an autumn visit. Both are illustrated with sketches of figures in eighteenth century costume.

18   The NNT had 786 members in 1952 (the NNNS had 461 in 1953), 900 in 1962, 4104 in 1972, 7170 in 1982 and 11,708 in 1990. Russell Colman was President 1926–45,

Sir Henry Upcher 1945–54, Colin McLean 1955–62, followed by Timothy Colman. Royal patronage was granted in January 1945.

19 Norfolk Naturalists Trust Council Minute Book 1 1944–1952, 14 June 1950; Norfolk Wildlife Trust archive, Norwich.

20 The 'marsh cowboy' phraseology appeared again in shooting disputes at Rockland Broad in 1964 (EDP 1964f; 1964g).

21 Norfolk Naturalists Trust Council Minute Book 1 1944–1952, 2 February 1944; Norfolk Wildlife Trust archive, Norwich.

22 Norfolk Naturalists Trust Council Minute Book 1 1944–1952, 23 March/20 April 1945; Norfolk Wildlife Trust archive, Norwich.

23 Norfolk Naturalists Trust Council Minute Book 1 1944–1952, 16 October 1948; NNT Annual report for 1948; Norfolk Wildlife Trust archive, Norwich.

24 Norfolk Naturalists Trust Council Minute Book 1 1944–1952, 9 March 1945; Hickling Endowment 1945 file; Norfolk Wildlife Trust archive, Norwich.

25 Norfolk Naturalists Trust Council Minute Book 1 1944–1952, 5 July 1952; Norfolk Wildlife Trust archive, Norwich.

26 Whiteslea Estate 1950/1989 file, undated, likely late 1951; Norfolk Wildlife Trust archive, Norwich.

27 Water Skiing, Hickling 1959–60 file; Hickling Management Committee file, 18 October 1963; Norfolk Wildlife Trust archive, Norwich. The NNT wrote to all boathouse tenants to rally support against water skiing; these included Piggin's son, who owned two fast motor-boats.

28 Hickling Management Committee file, 23 October 1969 / 26 August 1970; Norfolk Wildlife Trust archive, Norwich.

29 Letter, G Carmichael Low (acting RSPB Chair)–Constance Gay (NNT Secretary), 4 July 1946, Hickling Appeal 1945 file; Norfolk Wildlife Trust archive, Norwich.

30 *Broadland Summer* and *Broadland Winter*, JJB Productions, 1965, available online through the East Anglian Film Archive, Norwich; www.eafa.org.uk (accessed January 2014).

31 Norfolk Naturalists Trust Council Minute Book 2 1954–1962, 25 February 1958; Norfolk Wildlife Trust archive, Norwich.

32 Whiteslea Estate 1950/1989 file; Norfolk Wildlife Trust archive, Norwich.

33 Letter, Constance Gay–Aubrey Buxton, 16 January 1958, Whiteslea Estate 1950/1989 file; Norfolk Wildlife Trust archive, Norwich.

34 Letter, Aubrey Buxton–Constance Gay, 20 January 1958, Whiteslea Estate 1950/1989 file; Norfolk Wildlife Trust archive, Norwich.

35 Norfolk Naturalists Trust Council Minute Book 2 1954–1962, 22 March 1958; Norfolk Wildlife Trust archive, Norwich.

36 Whittle takes the name from that of an actual sixteenth century Norwich Cathedral composer: 'it's fairly probable that Osbert Parsley of Irstead is a descendant of his' (Whittle 1955: 27; Payne 2004).

37 *Woman's Hour* (East Anglian Edition), BBC Light Programme, 6 November 1959, 14.00–15.00; BBC Written Archives, Caversham Park, Reading.

38 'A Story About the Norfolk Broads', by 'BB', *Country Schools*, BBC Home Service, 21 November 1955, 13.40–14.00; BBC Written Archives, Caversham Park, Reading.

39 'Broadlands Journey', *Children's Hour*, BBC Home Service, 28 June 1952, 17.15–17.55; BBC Written Archives, Caversham Park, Reading.

40 Norwich City football club then played in the lowest tier of the Football League, the Third Division (South). Curator Rainbird Clarke had wanted Ellis to concentrate on museum

work, dropping 'spare-time responsibilities' and 'any future broadcasting'. Ellis told NNNS Secretary Gay he was minded to 'cut adrift from the museum and face the prospect of free-lancing': 'I must warn you that Mr. Clarke has instructed me not to answer the telephone if you ring me up at the museum'; letter, EA Ellis–Constance Gay, 19 March 1956, Box 65, EA Ellis Archive, Castle Museum, Norwich.

41 'Night Excursions and Alarms on the Broads', *Through East Anglian Eyes*, BBC Midland Home Service, 23 July 1957, 18.50–19.00; BBC Written Archives, Caversham Park, Reading.

42 Letter, EA Ellis–Philippa Pearce, 28 March 1946, Box 65, EA Ellis Archive, Castle Museum, Norwich.

43 Letter, EA Ellis–Philippa Pearce, 2 April 1946, Box 65, EA Ellis Archive, Castle Museum, Norwich.

44 *Norfolk Broads*, Rural Schools, BBC Home Service, 15 May 1946, 1.50-2.10; BBC Written Archives, Caversham Park, Reading.

45 I am grateful to Peter Marren for providing details on the production of *The Broads* further to those included in his book on *The New Naturalists* (Marren 1995).

46 Clifford and Rosemary Ellis's draft dust jackets included a grebe's head and a windmill reflected in rippling water (Marren 1995: plate 13).

47 Norfolk Naturalists Trust Council Minute Book 1 1944–1952, 5 May 1949; Norfolk Wildlife Trust archive, Norwich.

48 *The Parliamentary Debates* (Hansard) Fifth Series, vol. 444, 1952–1960.

49 Minutes of General Meeting, 18 January 1912, Linnean Society archive, London.

50 Walter Higham, *The Bittern*, made as part of the 'Secrets of Nature' series, and available on the 2011 BFI DVD collection *Secrets of Nature: Pioneering Natural History Films*; the film, lasting nine minutes, is likely to have been shot at either Horsey (where Higham also filmed post-war) or Hickling.

51 *Hickling Broad*, BBC Midland Home Service, 9 May 1947, 21.30–22.00; BBC Written Archives, Caversham Park, Reading.

52 *Norfolk Broads*, Rural Schools, BBC Home Service, 15 May 1946, 1.50–2.10; BBC Written Archives, Caversham Park, Reading.

53 '*The Naturalist*' in East Anglia, BBC Home Service, 24 May 1959, 13.10–13.40; BBC Written Archives, Caversham Park, Reading.

54 *Saturday Review*, BBC London Calling Area, 29 November 1952, 13.25–13.45; BBC Written Archives, Caversham Park, Reading.

55 Detail on the Norfolk Non-native Species Initiative, within the Norfolk Biodiversity Partnership, can be found at www.norfolkbiodiversity.org (accessed July 2012). In October 2011 the Broads Authority published a substantial Biodiversity Audit and Tolerance Sensitivity Mapping for the Broads, available at www.broads-authority.gov.uk/managing/broads-biodiversity-action-plan.html (accessed July 2012).

56 Norfolk Naturalists Trust Council Minute Book 2 1954–1962, 5 March 1955, 17 December 1955; Norfolk Wildlife Trust archive, Norwich.

57 *Broadland Winter*, JJB Productions, 1965, available online through the East Anglian Film Archive, Norwich; www.eafa.org.uk (accessed January 2014).

58 'Night Excursions and Alarms on the Broads', *Through East Anglian Eyes*, BBC Midland Home Service, 23 July 1957, 18.50–19.00; BBC Written Archives, Caversham Park, Reading.

59 Norfolk Naturalists Trust Council Minute Book 1 1944–1952, 6 December 1952; Norfolk Wildlife Trust archive, Norwich.

Chapter Five: Plant Landscapes

1   Purdy observed with Walter Rye, who WG Clarke would recall having 'stayed out night after night with the late R. J.W. Purdy to see the luminous owl' (TNNNS 1929: 723).

2   A 1916 *TNNNS* obituary also notes that CH Martin of Abergavenny, killed in action 3 May 1915, 'was for a short time in charge of the Gurney Laboratory at Sutton Broad', before moving to Glasgow University. Like Robert Gurney, Martin attended Eton and Oxford, studying zoology, and had worked at Naples Zoological Station, where Eustace had also worked (TNNNS 1916).

3   *Eastern Daily Press* cutting, 8 August 1931; Item 91, Robert Gurney archive, Castle Museum, Norwich.

4   Item 1.990.1.019, 'Fauna and Flora of Norfolk (Norfolk Room)', EA Ellis Archive, Castle Museum, Norwich.

5   Robert Gurney provided a counter-view, a Netherlands visit with Sydney Long in 1920 showing *Azolla* as 'a troublesome pest' in ditches and canals (Gurney 1923: 371).

6   Robert Gurney diary, 31 December 1928; Item 91, Robert Gurney archive, Castle Museum, Norwich.

7   I am grateful to Rob Driscoll for this information.

8   Letter, 15 August 1911, Letter No. 6, 'E.S.C. – Family – European Trip 1911'; Edith S. and Frederic E. Clements Collection, Acc. 1678, Box 21, folder 8, University of Wyoming American Heritage Center. 'Fred' is Edith's ecologist husband Frederic Clements.

9   Letter, 15 August 1911, Letter No. 6, 'E.S.C. – Family – European Trip 1911'; Edith S. and Frederic E. Clements Collection, Acc. 1678, Box 21, folder 8, University of Wyoming American Heritage Center.

10   BVC Preliminary Meeting, 3 December 1904, item 6, 'Local Workers'; BVC 1904–1912 minute book, BES/MB/2, British Ecological Society Archive, London.

11   I am grateful to Rob Driscoll for this information.

12   Interview with Joyce Lambert, 27 June 2001.

13   The biographies of Ted Ellis by Eugene Stone, and Phyllis Ellis by Pete Kelley, are by the same person; Stone was a pseudonym used for the writing of the 1988 EA Ellis biography, with the 2011 biography of Phyllis Ellis published under Kelley's own name.

14   EA Ellis, form of recommendation for a Fellow of the Linnean Society of London; Linnean Society archive, London.

15   'Night Excursions and Alarms on the Broads', *Through East Anglian Eyes*, BBC Midland Home Service, 23 July 1957, 18.50–19.00; BBC Written Archives, Caversham Park, Reading. Also Ellis 1952f: 206; 1965: 96.

16   'Probing the Undergrowth', *Through East Anglian Eyes*, BBC Midland Home Service, 6 August 1957, 18.50–19.00; BBC Written Archives, Caversham Park, Reading.

17   From Joyce Lambert, 'Broadland', in letter, Joyce Lambert–David Matless, 27 November 2000. This is the opening verse of a seven verse poem. Lambert also published a poem, 'A Tale of Two Boats', on Puffin and Mudlark, a dinghy and punt she owned 50 years ago, in 2000 in the Broads Society magazine *The Harnser* (Lambert 2000).

18   Norfolk Naturalists Trust annual report 1948; Norfolk Naturalists Trust Council Minute Book 1 1944–1952, 16 October 1948; Norfolk Wildlife Trust archive, Norwich.

19   Norfolk Naturalists Trust Council Minute Book 1 1944–1952, 5 July 1952; Norfolk Wildlife Trust archive, Norwich.

20   Letter, Eric Duffey–Paul Merchant, 27 October 2002.

21 The Nature Conservancy also encouraged Broadland visits with one of its contributions to European Conservation Year in 1970, *A Country Drive for Motorists in the Norwich District*, which took in Ranworth church, Horning ferry and Wroxham Broad, explaining conservation management and ecology along the way (Nature Conservancy 1970).

22 Interview with Humphrey Boardman, Norwich, 1988.

23 Ellis recorded a similar excursion in June 1954 (Ellis 1954). Humphrey Boardman was elected to the NNNS in 1939, living by then at Horning.

24 Botanical ornament was paralleled in Boardman's converted wherry Hathor, designed for the Colman sisters, named for an Egyptian god and decorated in Egyptian style (and later owned by Claud Hamilton of Hamilton's Guides), a vernacular vessel gone exotic (EDP 1954a; Bower 1989).

25 Details are given on the Trust's website at howhilltrust.org.uk (accessed July 2012).

26 Likewise it leads to a 2010 bottle of 'Reedcutter' ale from Humpty Dumpty Brewery of Reedham, its label design with a cutter working by a stack and mill.

27 *Inside Out*, BBC East, broadcast Wednesday 18 February 2009, 7.30 pm, BBC1.

28 *Points of View: Capturing the 19ᵗʰ Century in Photographs*, British Library exhibition, 30 October 2009–7 March 2010.

29 In the 1938 *Norfolk* volume of the interwar Land Utilisation Survey of Britain John Mosby had recorded 20 'hardy and independent' Broadland marshmen, most having inherited their job (Mosby 1938a: 196).

30 The same photograph had however appeared, attributed to Emerson, in the 1957 *Norfolk Shell Guide* (Harrod and Linnell 1957: 7).

31 The quotation is from Evans' 1966 book *The Pattern Under the Plough* (Evans 1966: 17).

32 Quotations from exhibition captions at Justin Partyka, *The East Anglians*, Sainsbury Centre for Visual Arts, University of East Anglia, Norwich, 29 September–13 December 2009. Partyka was born in Norfolk in 1972, becoming a photographer and folklorist in Canada.

33 Justin Partyka, preface to exhibition brochure for *The East Anglians*.

34 Partyka also provided photographs for an updated 2011 edition of Mary Chamberlain's 1975 Fenland oral and social history *Fenwomen* (Chamberlain 1975; 2011).

35 Letter, Marietta Pallis–Joan Wake, 22 June 1935; Box 36, Joan Wake Collection, Northamptonshire Record Office.

36 Letter, Marietta Pallis–Joan Wake, 7 June 1935; Box 36, Joan Wake Collection, Northamptonshire Record Office.

37 Letter, Marietta Pallis–Catherine Gurney, 7 September 1955, Pallis/Vlasto Archive, Hickling.

38 Arthur Rimbaud, last verse of 'Les Amis', from 'Comédie de la Soif', in *Derniers Vers* (1872) (Bernard 1962: 210).

39 Letter, Marietta Pallis–EA Ellis, 14 July 1960; found in 2002 by David Nobbs, Warden of the Ted Ellis Trust Nature Reserve, in the Wheatfen apple store.

40 Interview with Phyllis Ellis, 28 June 2001.

41 Minutes of the Smallburgh Internal Drainage Board, 7 January 1941; King's Lynn Consortium of Internal Drainage Boards archive, King's Lynn.

42 Minutes of the Smallburgh Internal Drainage Board, 28 November 1941; King's Lynn Consortium of Internal Drainage Boards archive, King's Lynn.

43 Minutes of the Smallburgh Internal Drainage Board, 10 April 1942; King's Lynn Consortium of Internal Drainage Boards archive, King's Lynn.

Icon II: Windmill

1   Wailes undertook a post-war survey for the Society for the Protection of Ancient Buildings, having been honorary technical adviser to their Windmill Section since its 1931 formation.
2   A new Norfolk Mills and Pumps Trust was formed in 1992, including nominees from the BA and County Council; the old Windmills Trust was wound up in 1995, but the name was taken on by the new Trust for public continuity.
3   Nicholas Ward RE has produced notable etchings of Norfolk landscape, with a particular interest in ramshackle, ephemeral and workaday features; a selection of work can be seen on the website of the Bircham Gallery, Holt: www.birchamgallery.co.uk (accessed November 2013). I am grateful to him for discussing his Somerton drawings with me.
4   Ecotricity website: www.ecotricity.co.uk/our-green-energy/our-green-electricity/from-the-wind/wind-parks-gallery/somerton-norfolk (accessed July 2012). The ten Blood Hills turbines were removed in early 2014, to be replaced by two new, larger turbines.

Chapter Six: The Ends of Landscape

1   EA Ellis, 'The Norfolk Broads', five minute talk delivered on *Woman's Hour*, BBC Light Programme, 11 May 1964, 14.00–15.00; BBC Written Archives, Caversham Park, Reading. See also Ellis 1961; 1962.
2   Hickling Management Committee minutes, 14 May 1962; Norfolk Wildlife Trust archive, Norwich. The Committee included Ellis and Martin George.
3   *No Lullaby for Broadland*, Anglia Television, 1979, viewed at the East Anglian Film Archive, Norwich.
4   While *No Lullaby for Broadland* was not shown until 20 November, it had been scheduled for broadcast on 22 August, but postponed due to industrial action; Breydon Friends of the Earth would appear to have seen the film in advance of broadcast, or at least have been aware of its content.
5   The article was 'based partly on information from John Barkham, David Dent, Richard Hey, Timothy O'Riordan, and Kerry Turner of the School of Environmental Sciences, University of East Anglia'. The acknowledgement to UEA staff was omitted from the published article, and appeared with an apology in the 15 January 1981 issue of *New Scientist*, 170.
6   Letter, Marietta Pallis–Joyce Lambert, 11 October 1953, Pallis/Vlasto Archive, Hickling.
7   Photographs of Eccles church tower standing on the beach in 1887, and its low remains in 1938, are shown in Mosby 1939: facing 415.
8   Collins 1995: 248.
9   Algernon Swinburne, 'Evening on the Broads', from *Studies in Song* (1880); Swinburne 1927: 473–81.
10 PH Emerson, 'A Study in Gold and Blue'; Emerson 1898b: 54.
11 Davies 1891: 106.
12 Turner 1924: 96.
13 EHT Jukes, 'The Cruise of the Scamp', Norfolk Heritage Centre, Millennium Library, Norwich; from a diary passage set at Martham, May 1935.
14 This was the 1938 issue of *TNNNS*, though not published until March 1939. Buxton also noted Vincent's role, raising banks to secure Hickling and writing the effort up for *Country Life* (Buxton 1938a: 351–2/359).

15   *The Horsey Mail* is included on the 2009 British Film Institute DVD set *We Live in Two Worlds*, volume two of their GPO Film Unit Collection, BFIVD759, and is also available via the BFI's Screenonline website: www.screenonline.org.uk (accessed January 2014).

16   The spelling of O'Brian and Simmonds' names varies; *Post Office Magazine* coverage of the flood has them as RJ O'Brien and Claude Simmons; I have followed the spelling on the film credits here. One other source might suggest Simmons rather than Simmonds; a website describing the building of military defences at nearby Winterton in World War II includes a photograph of the local Home Guard c.1943, listing Claude Simmons, with his home village as Somerton: http://www.pillbox-study-group.org.uk/index.php/defence-articles/building-winterton-on-seas-defences/ (accessed January 2014).

17   Happisburgh to Winterton Internal Drainage Board Minute Book, 1943–1967: 200; King's Lynn Consortium of Internal Drainage Boards archive, King's Lynn.

18   The 1962 edition of Steers' *The Sea Coast* would include an appendix on 'The Storm of 1953 and its Aftermath' (Steers 1962).

19   Conversation with Brian Moss, Nottingham, 1 October 2009.

Chapter Seven: Concluding

1   'Bibliography', including its title, is formed of terms from regional book titles which also include the words Broads, Broadland or Broad, found in this book's References.

2   If similarities and links are found across the North Sea, however, inland to the west of Norfolk brings contrast between the Broads and the Fens, the flat East Anglian agricultural land which includes that district of Lincolnshire named 'Holland'. Broadland may echo the Netherlands, but it departs from Holland, Lincs.

3   Richard Denyer, *Neither Land nor Water*, photographic exhibition, 10 February–11 March 2012, Norwich Arts Centre.

4   Letter, Marietta Pallis–Joan Strachey, Box X85, Joan Wake Collection, Northamptonshire Record Office. The quotation is from the page of the draft text numbered 125–6.

References

Adams, M (1914) *In the Footsteps of Borrow and Fitzgerald*. London: Jarrold.

Adderson, R and Kenworthy, G (2007) *Melton Constable to Yarmouth Beach*. Midhurst: Middleton Press.

Adderson, R and Kenworthy, G (2010) *Branch Lines East of Norwich: The Wherry Lines*. Midhurst: Middleton Press.

Adeney, M (2010) Aitken Clark (Obituary). *Guardian*, 20 April.

Agricultural Development and Advisory Service (1984) *ADAS Annual Report 1983*. London: HMSO.

Alberti, S (2001) Amateurs and Professionals in One County: Biology and Natural History in Late Victorian Yorkshire. *Journal of the History of Biology* 34: 115–147.

Alberti, S (2003) Conversaziones and the Experience of Science in Victorian England. *Journal of Victorian Culture* 8: 208–230.

Allen, HE (1923a) Ancient Grief. *Atlantic Monthly* 131: 177–187.

Allen, HE (1923b) A Glut of Fruit. *Atlantic Monthly* 132: 343–352.

Allen, HE (1927) The Fanciful Countryman. *The Dial* 83: 477–500.

Allen, HE and Meech, SB (eds) (1940) *The Book of Margery Kempe*. Oxford: Oxford University Press.

Allthorpe-Guyton, M (1977) *John Thirtle*. Norwich: Norfolk Museums Service.

Allthorpe-Guyton, M (1986) *Henry Bright 1810–1873*. Norwich: Norfolk Museums Service.

Alpers, S (1983) *The Art of Describing: Dutch Art in the Seventeenth Century*. London: John Murray.

Anderson, B and Harrison, P (eds) (2010) *Taking Place: Non-Representational Theories and Geography*. Farnham: Ashgate.

'Anonymously' (1960) Origin of the Broads. *Eastern Daily Press*, 18 March.

Apling, H (1984) *Norfolk Corn Windmills*. Norwich: Norfolk Windmills Trust.

Appleton, J (1975) *The Experience of Landscape*. Chichester: John Wiley & Sons, Ltd.

Appleton, J (1994) *How I Made the World: Shaping a View of Landscape*. Hull: University of Hull Press.

In the Nature of Landscape: Cultural Geography on the Norfolk Broads, First Edition. David Matless.
© 2014 David Matless. Published 2014 by John Wiley & Sons, Ltd.

Arrow, J (1930) *JC Squire v DH Lawrence: A Reply to Mr Squire's Article in 'The Observer' of March 9ᵗʰ, 1930*. London: Blue Moon Booklets.

Arrow, J (1932) *Young Man's Testament*. London: Putnam.

Arrow, J (1948) A Study in Neglect. *Norfolk Magazine*: July–September, 50–57.

Arrow, J (1949) The Broads as a National Park. *Architectural Review* 106 (632): 86–100.

Arrow, J (ed) (1951) *The Pleasures of Sailing: An Anthology*. London: Art and Technics.

Baker, JA (1967) *The Peregrine*. London: Collins.

Baker, JA (1970) *The Hill of Summer*. London: Collins.

Balfour-Browne, F (1905) A Study of the Aquatic Coleoptera and their Surroundings in the Norfolk Broads District (Part 1). *Transactions of the Norfolk and Norwich Naturalists' Society* 8 (1): 58–82.

Balfour-Browne, F (1906) A Study of the Aquatic Coleoptera and their Surroundings in the Norfolk Broads District (Part 2). *Transactions of the Norfolk and Norwich Naturalists' Society* 8 (2): 290–306.

Balfour-Browne, F (1940–58) *British Water Beetles* (Three Volumes). London: Ray Society.

Barclay, FH (1909) *Azolla Caroliniana*. *Transactions of the Norfolk and Norwich Naturalists' Society* 8 (5): 856–858.

Barker, D (2004) Smethurst, Allan Francis. *Oxford Dictionary of National Biography*.

Barrell, J (1982) Geographies of Hardy's Wessex. *Journal of Historical Geography* 8: 347–361.

Barthes, R (1972) *Mythologies*. London: Jonathan Cape (first published 1957).

The Battle of the Floods (1953). Amsterdam: Netherlands Booksellers and Publishers Association.

Beach, Thomas W (1907) Three Views of East Anglia. *Times Literary Supplement*: 31 October, 331.

Beadle, J (2008) A Pleasing Effect upon the Mind: James Stark's *Scenery of the Rivers of Norfolk*. *Norfolk Archaeology* 45: 277–292.

Bearman, CJ (1999) Kate Lee and the Foundation of the Folk-Song Society. *Folk Music Journal* 7: 627–643.

Beckett, J and Watkins, C (2011) Natural History and Local History in Late Victorian and Edwardian England: The Contribution of the Victoria County History. *Rural History* 22: 59–87.

Bellamy, D and Quayle, B (1990) *Wetlands: An Exploration of the Lost Wilderness of East Anglia*. London: Sidgwick and Jackson.

Benham, H, Boardman, H, Clark, GR, Ffiske, GE, Glendenning, SE, Miller, M, Morrison, V, Perks, CGH, Pollen, WMH, Poyser, FC, Read, HL and Storey L (1949) The Norfolk Wherry (Letter). *Eastern Daily Press*, 14 January.

Bennett, A (1910) *Naias Marina*, L., and *Chara Stelligera*, Bauer, as Norfolk Plants. *Transactions of the Norfolk and Norwich Naturalists' Society* 9 (1): 47–50.

Bennett, A (1916) Notes on Mr. Nicholson's Flora of Norfolk. *Transactions of the Norfolk and Norwich Naturalists' Society* 10 (2): 126–137.

Bernard, O (ed) (1962) *Rimbaud*. Harmondsworth: Penguin.

Betjeman, J (1974) *Collected Poems*. London: John Murray.

Betts, AR (1956) The Commercialised Broads. *Eastern Daily Press*, 12 May.

Binyon, L (1921) Appreciation of John Crome, in Castle Museum, *Souvenir Catalogue of the Crome Centenary Exhibition*. Norwich: Castle Museum, 7–19.

Bird, MCH (1913) Attempted Acclimatisation of Wild Rice (*Zizania Aquatica*) in East Norfolk, *Transactions of the Norfolk and Norwich Naturalists' Society* 9 (4): 603–606.

Bird, MCH (1922) The Drought of 1921. *Transactions of the Norfolk and Norwich Naturalists' Society* 11 (3): 241–245.

Blackbourn, D (2006) *The Conquest of Nature: Water, Landscape and the Making of Modern Germany*. London: Jonathan Cape.

Blake, H (1908) *Catalogue of Yachts, Wherries and Boats*. London: Blake.

Blake, H (1916) *Yachting List*. London: Blake.

Blake, P, Bull, J, Cartwright, A and Fitch, A (1958) *The Norfolk We Live In*. Norwich: Jarrold.

Blake, P, Bull, J, Cartwright, A and Fitch, A (1974) *the Norfolk We Live In*. Norwich: George Nobbs.

Blakes (1922) *Broad Smiles. Or How Not to Do a Norfolk Broads Holiday*. London: Blakes.

Blayney Brown, D, Hemingway, A and Lyles, A (2000) *Romantic Landscape: The Norwich School of Painters*. London: Tate Gallery.

Blyth, J (1903) *Juicy Joe: A Romance of the Norfolk Marshlands*. London: Grant Richards.

Boardman, C (1956) A Broads Society. *Eastern Daily Press*, 8 May.

Boardman, ET (1939) The Development of a Broadland Estate at How Hill, Ludham, Norfolk. *Transactions of the Norfolk and Norwich Naturalists' Society* 15 (1): 5–21.

Boardman, H (1926) The Bittern at its Nest. *Transactions of the Norfolk and Norwich Naturalists' Society* 12 (2): 203–206.

Boardman, H (1933) Reed Thatching in Norfolk. *The Architect's Journal* 26 April: 563–567.

Borrow, G (1924) *Lavengro*. London: T Nelson (first published 1851).

Bourdieu, P (1991) *Language and Symbolic Power*. Cambridge: Polity.

Bower, P (1989) *Hathor*. Norwich: Broads Authority.

Boyes, G (1993) *The Imagined Village: Culture, Ideology and the English Folk Revival*. Manchester: Manchester University Press.

Boyes, J and Russell, R (1977) *The Canals of Eastern England*. Newton Abbot: David and Charles.

Bracewell, M and Linder (2003) *I Know Where I'm Going: A Guide to Morecambe and Heysham*. London: BookWorks.

Brady, GS and Robertson, D (1870) The Ostraccoda and Foraminifera of Tidal Rivers. *The Annals and Magazine of Natural History* 31: 1–33.

Brassley, P (2004) Industries in the Early Twentieth-Century Countryside: The Oxford Rural Industries Survey of 1926/7, in RW Hoyle (ed) *People, Landscape and Alternative Agriculture: Essays for Joan Thirsk*. London: British Agricultural History Society, 133–148.

Brassley, P (2006) The Wheelwright, the Carpenter, Two Ladies from Oxford, and the Construction of Socio-Economic Change in the Countryside Between the Wars, in P Brassley, J Burchardt and L Thompson (eds) *The English Countryside Between the Wars: Regeneration or Decline?* Woodbridge: Boydell Press, 212–234.

Breydon Friends of the Earth (1979) *Eastern Mourning News*. Great Yarmouth: Breydon Friends of the Earth.

British Association for the Advancement of Science (1935) *General Excursions Saturday, Sept. 7th*. London: British Association for the Advancement of Science.

British Association for the Advancement of Science (1961a) *Norwich and Its Region*. Norwich: Norwich Local Executive Committee of the British Association.

British Association for the Advancement of Science (1961b) *Programme of the 123rd Annual Meeting*. Norwich: Norwich Local Executive Committee of the British Association.

British Association for the Advancement of Science (1961c) *Programme of Excursions*. Norwich: Norwich Local Executive Committee of the British Association.

Broads Authority (1982) *What Future for Broadland?* Norwich: Broads Authority.

Broads Authority (1987) *Cockshoot Broad: An Experiment in the Restoration of a Dead Broad* (Leaflet). Norwich: Broads Authority.

Broads Authority (1989) *The Broads … Last Enchanted Land (Leaflet)*. Norwich: Broads Authority.

Broads Authority (c.1990) *Broad Walks: Bure Valley*. Norwich: Broads Authority.

Broads Authority (1993) *No Easy Answers: Draft Broads Plan 1993*. Norwich: Broads Authority.

Broads Authority (1997) *Broad Sheet* (January). Norwich: Broads Authority.

Broads Authority (2005) *From Darkness to Light: The Restoration of Barton Broad.* Norwich: Broads Authority.

Broads Authority (2011) *Broads Plan 2011.* Norwich: Broads Authority.

Broads Conference (1947) *Report on the Preservation and Control of the Broads Area.* Norwich: Norfolk County Council.

Broads Consortium (1971) *Broadland Study and Plan.* Norwich: Norfolk County Council.

Broads Reed and Sedge Cutters Association (2009) *Marsh Workers of the Broads* (Booklet and three DVDs). Norwich: Broads Reed and Sedge Cutters Association.

Broadwood, L, Vaughan Williams, R and Gilchrist, AG (1910) Songs From Norfolk. *Journal of the Folk-Song Society* 4: 84–91.

Brogan, H (1984) *The Life of Arthur Ransome.* London: Jonathan Cape.

Browne, T (1902) *Notes and Letters on the Natural History of Norfolk More Especially on the Birds and Fishes* (ed T Southwell). London: Jarrold.

Bryan, PW (1933) *Man's Adaptation of Nature: Studies of the Cultural Landscape.* London: University of London Press.

Bulmer, J (1987) *The Norfolk Broads in Colour.* Norwich: Jarrold.

Bunting, B (1968) *Collected Poems.* London: Fulcrum Press.

Bunting, B (2009) *Briggflatts.* Tarset: Bloodaxe Books (with DVD of 1982 film by Peter Bell).

Burrell, WH (1914) *Azolla Filiculoides* Lam. *Transactions of the Norfolk and Norwich Naturalists' Society* 9 (5): 734–742.

Burton, H (1960) *The Great Gale.* Oxford: Oxford University Press.

Buxton, A (Anthony) (1916) Birds of the Western Front. *Transactions of the Norfolk and Norwich Naturalists' Society* 10 (2): 148–154.

Buxton, A (1920) *Sport in Peace and War.* London: Arthur L Humphreys.

Buxton, A (1921) Spring Birds At Geneva. *Transactions of the Norfolk and Norwich Naturalists' Society* 11 (2): 162–167.

Buxton, A (1932) *Sporting Interludes at Geneva.* London: Country Life.

Buxton, A (1933) Birds of the Broads. *South-Eastern Naturalist* 38: 84–86.

Buxton, A (1938a) General Effects of the Flood. *Transactions of the Norfolk and Norwich Naturalists' Society* 14 (4): 349–373.

Buxton, A (1938b) The Norfolk Floods (Letter). *The Times*, 24 February.

Buxton, A (1939a) The Frost of January–February, 1940. *Transactions of the Norfolk and Norwich Naturalists' Society* 15 (1): 102–105.

Buxton, A (1939b) General Effects of the Flood Seen in 1939. *Transactions of the Norfolk and Norwich Naturalists' Society* 15 (1): 22–40.

Buxton, A (1940) General Effects of the Flood Seen in 1940. *Transactions of the Norfolk and Norwich Naturalists' Society* 15 (2): 150–159.

Buxton, A (1941) General Effects of the February 1938 Flood Seen in 1941. *Transactions of the Norfolk and Norwich Naturalists' Society* 15 (3): 259–267.

Buxton, A (1942) General Effects of the February 1938 Flood Seen in 1942. *Transactions of the Norfolk and Norwich Naturalists' Society* 15 (4): 332–341.

Buxton, A (1943) General Effects of the February, 1938 Flood Seen in 1943. *Transactions of the Norfolk and Norwich Naturalists' Society* 15 (5): 410–419.

Buxton, A (1944) A Man of the Broads. *Country Life*, 24 November: 908.

Buxton, A (1946) *Fisherman Naturalist.* London: Collins.

Buxton, A (1948) *Travelling Naturalist.* London: Collins.

Buxton, A (1950) *Happy Year: The Days of a Fisherman-Naturalist.* London: Collins.

Buxton, A (1953) Effects of Sea Flooding on Wild Life. *Eastern Daily Press*, 12 November.

Buxton, A (1955a) Horsey Warren (Letter). *Eastern Daily Press*, 3 August.

Buxton, A (1955b) Sea Defence by Planting. *Eastern Daily Press*, 27 July.

Buxton, A, Hornor, B and Mann, JC (1938) *East Norfolk Floods*. Norwich: Eastern Daily Press (Pamphlet from articles published 2–3 December 1938).

Buxton, A (Aubrey) (1955) *The King in His Country*. London: Longmans, Green.

Buxton, J and Durdin, C (2011) *The Norfolk Cranes' Story*. Sheringham: Wren.

Cable, T (1991) *Broadland Tom: The Trials of a Norfolk Water Bailiff 1952–1976*. Wymondham: Reeve.

Cage, J (1968) *Silence*. London: Calder and Boyars.

Calman, WT (1950) Dr Robert Gurney (Obituary). *Nature* 165 (4198): 587–588, 15 April.

Cameron, L (1997) *Openings: A Meditation on History, Method, and Sumas Lake*. Kingston: McGill-Queen's University Press.

Cameron, L and Matless, D (2003) Benign Ecology: Marietta Pallis and the Floating Fen of the Delta of the Danube, 1912–1916. *Cultural Geographies* 10: 253–277.

Cameron, L and Matless, D (2011) Translocal Ecologies: The Norfolk Broads, the 'Natural,' and the International Phytogeographical Excursion, 1911. *Journal of the History of Biology* 44: 15–41.

Campling, CA (1871) *The Log of 'The Stranger': A Cruise on the Broads of Norfolk*. Beccles: William Moore.

Carill-Worsley, PET (1931) A Fur Farm in Norfolk. *Transactions of the Norfolk and Norwich Naturalists' Society* 13 (1): 105–115.

Castell, J (1960) Origin of the Broads. *Eastern Daily Press*, 23 March.

Castle Museum (1949) *Guide to the Norwich Museums*. Norwich: Norwich Museums Committee.

Castle Museum (1968) *Guide to the Castle Museum*. Norwich: Norwich Museums Committee.

Catling, G (1958) *The Gang on the Broads*. London: Ernest Benn.

Cator, HJ (1952) The Case for Coypu (Letter). *Eastern Daily Press*, 6 December.

Caulfield, C (1981) What Can We Save of the Broadlands? *New Scientist* 1 January: 28–31.

Chadwin, WT (1913) The Cantley Sugar Beet Factory. *Journal of the Board of Agriculture* 20: 569–582.

Chamberlain, M (1975) *Fenwomen: A Portrait of Women in an English Village*. London: Virago.

Chamberlain, M (2011) *Fenwomen: A Portrait of Women in an English Village*. Framlingham: Full Circle Editions.

Cherry, D (1993) *Painting Women: Victorian Women Artists*. London: Routledge.

Clark, MA (1989) Britain's Newest and Very Special National Park, in R Denyer, *Still Waters*. Norwich: Still Waters Press, 2–3.

Clark, R (1949) Norfolk Marsh Mills (Letter). *Eastern Daily Press*, 6 January.

Clark, R (1961) *Black-Sailed Traders: The Keels and Wherries of Norfolk and Suffolk*. Newton Abbot: David and Charles.

Clarke, D (2010) *The Broads in Print: The Days of Discovery: Early 1800s to 1920*. Norwich: Joy and David Clarke.

Clarke, RR (1960) The Broads and the Medieval Peat Industry. *Norfolk Archaeology* 32: 209–210.

Clarke, RR and Ellis EA (1955) Origin of the Broads. *Eastern Daily Press*, 14 April.

Clarke, RR, Ellis, EA and Lambert, JM (1955) Origin of the Broads. *Eastern Daily Press*, 26 April.

Clarke, WG (1894) A Breckland Ramble. *The Naturalists' Journal* 3: 90–92 / 105–107.

Clarke, WG (1918) The Natural History of Norfolk Commons. *Transactions of the Norfolk and Norwich Naturalists' Society* 10 (4): 293–318.

Clarke, WG (1921) *Norfolk and Suffolk*. London: A and C Black.

Clarke, WG (1923) A Novice on the Norfolk Broads. *Open Air* 1: 35–37.

Clarke, WG with Clarke, RR (1937) *In Breckland Wilds*. Cambridge: Heffer.

Clements, FE (1916) *Plant Succession*. Washington: Carnegie Institute of Washington.

Cleveland, D (2009) *Films Were Made*. Manningtree: David Cleveland.

Clout, H (2009) *Patronage and the Production of Geographical Knowledge in France: The Testimony of the First Hundred Regional Monographs, 1905–1966*. London: Historical Geography Research Group, Research Series Number 41.

Cocker, M (2007) *Crow Country*. London: Jonathan Cape.

Cocker, M (2008) Arthur Patterson: The Life of a Great Norfolk Naturalist. *Transactions of the Norfolk and Norwich Naturalists' Society* 41 (1): 1–8.

Cocker, M (2010) Introduction, in JA Baker *The Peregrine, the Hill of Summer and Diaries: The Complete Works*. London: Collins, 4–15.

Collett, A (1926) *The Changing Face of England*. London: Nisbet.

Collins, I (1990) *A Broad Canvas: Art in East Anglia Since 1880*. Norwich: Parke Sutton.

Collins, W (1995) *Armadale*. Harmondsworth: Penguin (first published 1866).

Colls, R (2002) *The Identity of England*. Oxford: Oxford University Press.

Colls, R (2007) The New Northumbrians, in R Colls (ed) *Northumbria: History and Identity 547–2000*. Chichester: Phillimore, 151–177.

Committee of Inquiry into Inland Waterways (1958) *Report of the Committee of Inquiry into Inland Waterways*. London: HMSO, Cmnd.486.

Conan Doyle, A (1988) The 'Gloria Scott', in *The Complete Sherlock Holmes*. New York: Doubleday: 373–385.

Cornell, D (2008) *A History of the River Thurne Bungalows*. Potter Heigham: River Thurne Tenants Association.

Cornford, B (1979) The Sea Breach Commission in East Norfolk 1609–1743. *Norfolk Archaeology* 37: 137–145.

Cornford, B (1982) Past Water Levels in Broadland. *Norfolk Research Committee Bulletin* 28: 14–19.

Cornish, V (1930) *National Parks, and the Heritage of Scenery*. London: Sifton Praed.

Cornish, V (1937a) *The Preservation of Our Scenery*. Cambridge: Cambridge University Press.

Cornish, V (1937b) *The Scenery of England* (Second Edition). London: Alexander Maclehose.

Cosgrove, D (1984) *Social Formation and Symbolic Landscape*. London: Croom Helm.

Cosgrove, D (1993) *The Palladian Landscape*. Leicester: Leicester University Press.

Cosgrove, D (2001) *Apollo's Eye*. Baltimore: Johns Hopkins University Press.

Cosgrove, D and Petts, G (eds) (1990) *Water, Engineering and Landscape*. London: Belhaven.

Cosgrove, I and Jackson, R (1972) *The Geography of Recreation and Leisure*. London: Hutchinson.

Court, AN (1970) *The Norfolk Broads*. Norwich: Jarrold.

Crompton, J (1870) Address. *Transactions of the Norfolk and Norwich Naturalists' Society* 1 (1): 3– 8.

Crowfoot, WM (1911) President's Address. *Transactions of the Norfolk and Norwich Naturalists' Society* 9 (2): 143–146.

Cultural Geographies (2009) Reflections on the Career of Denis Cosgrove 1948–2008. *Cultural Geographies* 16: 5–28.

Daily Express (1947) *The Laski Libel Action: Verbatim Report*. London: Daily Express.

Daley, PV (1935) *The Broadland Guide*. Norwich: Jarrold.

Daley, PV (1938) The Norfolk Floods: Their Effect on Broadland. *East Anglian Magazine* 3: 268–272.

Daley, PV (1949) *Broadland in Pictures*. Norwich: Jarrold.

Daniels, S (1989) Marxism, Culture and the Duplicity of Landscape, in R Peet and N Thrift (eds) *New Models in Geography, Vol. II*. London: Unwin Hyman, 196–220.

Daniels, S and Endfield, G (2009) Narratives of Climate Change: Introduction. *Journal of Historical Geography* 35: 215–222.

Daniels, S and Lorimer, H (2012) Until the End of Days: Narrating Landscape and Environment. *Cultural Geographies* 19: 3–9.

Daniels, S, Pearson, M and Roms, H (2010) Editorial. *Performance Research* 15: 1–4 (introduction to issue on 'Fieldworks').

Darby, HC (1962) The Problem of Geographical Description. *Transactions of the Institute of British Geographers* 30: 1–14.

Davenport-Hines, R (2008) *Ettie: The Intimate Life and Dauntless Spirit of Lady Desborough.* London: Weidenfeld and Nicolson.

Davidson, J, Bondi, L and Smith, M (eds) (2007) *Emotional Geographies.* Farnham: Ashgate.

Davies, GC (1873) *Mountain, Meadow and Mere.* London: Henry King.

Davies, GC (1876) *The Swan and Her Crew.* London: Frederick Warne.

Davies, GC (1882) *Handbook to the Rivers and Broads of Norfolk and Suffolk.* London: Jarrold.

Davies, GC (1883a) *The Scenery of the Broads and Rivers of Norfolk and Suffolk.* London: Jarrold.

Davies, GC (1883b) *Norfolk Broads and Rivers.* London: Blackwood.

Davies, GC (1891) *Handbook to the Rivers and Broads of Norfolk and Suffolk.* Norwich: Jarrold (18th Edition).

Davies, GC (1894) *Cruising in the Netherlands.* Norwich: Jarrold.

Davis, R (1956) The Coypu. *Agriculture* 63: 127–129.

Davis, R (1963) Feral Coypus in Britain. *Annals of Applied Biology* 51: 345–348.

Davis, S (2011) Militarised Natural History: Tales of the Avocet's Return to Postwar Britain. *Studies in the History and Philosophy of Biological and Biomedical Sciences* 42: 226–232.

Dawney, M (1976) George Butterworth's Folk Manuscripts. *Folk Music Journal* 3: 99–113.

Day, JW (1931) *Speed: The Authentic Life of Sir Malcolm Campbell.* London: Hutchinson.

Day, JW (1935) *King George V as a Sportsman: An Informal Study of the First Country Gentleman in Europe.* London: Cassell.

Day, JW (1943) *Farming Adventure.* London: Harrap.

Day, JW (1946) *Harvest Adventure.* London: Harrap.

Day, JW (1948a) *Gamblers' Gallery.* London: Background Books.

Day, JW (1948b) *Wild Wings and Some Footsteps.* London: Blandford Press.

Day, JW (1949) *Coastal Adventure.* London: Harrap.

Day, JW (1950) *Marshland Adventure.* London: Harrap.

Day, JW (1951) *Broadland Adventure.* London: Country Life.

Day, JW (1953) *Norwich and the Broads.* London: Batsford.

Day, JW (1957) *Poison on the Land: The War on Wild Life, and Some Remedies.* London: Eyre and Spottiswoode.

Day, JW (1958) *Lady Houston OBE: The Woman Who Won the War.* London: Allan Wingate.

Day, JW (1960) *British Animals of the Wild Places.* London: Blandford Press.

Day, JW (1967) *Portrait of the Broads.* London: Robert Hale.

Delyser, D (2005) *Ramona Memories: Tourism and the Shaping of Southern California.* Minneapolis: University of Minnesota Press.

Demangeon, A (1939) *The British Isles.* London: Heinemann.

Denton, FH (1956) Theory or Fact? *Eastern Daily Press*, 23 March.

Denyer, R (1989) *Still Waters.* Norwich: Still Waters Press.

Departmental Committee on Coastal Flooding (1954) *Report of the Departmental Committee on Coastal Flooding.* London: HMSO, Cmd.9165.

Derham, M (1952) *The Cruise of the Clipper.* London: Children's Special Service Mission.

Desilvey, C (2007) Salvage Memory: Constellating Material Histories on a Hardscrabble Homestead. *Cultural Geographies* 14: 401–424.

Desilvey, C (2012) Making Sense of Transience: An Anticipatory History. *Cultural Geographies* 19: 31–54.

Dicker, G (1998) Obituary: Lady Mayhew. *The Harnser* (October–December): 18–19.

Dixon, ES (1851) *The Dovecote and the Aviary*. London: WS Orr.

Dodd, AB (1896) *On the Broads*. London: Macmillan.

Dolar, M (2006) *A Voice and Nothing More*. Cambridge, Mass.: MIT Press.

Doughty, H (1889a) *On Friesland Meres*. Norwich: Jarrold.

Doughty, H (1889b) *Summer in Broadland*. Norwich: Jarrold.

Doughty, H (1891) *Our Wherry in Wendish Lands*. Norwich: Jarrold.

Dower, J (1945) *National Parks in England and Wales*. London: HMSO, Cmd.6628.

Dower, M (1965) *The Challenge of Leisure*. London: Civic Trust.

Doyle, B (1995) Urban Liberalism and the 'Lost Generation': Politics and Middle Class Culture in Norwich, 1900–1935. *The Historical Journal* 38: 617–634.

Driscoll, RJ (1982) The Dyke Vegetation at Oby and Thurne: A Comparison of Late 19th Century and Recent Records. *Transactions of the Norfolk and Norwich Naturalists' Society* 26 (1): 43–49.

Driscoll, RJ (1983) Broadland Dykes: The Loss of an Important Wildlife Habitat. *Transactions of the Norfolk and Norwich Naturalists' Society* 26 (3): 170–172.

Driscoll, RJ (1985) The Effect of Changes in Land Management on the Dyke Flora At Somerton and Winterton. *Transactions of the Norfolk and Norwich Naturalists' Society* 27 (1): 33–41.

Driscoll, RJ and Parmenter, JM (1994) Robert Gurney's 1908/1909 Vegetation Survey of Broadland. *Transactions of the Norfolk and Norwich Naturalists' Society* 30 (1): 71–88.

Du Maurier, G (1995) *Trilby*. Ware: Wordsworth (first published 1894).

Duffey, E (1958) *Dolomedes Plantarius*. *Transactions of the Norfolk and Norwich Naturalists' Society* 18 (7): 1–5.

Duffey, E (1964) The Norfolk Broads. A Regional Study of Wild Life Conservation in a Wetland Area with High Tourist Attraction, in *Proceedings of the MAR Conference*. International Union for the Conservation of Nature Publications: 290–301.

Dutt, WA (1901) *Highways and Byways in East Anglia*. London: Macmillan.

Dutt, WA (1903) *The Norfolk Broads*. London: Methuen.

Dutt, WA (1906) *Wild Life in East Anglia*. London: Methuen.

Dutt, WA (1907) *Some Literary Associations of East Anglia*. London: Methuen.

Dutt, WA (1923) *A Guide to the Norfolk Broads*. London: Methuen.

Dye, K, Fiszer, M and Allard, P (2009) *Birds New to Norfolk: The Accounts of their Discovery and Identification*. Sheringham: Wren.

Dyer, AER (1924) The Cruise of the Valentine. *Open Air* 2: 351–354.

Earwaker, J and Becker, K (1998) *Literary Norfolk*. Ipswich: Chapter 6 Publishing.

East Anglian Magazine (1936) *East Anglian Authors: No.2 – AG Miller ('Drew' Miller)*. 1: 514, June.

Eastern Daily Press (1938a) Horsey is Not Yet Deserted. 21 February.

Eastern Daily Press (1938b) Motorists in East Norfolk. 22 February.

Eastern Daily Press (1938c) More Water Through Horsey Breach. 3 March.

Eastern Daily Press (1938d) Post Office Magic. 25 October.

Eastern Daily Press (1938e) Post Office Film Week. 9 November.

Eastern Daily Press (1947a) Eccles Beach Erosion Brings Ruined Church to Light. 17 March.

Eastern Daily Press (1947b) Problems of Defending Norfolk Against Floods. 12 April.

Eastern Daily Press (1947c) Future of the Broads: East Anglian MPs Concerned. 26 November.

Eastern Daily Press (1947d) Sea Level (Editorial). 4 December.

Eastern Daily Press (1948a) River Access (Editorial). 10 January.

Eastern Daily Press (1948b) Problem of Ending River Bank Tenancies. 13 March.

Eastern Daily Press (1948c) Yare Barrage. 19 June.

Eastern Daily Press (1948d) Restoring the Broads / The Broads (Editorial). 10 July.

Eastern Daily Press (1948e) Rand Tenants. 9 October.

Eastern Daily Press (1949a) Cost of Restoring Broads as They Were in 1929 / Threat to the Broads (Editorial). 13 January.

Eastern Daily Press (1949b) Plea for Priority for the Broads. 2 April.

Eastern Daily Press (1949c) Action Over a Norfolk Broad. 27 April.

Eastern Daily Press (1949d) Action Over a Norfolk Broad. 7 May.

Eastern Daily Press (1949e) Public Access Lost to Broad / A Trust Abused (Editorial). 27 June

Eastern Daily Press (1949f) Mr Silkin Spends Day Cruising on Norfolk Waterways. 13 October.

Eastern Daily Press (1949g) Mr Silkin Afloat (Editorial). 13 October.

Eastern Daily Press (1950) Preserving Reed Thatching Craft. 11 January.

Eastern Daily Press (1951a) Exhibitions with East Anglian Background. 11 June.

Eastern Daily Press (1951b) Rate Appeal by Norfolk Naturalists Trust. 14 February.

Eastern Daily Press (1951c) Formation of Norfolk Broads Began About 500 B.C. 4 December.

Eastern Daily Press (1952a) Rockland Broad Again Open to River Traffic / Rockland Broad (Editorial). 6 August.

Eastern Daily Press (1952b) Surlingham Broad for Naturalists' Trust. 12 August.

Eastern Daily Press (1952d) Work of Broads Survey Vessel Shown on Television. 13 December.

Eastern Daily Press (1953a) Reed Harvest in Progress At Surlingham Broad. 3 January.

Eastern Daily Press (1953b) Floodwater Still Flows Through Breydon Gap. 4 February.

Eastern Daily Press (1953c) 300 Tons of Sandbags for 85-Yard Rockland Breach. 7 February.

Eastern Daily Press (1953d) Dutch Experience in Recovery of Land Flooded By Sea. 11 February.

Eastern Daily Press (1953e) Potter Heigham Bridge is to be Replaced. 2 March.

Eastern Daily Press (1953f) Potter Heigham (Editorial). 3 March.

Eastern Daily Press (1953g) Potter Heigham Bridge: Anglers' Advice to 'Floating Charabancs'. 17 March.

Eastern Daily Press (1953h) Naturalist Trust's Intention on Surlingham Broad Explained. 31 March.

Eastern Daily Press (1953i) Wherry Albion to Cease Trading: Heavy Loss. 2 May.

Eastern Daily Press (1953j) New Theory on Origin of the Norfolk Broads. 7 July.

Eastern Daily Press (1953k) The Broads (Editorial). 7 July.

Eastern Daily Press (1953l) Windmills (Editorial). 15 July.

Eastern Daily Press (1953m) Peat Cutting Theory on Broads Challenged by Norfolk Man. 22 August.

Eastern Daily Press (1953n) Origin of the Broads (Editorial). 22 August.

Eastern Daily Press (1953o) Windmills (Editorial). 12 October.

Eastern Daily Press (1954a) Winter Quarters. 6 March.

Eastern Daily Press (1954b) Girls Are Apple Graders – and Rush Mat Weavers. 9 March.

Eastern Daily Press (1954c) Evidence Produced in Support of Man-Made Broads Theory. 21 June.

Eastern Daily Press (1954d) R.D.C. Views on Second Line of Sea Defence. 1 December.

Eastern Daily Press (1955a) EA Ellis, Naturalist and Writer. 24 February.

Eastern Daily Press (1955b) Potter Heigham Will Have New Bridge: But Conference First / Potter Heigham (Editorial). 28 February.

Eastern Daily Press (1955c) The Origins of the Norfolk Broads. 12 April.

Eastern Daily Press (1955d) Bookshop and Workshop. 3 June.

Eastern Daily Press (1955e) Alarm at Horsey Over 'Gap' in Sea Defences. 5 July.

Eastern Daily Press (1955f) Yarmouth Barrage Could be Built in Two Years. 19 November.

Eastern Daily Press (1955g) Yachtsmen Oppose New Potter Heigham Bridge. 3 December.

Eastern Daily Press (1956a) E.A.E. (Editorial). 5 April.

Eastern Daily Press (1956b) Decision to Form Broads Society. 5 June.

Eastern Daily Press (1956c) The Broads Society (Advertisement). 29 June.

Eastern Daily Press (1956d) High Level Bridge for Haddiscoe Cut / The Cut Reprieved (Editorial). 26 November.

Eastern Daily Press (1957a) Tributes to 'John Knowlittle'. 21 October.

Eastern Daily Press (1957b) 'Broadside' on the Broads. 29 October.

Eastern Daily Press (1957c) Cheap Peeps. 29 October.

Eastern Daily Press (1958a) Loddon Back 'Within Framework of the Broads'. 22 May.

Eastern Daily Press (1958b) The Broads (Editorial). 24 May.

Eastern Daily Press (1958c) Ah! Wilderness! 28 June.

Eastern Daily Press (1958d) Coypu Country (Editorial). 11 August.

Eastern Daily Press (1959a) Winter Broads Tour. 6 January.

Eastern Daily Press (1959b) Sea Defence in East Anglia (Supplement), 19 February.

Eastern Daily Press (1959c) Coypu Threat to Farm Crops. 20 March.

Eastern Daily Press (1960a) Broads Petition Goes in Today. 7 January.

Eastern Daily Press (1960b) Mr Roberts Baden Gladden. 11 February.

Eastern Daily Press (1960c) Vice-Chancellor of University to be Chosen Soon. 1 October.

Eastern Daily Press (1960d) Hoseason's Panorama. 11 November.

Eastern Daily Press (1961a) Norfolk Broads Not to Become National Park. 2 August.

Eastern Daily Press (1961b) Brighter Science (Editorial). 11 August.

Eastern Daily Press (1962a) Broadland Council? (Editorial). 30 May.

Eastern Daily Press (1962b) Coypu Toll in Broads Area 100,000. 9 June.

Eastern Daily Press (1962c) Now Floating Housing Estates? 9 November.

Eastern Daily Press (1963a) Battle of Broads 'Getting Worse'. 14 March.

Eastern Daily Press (1963b) The Broads (Editorial). 15 March.

Eastern Daily Press (1963c) Report of Broads Conflict is 'Bilge' Says Oulton Man / Letting Agents Criticise Conservancy View. 15 March.

Eastern Daily Press (1963d) Extend Broads Says Naturalist. 18 March.

Eastern Daily Press (1963e) Plan to Stop 'Marsh Cowboys' Shooting Over Breydon Water. 30 September.

Eastern Daily Press (1963f) Broads Society Hits At Report. 29 October.

Eastern Daily Press (1963g) Buchanan Afloat. 13 December.

Eastern Daily Press (1963h) Broads Area Needs 'A New Form of Authority'. 16 December.

Eastern Daily Press (1963i) TVA for Broads? (Editorial). 17 December.

Eastern Daily Press (1964a) Broadland Conservation Talks. 28 January.

Eastern Daily Press (1964b) Broads Policy (Editorial). 29 January.

Eastern Daily Press (1964c) Squall on the Broads (Editorial). 17 February.

Eastern Daily Press (1964d) Broads Estate Makes £13,000. 8 June.

Eastern Daily Press (1964e) In Defence of Nature (Editorial). 3 December.

Eastern Daily Press (1964f) Rockland Broad: Peaceful Backwater or Just Another Boating Lake? 3 December.

Eastern Daily Press (1964g) Marsh Cowboys (Editorial). 4 December.

Eastern Daily Press (1965a) Search Goes on for Broads Area Youth Hostel. 12 January.

Eastern Daily Press (1965b) Blitz Starts on Coypu's Last Redoubt. 17 February.

Eastern Daily Press (1965c) Scientist Clears Boats as Pollution Suspects. 1 May.

Eastern Daily Press (1965d) New Light on Man-Made Broads Theory. 14 May.

Eastern Daily Press (1965e) Broads Claim Criticised. 15 May.

Eastern Daily Press (1965f) Broads Formed Naturally Says River Engineer. 29 April.

Eastern Daily Press (1965g) Speed on the Broads (Editorial). 29 April.

Eastern Daily Press (1965h) Broads Centre Plan Biggest Since War / An Acle Marina (Editorial). 18 May.

Eastern Daily Press (1965i) Nature Conservancy Urges Broadland Council / The Broads (Editorial). 3 July.

Eastern Daily Press (1965j) Future of Broadland (Letters) 6 July.

Eastern Daily Press (1965k) Broadland Report Welcomed. 16 July.

Eastern Daily Press (1965l) Commissioners Back Broadland Report. 24 July.

Eastern Daily Press (1965m) Yare's Biggest Holiday Plan for Thorpe 'Island'. 16 September.

Eastern Daily Press (1965n) £1 m Scheme for 'Tragic' Breydon. 16 September.

Eastern Daily Press (1965o) Yacht Centres (Editorial). 16 September.

Eastern Daily Press (1966) Coypu with Tiny Radio on its Body Aids Science. 5 February.

Eastern Daily Press (1967a) Revolutionary Reed Cutter Can Work on Land or Water. 18 February.

Eastern Daily Press (1967b) Cleansing the Broads (Editorial). 16 March.

Eastern Daily Press (1972) 'Survival' Film Handed to Trust. 17 May.

Eastern Daily Press (1975) Broads in Danger. 20 May.

Eastern Daily Press (1979) Chemical Cocktail. 20 November.

Eastern Daily Press (1984a) Halvergate Drainage is Halted by Agreement. 16 June.

Eastern Daily Press (1984b) Marsh Deal (Editorial). 18 June.

Eastern Daily Press (1988) The Broads: Dawn of a New Era. 29 March.

Eastern Daily Press (2006) Soap Star Roy Launched Verbal Tirade on Woman. 27 May.

Eastern Daily Press (2008a) Villagers Air Broads Flood Plan Anger. 9 April.

Eastern Daily Press (2008b) Broads Surrender 'Unacceptable'. 17 April.

Eastern Daily Press (2009) Victory for the People. 31 March.

Eastern Daily Press (2010) Anglers Vowing to Step Up Protests Over Broads Plan. 28 August.

Eastern Daily Press (2011a) Yare Flood Barrier Would Also Harvest Tidal Power. 2 February.

Eastern Daily Press (2011b) A Wherry Worthwhile Waterway Project. 5 February.

Eastern Daily Press (2011c) Anglers Threaten Blockade on River Thurne. 1 March.

Eastern Evening News (1951) Two Good Exhibitions. 12 June.

Eastern Evening News (1991) Fans Turn Out Nice Again for George. 10 June.

Eastern Evening News (1994) Rejection for Broads Wind Turbines Idea. 1 October.

Eastern Evening News (1996) Cash Bids May Mean Trouble at Mills. 24 February.

Ecclestone, AW and Ecclestone, JL (1959) *The Rise of Great Yarmouth: The Story of a Sandbank.* Great Yarmouth: AW Ecclestone.

Edwards, JK (1965) Communications and the Economic Development of Norwich 1750–1850. *Journal of Transport History* 7: 96–108.

Edwards, LA (1955) The Shrinking Broads. *Eastern Daily Press*, 5 August.

Edwards, P (1948) *Call Me At Dawn.* Ipswich: East Anglian Daily Times.

Edwards, WF (1984) The First Fifty Years. *Norfolk Research Committee Bulletin* 32: 1–3.

Ellis, AE (1926) *British Snails.* Oxford: Oxford University Press.

Ellis, AE (1941a) The Natural History of Wheatfen Broad Surlingham. Part IV: The Woodlice and Harvestmen. *Transactions of the Norfolk and Norwich Naturalists' Society* 15 (3): 291–300.

Ellis, AE (1941b) The Mollusca of a Norfolk Broad. *Journal of Conchology* 21: 224–243.

Ellis, EA (1934a) Flora of Norfolk: Rust Fungi (*Uredinales*). *Transactions of the Norfolk and Norwich Naturalists' Society* 13 (5): 489–505.

Ellis, EA (1934b) Wheatfen Broad, Surlingham. *Transactions of the Norfolk and Norwich Naturalists' Society* 13 (5): 422–451.

Ellis, EA (1938) Detailed Observations. *Transactions of the Norfolk and Norwich Naturalists' Society* 14 (4): 373–390.

Ellis, EA (1939) The Natural History of Wheatfen Broad Surlingham. Part II. *Transactions of the Norfolk and Norwich Naturalists' Society* 15 (1): 115–128.

Ellis, EA (1940) The Natural History of Wheatfen Broad Surlingham. Part III: The Micro-Fungi and their Hosts. *Transactions of the Norfolk and Norwich Naturalists' Society* 15 (2): 191–219.

Ellis, EA (1941) The Natural History of Wheatfen Broad Surlingham. Part V: Bugs (Hemiptera-Heteroptera). *Transactions of the Norfolk and Norwich Naturalists' Society* 15 (3): 301–306.

Ellis, EA (1948) Low Tide on the Broad. *Eastern Daily Press*, 11 January.

Ellis, EA (1949) The Broads as a Relict Marsh. *The New Naturalist* 6: 28–32.

Ellis, EA (1951) Norfolk Peat. *Eastern Daily Press*, 3 October.

Ellis, EA (1952a) Old Shore Lines. *Eastern Daily Press*, 11 January.

Ellis, EA (1952b) Surlingham Broad. *Eastern Daily Press*, 12 August.

Ellis, EA (1952c) Submerged Forest. *Eastern Daily Press*, 28 August.

Ellis, EA (1952d) Origin of the Broads. *Eastern Daily Press*, 29 October.

Ellis, EA (1952e) Land Drowning. *Eastern Daily Press*, 15 December.

Ellis, EA (1952f) Wild Life in Broadland, in RH Mottram, *The Broads*. London: Robert Hale, 185–217.

Ellis, EA (1953a) In the Gale's Wake. *Eastern Daily Press*, 2 February.

Ellis, EA (1953b) Lesson From Horsey. *Eastern Daily Press*, 3 February.

Ellis, EA (1953c) Is Britain Tilting? *Eastern Daily Press*, 24 February.

Ellis, EA (1954) The Ant Valley. *Eastern Daily Press*, 16 June.

Ellis, EA (1955a) Coast Protection. *Eastern Daily Press*, 27 May.

Ellis, EA (1955b) Broadland Bird Renowned for its Courage. *Eastern Daily Press*, 18 June.

Ellis, EA (1957) Arthur Henry Patterson. *Eastern Daily Press*, 21 October.

Ellis, EA (1958a) A Water Weed Mystery. *Eastern Daily Press*, 21 July.

Ellis, EA (1958b) Getting Out of Hand. *Eastern Daily Press*, 10 December.

Ellis, EA (1959) Bottle's Voyage in Broadland. *Eastern Daily Press*, 4 February.

Ellis, EA (1960a) The Coypu Threat in East Anglia. *Country Life*, 29 December: 1590–1591.

Ellis, EA (1960b) Spoiling the River. *Eastern Daily Press*, 3 August.

Ellis, EA (1960c) Coypu Make Way for Snipe. *Eastern Daily Press*, 12 October.

Ellis, EA (1961) Changes in a Broad. *Eastern Daily Press*, 1 August.

Ellis, EA (1962) Disappearing Reed Beds. *Eastern Daily Press*, 27 March.

Ellis, EA (1963) A Forgotten Broad. *Eastern Daily Press*, 16 December.

Ellis, EA (1965) *The Broads*. London: Collins.

Ellis, EA (1967) Are Research Activities Always Compatible with Conservation?, in E Duffey (ed) *The Biotic Effects of Public Pressures on the Environment*. London: Nature Conservancy, 35–37.

Ellis, EA (1973) *Wild Flowers of the Waterways and Marshes*. Norwich: Jarrold.

Ellis, EA (1976a) *British Fungi: Book 1*. Norwich: Jarrold.

Ellis, EA (1976b) *British Fungi: Book 2*. Norwich: Jarrold.

Ellis, EA (1982) *Ted Ellis's Countryside Reflections*. Sprowston: Wilson-Poole.

Ellis, EA (1998) *Tapestry of Nature*. Norwich: Eastern Counties Newspapers.

Ellis, EA and Cooke, GJ (1943) Fungus Forays in 1943. *Transactions of the Norfolk and Norwich Naturalists' Society* 15 (5): 428–429.

Ellis, EA, Ellis, M and Ellis, JP (1951) British Marsh and Fen Fungi. *Transactions of the British Mycological Society* 34: 147–169 / 497–514.

Emerson, PH (1887a) *Pictures From Life in Field and Fen*. London: George Bell.

Emerson, PH (1887b) *Idyls of the Norfolk Broads*. London: Autotype.

Emerson, PH (1888) *Pictures of East Anglian Life*. London: Sampson Low, Marston, Searle and Rivington.

Emerson, PH (1889a) *Naturalistic Photography for Students of the Art*. London: Sampson Low, Marston, Searle and Rivington.

Emerson, PH (1889b) *English Idyls*. London: Sampson Low.

Emerson, PH (1890a) *Wild Life on a Tidal Water*. London: Sampson Low, Marston, Searle and Rivington.

Emerson, PH (1890b) *Naturalistic Photography for Students of the Art*. London: Sampson Low, Marston, Searle and Rivington (Second Edition, revised).

Emerson, PH (1890c) *The Death of Naturalistic Photography*. London: The Author.

Emerson, PH (1892) *A Son of the Fens*. London: Sampson Low, Marston.

Emerson, PH (1893) *On English Lagoons*. London: David Nutt.

Emerson, PH (1894) *Tales from Welsh Wales*. London: David Nutt.

Emerson, PH (1895a) *Marsh Leaves From the Norfolk Broad-land*. London: David Nutt.

Emerson, PH (1895b) *Birds, Beasts and Fishes of the Norfolk Broadland*. London: David Nutt

Emerson, PH (1898a) *The English Emersons*. London: David Nutt.

Emerson, PH (1898b) *Marsh Leaves From the Norfolk Broadland* (Popular Edition). London: David Nutt.

Emerson, PH (1925) *The Blood Eagle and Other Tales*. London: Andrew Melrose.

Emerson, PH and Goodall, TF (1886) *Life and Landscape on the Norfolk Broads*. London: Sampson Low, Marston, Searle and Rivington.

Evans, EE (1996) *Ireland and the Atlantic Heritage*. Dublin: Lilliput Press.

Evans, GE (1966) *The Pattern under the Plough*. London: Faber and Faber.

Evans, GE (1970) *Where Beards Wag All: The Relevance of the Oral Tradition*. London: Faber and Faber.

Evans, HM (1929) The Sandbanks of Yarmouth and Lowestoft. *The Mariner's Mirror* 15: 251–270.

Everett, N (2007) Basil Bunting's *Briggflatts*, in R Colls (ed) *Northumbria: History and Identity 547–2000*. Chichester: Phillimore, 314–333.

Everitt, N (1902) *Broadland Sport*. London: Everitt.

Everitt, N (1903) Wild-Fowling, in WA Dutt, *The Broads*. London: Methuen, 323–330.

Ewans, M (1992) *The Battle for the Broads*. Lavenham: Terence Dalton.

Fagan, B (2001) *Grahame Clark: An Intellectual Life of an Archaeologist*. Boulder: Westview.

Fawcett, CB (1919) *Provinces of England*. London: Williams and Norgate.

Fawcett, T (1977) Thorpe Water Frolic. *Norfolk Archaeology* 36: 393–398.

Fidler, K (1955) *Brydons on the Broads*. London: Lutterworth Press.

Field (1944) Front Cover Tribute to Jim Vincent. *The Field*: 18 November.

Finnegan, D (2009) *Natural History Societies and Civic Culture in Victorian Scotland*. London: Pickering and Chatto.

Fitter, RSR (1959) *The Ark in Our Midst: The Story of the Introduced Animals of Britain*. London: Collins.

FitzRandolph, HE and Hay, MD (1926) *The Rural Industries of England and Wales II: Osier-Growing and Basketry and Some Rural Factories*. Oxford: Oxford University Press.

Forman, B (1966) Broadland, in R Jellis (ed) *Land and People: The Countryside for Use and Leisure*. London: British Broadcasting Corporation, 56–59.

Foucault, M (1986) Preface to the History of Sexuality, Volume II, in P Rabinow (ed) *The Foucault Reader*. Harmondsworth: Penguin, 333–339.

Foucault, M (2000) *Ethics: Subjectivity and Truth*. Harmondsworth: Penguin.

Fowler, E (1947) The Future of the Broads. *Transactions of the Norfolk and Norwich Naturalists' Society* 16 (4): 323–327.

Fowler, E (1961) An English Province and its Capital, in British Association for the Advancement of Science *Norwich and Its Region*. Norwich: Norwich Local Executive Committee of the British Association, 10–17.

Fowler, E (1970) *Broadland in Colour*. Norwich: Jarrold.

Freeman, A (1905) Fauna and Flora of Norfolk. Rotifera. *Transactions of the Norfolk and Norwich Naturalists' Society* 8 (1): 137–147.

Friday, L (ed) (1997) *Wicken Fen: The Making of a Wetland Nature Reserve*. Colchester: Harley Books.

Funnell, BM (1979) History and Prognosis of Subsidence and Sea-Level Change in the Lower Yare Valley, Norfolk. *Bulletin of the Geological Society of Norfolk* 31: 35–44.

Gage, AT and Stearn, WT (1988) *A Bicentenary History of the Linnean Society of London*. London: Academic Press.

Gardiner, JS (ed) (1932) *The Natural History of Wicken Fen*. Cambridge: Bowes and Bowes.

Geldart, A (1906) *Stratiotes Aloides*, L. *Transactions of the Norfolk and Norwich Naturalists' Society* 8 (2): 181–200.

Geldart, H (1870) On the Division of the County for Botanical Purposes. *Transactions of the Norfolk and Norwich Naturalists' Society* 1 (1): 19–22.

Geldart, H (ed) (1901) Botany, in W Rye (ed) *The Victoria History of the Counties of England: Norfolk (Volume 1)*. London: Constable, 39–76.

George, M (1976) Land Use and Nature Conservation in Broadland. *Geography* 61: 137–142.

George, M (1977) The Decline in Broadland's Aquatic Fauna and Flora: A Review of the Present Position. *Transactions of the Norfolk and Norwich Naturalists' Society* 24 (2): 41–53.

George, M (1978) Recent Research and Management Problems in Broadland. *Norfolk Research Committee Bulletin* 20: 4–6.

George, M (1992) *The Land Use, Ecology and Conservation of Broadland*. Chichester: Packard.

George, M (2000a) *Birds in Norfolk and the Law, Past and Present*. Norwich: Norfolk and Norwich Naturalists' Society.

George, M (2000b) Nature Conservation in Norfolk, in M Taylor, P Allard, M Seago and D Dorling, *The Birds of Norfolk*. Mountfield: Pica Press, 67–91.

George, M (2000c) Site Conservation in Broadland and its Terminology. *The Harnser*, January–March: 7–12

Giblett, R (1996) *Postmodern Wetlands: Culture, History, Ecology*. Edinburgh: Edinburgh University Press.

Gladden, RB (1953) Origin of the Broads. *Eastern Daily Press*, 3 September.

Gladden, RB (1955) Origin of the Broads. *Eastern Daily Press*, 25 April.

Gladden, RB (c.1956) *Map of the Norfolk Broad District During the Roman Occupation*. Privately published.

Godwin, GN (nd) Choir Outing. *Flegg Magazine*.

Godwin, H (1943) Coastal Peat Beds of the British Isles and North Sea. *Journal of Ecology* 31: 199–247.

Godwin, H and Turner, JS (1933) Soil Acidity in Relation to Vegetational Succession in Calthorpe Broad, Norfolk. *Journal of Ecology* 21: 235–262.

Gordon, DI (1968) *The Eastern Counties*. Newton Abbot: David and Charles.

Gosling, LM and Baker, SJ (1989) The Eradication of Muskrats and Coypus From Britain. *Biological Journal of the Linnean Society* 38: 39–51.

Graham, B (1994) The Search for the Common Ground: Estyn Evans's Ireland. *Transactions of the Institute of British Geographers* 19: 183–201.

Granta (2008) The New Nature Writing. *Granta* 102.

Grapes, MG (1960) Origin of the Broads. *Eastern Daily Press*, 16 March.

Grapes, S (1953) 'Our Bridge'. *Eastern Daily Press*, 19 March.

Grapes, S (1958a) *The Boy John*. Norwich: Norfolk News.

Grapes, S (1958b) *The Boy John Again*. Norwich: Norfolk News

Grapes, S (1974) *The Boy John Letters*. Norwich: Wensum Books.

The Graphic (1893) A Week in a Wherry. 28 October: 544–555.

Graves, AP (1913) *Irish Literary and Musical Studies*. London: Elkin Mathews.

Great Yarmouth Port and Haven Commissioners (1971) *Broadland* (Dissenting memorandum included in rear pocket of Broads Consortium (1971)). Great Yarmouth: Great Yarmouth Port and Haven Commissioners.

Green, C (1956a) The Birth of Broadland. *Eastern Daily Press*, 20 March.

Green, C (1956b) Conflict of Land and Sea in East Anglia. *The Times*, 14 March.

Green, C and Hutchinson, JN (1965) Relative Land and Sea Levels at Great Yarmouth, Norfolk. *Geographical Journal* 131: 86–90.

Green, GC (1953) *The Norfolk Wherry*. Wymondham: Reeve.

Gregory, ES (1912) *British Violets*. Cambridge: Heffer.

Gregory, J (1892) The Physical Features of the Norfolk Broads. *Natural Science* 1: 347–355.

Grey, E (1927) *The Charm of Birds*. London: Hodder and Stoughton.

Griffin, HFG (1953) *Quays Without Locks*. Lowestoft: Bradbeer.

Grimes, K (1953a) The Marsh Mills. *Eastern Daily Press*, 19 June.

Grimes, K (1953b) The Humber Keel. *Eastern Daily Press*, 15 December.

Grimes, K (1954) Problems of the Broads. *Journal of the Town Planning Institute* 40: 58–61, February.

Gurney, E (1906) Limnology, or, The Study of Fresh Waters. *Transactions of the Norfolk and Norwich Naturalists' Society* 8 (2): 155–174.

Gurney, E and Gurney, R (1908) *The Sutton Broad Fresh-Water Laboratory*. Sutton: Gurney.

Gurney, R (1905) The Life History of the Cladocera. *Transactions of the Norfolk and Norwich Naturalists' Society* 8 (1): 44–57.

Gurney, R (1907) The Crustacea of the East Norfolk Rivers. *Transactions of the Norfolk and Norwich Naturalists' Society* 8 (3): 410–438.

Gurney, R (1909) Estuarine Shells from Ludham. *Transactions of the Norfolk and Norwich Naturalists' Society* 8 (5): 855.

Gurney, R (1911a) Some Observations on the Waters of the River Bure and its Tributaries. *Geographical Journal* 37: 292–301.

Gurney, R (1911b) The Tides of the River Bure and its Tributaries. *Transactions of the Norfolk and Norwich Naturalists' Society* 9 (2): 216–243.

Gurney, R (1913) The Origins and Conditions of Existence of the Fauna of Fresh Water. *Transactions of the Norfolk and Norwich Naturalists' Society* 9 (4): 461–485.

Gurney, R (1922a) Utricularia in Norfolk in 1921: The Effects of Drought and Temperature. *Transactions of the Norfolk and Norwich Naturalists' Society* 11 (3): 260–266.

Gurney, R (1922b) A New British Prawn Found in Norfolk. *Transactions of the Norfolk and Norwich Naturalists' Society* 11 (3): 311–312.

Gurney, R (1923) Notes of a Naturalist in the Netherlands. *Transactions of the Norfolk and Norwich Naturalists' Society* 11 (4): 360–374.

Gurney, R (1931–33) *British Fresh-Water Copepoda* (Three Volumes). London: Ray Society.

Gurney, R (1934) The Hairy-Fisted Crab: A New Pest to Beware of. *Eastern Daily Press*, 15 January.

Gurney, R (1947) *Our Trees and Woodlands*. London: Medici Society.

Gurney, R (1949a) Aquatic Life in the Norfolk Broads. *The New Naturalist* 6: 22–25.

Gurney, R (1949b) Vegetational Changes. *Eastern Daily Press*, 26 August.

Gurney, R (1958) *Trees of Britain*. London: Faber and Faber.

Gurney, R and Gurney, SG (1920) *A Book About Plants and Trees*. London: C Arthur Pearson.

Gurney, R, Oliver, FW, Salisbury, EJ and Blofeld, TC (1927) The Functions of a Local Natural History Society. *Transactions of the Norfolk and Norwich Naturalists' Society* 12 (3): 307–316.

Hamilton's (1938) *Map and Chart of the Broads*. Norwich: Hamilton and Miller.

Hamilton's (1955) *Broads Navigation Charts and Index*. Oulton Broad: Hamilton Publications.

Hannaford, CA (c.1950) *The Charm of the Norfolk Broads*. Wroxham: Broads Tours.

Hardy, AC (1950) Dr Robert Gurney (Obituary). *Proceedings of the Linnean Society of London* 162: 118–120.

Hardy, D and Ward, C (2004) *Arcadia for All: The Legacy of a Makeshift Landscape*. Nottingham: Five Leaves Press (first published 1984).

Harrison, R (1959) *Scroby Sands*. Norwich: Jarrold.

Harrison, R (1960) *Breydon Water*. Norwich: Jarrold.

Harrod, W and Linnell, CLS (1957) *Shell Guide to Norfolk*. London: Faber and Faber.

Hauser, K (2007) *Shadow Sites: Photography, Archaeology, and the British Landscape 1927–1955*. Oxford: Oxford University Press.

Hauser, K (2008) *Bloody Old Britain: OGS Crawford and the Archaeology of Modern Life*. London: Granta.

Hawkes, J and Hawkes, J (1995) *The Spotters Book of Broads Hire Cruisers*. Presteigne: Black Hill Publications.

Haworth-Booth M (1988) *A Yarmouth Holiday: Photographs by Paul Martin*. London: Nishen.

Head, L and Atchison, J (2009) Cultural Ecology: Emerging Human-Plant Geographies. *Progress in Human Geography* 33: 236–245.

Hemingway, A (1979) *The Norwich School of Painters 1803–1833*. Oxford: Phaidon.

Hemingway, A (1988) Cultural Philanthropy and the Invention of the Norwich School. *Oxford Art Journal* 11: 17–39.

Hemingway, A (1992) *Landscape Imagery and Urban Culture in Early Nineteenth-Century Britain*. Cambridge: Cambridge University Press.

Hemingway, A (2000) 'Norwich School': Myth and Reality, in D Blayney Brown, A Hemingway and A Lyles *Romantic Landscape: The Norwich School of Painters*. London: Tate Gallery, 9–23.

Heppa, C (2005) Harry Cox and His Friends: Song Transmission in an East Norfolk Singing Community, c.1896–1960. *Folk Music Journal* 8: 569–593.

Herring, S (2009) Regional Modernism: A Reintroduction. *Modern Fiction Studies* 55: 1–10.

Hill, L (1985) *Lonely Waters: The Diary of a Friendship with EJ Moeran*. London: Thames.

Hillier, B (2004) Nerina Shute (Obituary). *Guardian*, 2 November: 27.

Hirsh, JC (1988) *Hope Emily Allen: Medieval Scholarship and Feminism*. Norman: Pilgrim Books.

Hirsh, JC (1989) *The Revelations of Margery Kempe: Paramystical Practices in Late Medieval England*. New York: EJ Brill.

Holmes, D (1988) *The How Hill Story*. Ludham: How Hill Trust.

Holmes, D (2011) *How Hill: Heart of the Norfolk Broads*. Ludham: How Hill Trust.

Holmes, PF and Pryor, MGM (1938) Barnacles in Horsey Mere. *Nature* 142 (3600): 795–796, 29 October.

Holtby, W (1936) *South Riding*. London: Collins.

Home, G (1925) *Through East Anglia*. London: Dent.

Hooper, J (1913) *Souvenir of the George Borrow Celebration*. Norwich: Jarrold.

Hopkins, H (1985) *The Long Affray: The Poaching Wars in Britain 1760–1914*. London: Secker and Warburg.

Howard, K (1928) *The Fast Gentleman*. London: Ernest Benn.

Howkins, A (1985) *Poor Labouring Men: Rural Radicalism in Norfolk 1872–1923*. London: Routledge.

Howkins, A (1986) The Discovery of Rural England, in R Colls and P Dodd (eds) *Englishness: Politics and Culture 1880–1920*. London: Croom Helm, 62–88.

Howkins, A (2004) Evans, George Ewart. *Oxford Dictionary of National Biography*.

Hunter, A (1944) *The Norwich Poems*. Norwich: Soman-Wherry Press.

Hunter, A (1957) *Gently Down the Stream*. London: Cassell.

Hunter, A (1965) A Jolt for the Turf-Cutting Theory. *Eastern Daily Press*, 7 May.

Hunter, A (1968) Broads Origins. *Eastern Daily Press*, 30 December.

Hurrell, HE (1911) Distribution of the Polyzoa in Norfolk Waters. *Transactions of the Norfolk and Norwich Naturalists' Society* 9 (2): 197–205.

Huxley, J and Koch, L (1938) *Animal Language*. London: Country Life.

Ingold, CT and Ellis, EA (1952) On Some Hyphomycete Spores, Including Those of *Tetracladium Maxilliformis*, From Wheatfen. *Transactions of the British Mycological Society* 35: 158–161.

Innes, AG (1911) Tidal Action in the Bure and its Tributaries. *Transactions of the Norfolk and Norwich Naturalists' Society* 9 (2): 244–262.

Insley, J (2008) Little Landscapes: Dioramas in Museum Displays. *Endeavour* 32: 27–31.

Jackson, I (1971) *The Provincial Press and the Community*. Manchester: Manchester University Press.

James, C, Pierce, S and Rowley, H (1945) *City of Norwich Plan*. Norwich: City of Norwich Corporation.

Jennings, JN (1950) The Origin of the Fenland Meres: Fenland Homologues of the Norfolk Broads. *Geological Magazine* 87: 217–225.

Jennings, JN (1952) *The Origin of the Broads*. London: Royal Geographical Society.

Jennings, JN (1955) Further Pollen Data from the Norfolk Broads. *The New Phytologist* 54: 199–207.

Jennings, JN and Lambert, JM (1949) The Shrinkage of the Broads. *The New Naturalist* 6: 26–27.

Jennings, JN and Lambert, JM (1951) Alluvial Stratigraphy and Vegetational Succession in the Region of the Bure Valley Broads. *Journal of Ecology* 39: 106–170.

Jennings, JN and Lambert, JM (1953) The Origin of the Broads. *Geographical Journal* 119: 91.

Jennings, JP (1892) *Sun Pictures of the Norfolk Broads*. Norwich: Jarrold (Text by E Suffling).

Johnson, M (2007) *Ideas of Landscape*. Oxford: Blackwell.

Jolly, HLP (1939) Supposed Land Subsidence in the South of England. *Geographical Journal* 93: 408–413.

Jones, M (2004) Social Justice and the Region: Grassroots Regional Movements and the 'English Question'. *Space and Polity* 8: 157–189.

Jones, M and Olwig, K (eds) (2008) *Nordic Landscapes: Region and Belonging on the Northern Edge of Europe*. Minneapolis: University of Minnesota Press.

Joyce, P (1991) *Visions of the People: Industrial England and the Question of Class 1848–1914*. Cambridge: Cambridge University Press.

Karpeles, M, Gilchrist, AG, Moeran, EJ, Freeman, AM and Howes, F (1931) Humorous and Disreputable Songs, and Ballads of Adventure. *Journal of the Folk-Song Society* 8: 270–279.

Kelley, P (ed) (2011) *From Osborne House to Wheatfen Broad: Memoirs of Phyllis Ellis*. Norwich: Wheatfen Books.

Kempe, M (1985) *The Book of Margery Kempe*. Harmondsworth: Penguin.

Kidson, C (1961) The Norfolk Broads. *Nature* 192 (4800): 314–315, 28 October.

Kirsch, S (2005) *Proving Grounds: Project Plowshare and the Unrealized Dream of Nuclear Earthmoving*. New Brunswick: Rutgers University Press.

Knights, S (1986) Change and Decay: Emerson's Social Order, in N McWilliam and V Sekules (eds) *Life and Landscape: PH Emerson Art and Photography in East Anglia 1885–1900*. Norwich: Sainsbury Centre for Visual Arts, 12–20.

Knights, S (1989) Representation and Reality: A View of the Broads, in R Denyer, *Still Waters*. Norwich: Still Waters Press, 4–5.

Koch, L (1945) The Bittern's Boom: Sound Recording in the Reeds of Horsey Mere. *The Times*, 4 August: 5.

Koch, L (1955) *Memoirs of a Birdman*. London: Phoenix House.

Kohler, R (2002) *Landscapes and Labscapes: Exploring the Lab–Field Border in Biology*. Chicago: Chicago University Press.

Lambert, JM (1946a) A Note on the Physiognomy of *Glyceria Maxima* Reedswamps in Norfolk. *Transactions of the Norfolk and Norwich Naturalists' Society* 16 (3): 246–259.

Lambert, JM (1946b) The Distribution and Status of *Glyceria Maxima* (Hartm.) Holmb. in the Region of Surlingham and Rockland Broads, Norfolk. *Journal of Ecology* 33: 230–267.

Lambert, JM (1947) Biological Flora of the British Isles. *Glyceria Maxima* (Hartm.) Holmb. *Journal of Ecology* 34: 310–344.

Lambert, JM (1948) A Survey of the Rockland-Claxton Level. *Journal of Ecology* 36: 120–135.

Lambert, JM (1950) The Ecological Status of the Bargate Nature Reserve. *Transactions of the Norfolk and Norwich Naturalists' Society* 17 (2): 123–135.

Lambert, JM (1951) Botanical Investigations in the Bure and Yare Valleys. *Norfolk Research Committee Bulletin* 3: 5–6.

Lambert, JM (1952a) Stratigraphical and Ecological Investigations in the Region of the Norfolk Broads – 1951. *Norfolk Research Committee Bulletin* 4: 7–8.

Lambert, JM (1952b) Summer Meeting in Norfolk, 6–13 July 1951. *Journal of Ecology* 40: 414–419.

Lambert, JM (1953a) The Past, Present and Future of the Norfolk Broads. *Transactions of the Norfolk and Norwich Naturalists' Society* 17: 223–258.

Lambert, JM (1953b) Stratigraphical Investigations in the Region of the Norfolk Broads in 1952. *Norfolk Research Committee Bulletin* 5: 5–6, August.

Lambert, JM (1953c) Origin of the Broads. *Eastern Daily Press*, 27 August.

Lambert, JM (1955) Origins of the Broads. *Eastern Daily Press*, 21 April.

Lambert, JM (1957a) The Danger to the Broads. *Eastern Daily Press*, 28 March.

Lambert, JM (1957b) Diminishing Broads. *The Times*, 11 July: 9.

Lambert, JM (1964a) The *Spartina* Story. *Nature* 204 (4964): 1136–1138, 19 December.

Lambert, JM (1964b) Ecology as a Biological Discipline. *The School Science Review* 157: 568–575.

Lambert, JM (ed) (1967) *The Teaching of Ecology*. Oxford: Blackwell.

Lambert, JM (2000) A Tale of Two Boats. *The Harnser*, January–March: 5.

Lambert, JM and Dale, MB (1964) The Use of Statistics in Phytosociology. *Advances in Ecological Research* 2: 59–99.

Lambert, JM and Davies, MR (1940) A Sandy Area in the Dovey Estuary. *Journal of Ecology* 28: 453–464.

Lambert, JM and Smith, CT (1960) The Norfolk Broads as Man-Made Features. *New Scientist* 7 (176): 775–777, 31 March.

Lambert, JM, Jennings, JN, Smith, CT, Green, C and Hutchinson, JN (1960) *The Making of the Broads: A Reconsideration of Their Origin in the Light of New Evidence*. London: Royal Geographical Society.

Landscape Research (2003) Special Issue on Landscape and Alien and Exotic Species. *Landscape Research* 28 (1): 5–137.

Langley, RF (2006) *Journals*. Exeter: Shearsman.

Latour, B (1993) *We Have Never Been Modern*. London: Harvard University Press.

Laurie, EMO (1946) The Coypu (*Myocastor Coypus*) in Great Britain. *Journal of Animal Ecology* 15: 22–34.

Lee, K (1899) Experiences of a Folk-Song Collector. *Journal of the Folk-Song Society* 1: 7–12 / 16–25.

Leney, F (1933) Progress of the Norfolk Room at the Norwich Castle Museum. *Transactions of the Norfolk and Norwich Naturalists' Society* 13 (4): 362–366.

Leney, F (1935) 'The Norfolk Room' at the Norwich Castle Museum. *East Anglian Magazine* 1 (3): 184–185.

Linsell, S (1990) *Hickling Broad and its Wildlife*. Lavenham: Terence Dalton.

Livingstone, D (1992) *The Geographical Tradition*. Oxford: Blackwell.

Livingstone, D (2003) *Putting Science in Its Place: Geographies of Scientific Knowledge*. Chicago: University of Chicago Press.

Löfgren, O (1985) Wish You Were Here! Holiday Images and Picture Postcards. *Ethnologia Scandinavica* 15: 90–107.

Long, S (1908) President's Address. *Transactions of the Norfolk and Norwich Naturalists' Society* 8 (4): 499–502.

Long, S (1925a) The Norwich Corporation Swanherd's Return for 1925. *Transactions of the Norfolk and Norwich Naturalists' Society* 12 (1): 111–113.

Long, S (1925b) A Supposed Winter Nest of a Harvest Mouse. *Transactions of the Norfolk and Norwich Naturalists' Society* 12 (1): 54–56.

Long, S (1936) Arthur H Patterson, ALS. *Proceedings of the Linnean Society of London* 148: 211.

Lorimer, H (2000) Guns, Game and the Grandee: The Cultural Politics of Deerstalking in the Scottish Highlands. *Cultural Geographies* 7: 403–431.

Lorimer, H (2003) Telling Small Stories: Spaces of Knowledge and the Practice of Geography. *Transactions of the Institute of British Geographers* 28: 197–217.

Lorimer, H (2005) Cultural Geography: The Busyness of Being 'More-Than-Representational'. *Progress in Human Geography* 29: 83–94.

Lorimer, H (2006) Herding Memories of Humans and Animals. *Environment and Planning D: Society and Space* 24: 497–518.

Lorimer, H and Lund, K (2003) Performing Facts: Finding a Way Over Scotland's Mountains, in B Szerszynski, W Heim and C Waterton (eds) *Nature Performed*. Oxford: Blackwell, 130–144.

Lorimer, H and Spedding, N (2005) Locating Field Science: A Geographical Family Expedition to Glen Roy, Scotland. *British Journal for the History of Science* 38: 13–33.

Lowe, P (1976) Amateurs and Professionals: The Institutional Emergence of British Plant Ecology. *Journal of the Society for the Bibliography of Natural History* 7: 517–535.

Lubbock, R (1879) *Observations on the Fauna of Norfolk, and More Particularly on the District of the Broads*. Norwich: Jarrold (first published 1845).

Lydekker, R (1895) Birds, Beasts and Fishes of the Norfolk Broadland (Review). *Nature* 52 (1339): 195–196, 27 June.

Mabey, R (1980) *The Common Ground*. London: Hutchinson.

MacDonald, F (2004) Paul Strand and the Atlanticist Cold War. *History of Photography* 28: 357–374.

MacDonald, F (2006a) The Last Outpost of Empire: Rockall and the Cold War. *Journal of Historical Geography* 32: 627–647.

MacDonald, F (2006b) Geopolitics and 'The Vision Thing': Regarding Britain and America's First Nuclear Missile. *Transactions of the Institute of British Geographers* 31: 53–71.

MacDonald, F (2011) Doomsday Fieldwork, or, How to Rescue Gaelic Culture? The Salvage Paradigm in Geography, Archaeology, and Folklore. *Environment and Planning D: Society and Space* 29: 309–335.

Macdonald, H (2002) 'What Makes You a Scientist is the Way You Look At Things': Ornithology and the Observer 1930–1955. *Studies in History and Philosophy of Biological and Biomedical Sciences* 33: 53–77.

Macdonald, H (2006) *Falcon*. London: Reaktion.

Macfarlane, R (2007) *The Wild Places*. Cambridge: Granta.

Macfarlane, R (2008) Ghost Species. *Granta* 102: 109–128.

MacKenzie, JM (1988) *The Empire of Nature: Hunting, Conservation and British Imperialism*. Manchester: Manchester University Press.

Mair, J and Delafons, J (2001) The Policy Origins of Britain's National Parks: The Addison Committee 1929–31. *Planning Perspectives* 16: 293–309.

Makin, P (1992) *Bunting: The Shaping of His Verse*. Oxford: Clarendon.

Malster, R (1963) The Northern River. *The Norfolk Sailor* 6: 21–27.

Malster, R (1966) Norfolk Navigations. *The Norfolk Sailor* 11: 19–27.

Malster, R (1971) *Wherries and Waterways*. Lavenham: Terence Dalton.

Malster, R (2003) *The Norfolk and Suffolk Broads*. Chichester: Phillimore.

Manning, SA (1943) The Natural History of Wheatfen Broad Surlingham. Part VI: Lichens. *Transactions of the Norfolk and Norwich Naturalists' Society* 15 (5): 420–422.

Manning, SA (1948) *Broadland Naturalist: The Life of Arthur H Patterson, ALS ('John Knowlittle')*. Norwich: Soman-Wherry Press.

Manning, SA (1980) *Portrait of Broadland*. London: Robert Hale.

Mardle, J (1949a) *Broad Norfolk*. Norwich: Norfolk News.

Mardle, J (1949b) Down-River with the Albion. *Eastern Daily Press*, 25 May.

Mardle, J (1950) A Threatened Coast. *Eastern Daily Press*, 3 May.

Mardle, J (1952) The Hungry Sea. *Eastern Daily Press*, 30 April.

Mardle, J (1953) The Sea's Triple Assault. *Eastern Daily Press*, 4 February.

Mardle, J (1955a) *Wednesday Mornings*. Norwich: Jarrold.

Mardle, J (1955b) Theorists and the Broads. *Eastern Daily Press*, 27 April.

Mardle, J (1955c) Emma's Island. *Eastern Daily Press*, 7 July.

Mardle, J (1960) The Making of the Broads. *Eastern Daily Press*, 17 February.

Mardle, J (1963) Autumn in a Nature Reserve. *Eastern Daily Press*, 20 November.

Mardle, J (1965) The Report on Broadland. *Eastern Daily Press*, 7 July.

Mardle, J (1966) Saving the Broads. *Eastern Daily Press*, 14 December.

Mardle, J (1967) Broad Norfolk. *Eastern Daily Press* 22 March.

Mardle, J (1970) Wardens of the Broads. *Eastern Daily Press*, 15 July.

Marren, P (1995) *The New Naturalists*. London: HarperCollins.

Marren, P and Gillmor, R (2009) *Art of the New Naturalists*. London: HarperCollins.

Mason, CF and Bryant, RJ (1975) Changes in the Ecology of the Norfolk Broads, *Freshwater Biology* 5: 257–270.

Mason, U (1944) Forays, 1943. *Transactions of the British Mycological Society* 27: 99.

Massingham, HJ (1923) *Untrodden Ways*. London: T Fisher Unwin.

Massingham, HJ (1924) *Sanctuaries for Birds and How to Make Them*. London: Bell.

Matless, D (1992) Regional Surveys and Local Knowledges: The Geographical Imagination in Britain 1918–39. *Transactions of the Institute of British Geographers* 17: 464–480.

Matless, D (1993) One Man's England: WG Hoskins and the English Culture of Landscape. *Rural History* 4: 187–207.

Matless, D (1994) Moral Geography in Broadland. *Ecumene* 1: 127–156.

Matless, D (1996) Visual Culture and Geographical Citizenship: England in the 1940s. *Journal of Historical Geography* 22: 424–439.

Matless, D (1998) *Landscape and Englishness*. London: Reaktion.

Matless, D (1999a) The Uses of Cartographic Literacy, in D Cosgrove (ed) *Mappings*. London: Reaktion, 193–212.

Matless, D (1999b) Review of Tom Williamson: *The Norfolk Broads. Journal of Historical Geography* 25: 411–413.

Matless, D (2000a) The Predicament of Englishness. *Scottish Geographical Journal* 116: 79–86.

Matless, D (2000b) Action and Noise over a Hundred Years: The Making of a Nature Region. *Body and Society* 6: 141–165.

Matless, D (2000c) Versions of Animal–Human: Broadland, 1945–70, in C Philo and C Wilbert (eds) *Animal Spaces, Beastly Places*. London: Routledge, 115–140.

Matless, D (2003a) Original Theories: Science and the Currency of the Local. *Cultural Geographies* 10: 354–378.

Matless, D (2003b) The Properties of Landscape, in K Anderson, M Domosh, S Pile and N Thrift (eds) *Handbook of Cultural Geography*. London: Sage, 227–232.

Matless, D (2005) Sonic Geography in a Nature Region. *Social and Cultural Geography* 6: 745–766.

Matless, D (2008a) Writing English Landscape History. *Anglia: Zeitschrift Für Englische Philologie* 126: 295–311.

Matless, D (2008b) Properties of Ancient Landscape: The Present Prehistoric in Twentieth-Century Breckland. *Journal of Historical Geography* 34: 68–93.

Matless, D (2009a) Nature Voices. *Journal of Historical Geography* 35: 178–188.

Matless, D (2009b) East Anglian Stones: Erratic Prehistories from the Early Twentieth Century, in J Parker (ed) *Written on Stone: The Cultural Reception of Prehistoric Monuments*. Newcastle: Cambridge Scholars Publishing, 66–81.

Matless, D (2010) Describing Landscape: Regional Sites. *Performance Research* 15: 72–82.

Matless, D (2011) *The Horsey Mail*: Documentary as Landscape, in S Anthony and J Mansell (eds) *The Projection of Britain: A History of the GPO Film Unit*. London: Palgrave Macmillan, 244–253.

Matless, D (2012) Accents of Landscape in GPO Country: *The Horsey Mail*, 1938. *Twentieth Century British History* 23: 57–79.

Matless, D and Cameron, L (2006) Experiment in Landscape: The Norfolk Excavations of Marietta Pallis. *Journal of Historical Geography* 32: 96–126.

Matless, D and Cameron, L (2007a) Geographies of Local Life: Marietta Pallis and Friends, Long Gores, Hickling, Norfolk. *Environment and Planning D: Society and Space* 25: 75–103.

Matless, D and Cameron, L (2007b) Devotional Landscape: Ecology and Orthodoxy in the Work of Marietta Pallis, in M Conan (ed) *Sacred Gardens and Landscapes: Ritual and Agency*, Washington: Dumbarton Oaks, 263–296.

Matless, D and Pearson, M (2012) A Regional Conversation. *Cultural Geographies* 19: 123–129.

Matless, D, Watkins, C and Merchant, P (2005) Animal Landscapes: Otters and Wildfowl in England 1945–1970. *Transactions of the Institute of British Geographers* 30: 191–205.

Matless, D, Watkins, C and Merchant, P (2010) Nature Trails: The Production of Instructive Landscapes in Britain, 1960–72. *Rural History* 21: 97–131.

Maxwell, D (1925) *Unknown Norfolk*. London: Bodley Head.

Maxwell, W (ed) (1982) *Sylvia Townsend Warner: Letters*. London: Chatto and Windus.

May, J (1952) *The Norfolk Broads Holiday Book and Pocket Pilot*. London: Hulton Press.

Mayes, W J (1956) A Broads Society? (Letter). *Eastern Daily Press*, 11 May.

McConkey, K (1986) Dr Emerson and the Sentiment of Nature, in N McWilliam and V Sekules (eds) *Life and Landscape: PH Emerson Art and Photography in East Anglia 1885–1900*. Norwich: Sainsbury Centre for Visual Arts, 48–56.

McGonigal, J and Price, R (eds) (2000) *The Star You Steer By: Basil Bunting and British Modernism*. Amsterdam: Editions Rodopi.

McGregor, J (2012) *This Isn't the Sort of Thing That Happens to Someone Like You*. London: Bloomsbury.

McLean, C (1936) The Status of Wild Duck in Our Area, With Some Observations on Their Breeding and Habits in Captivity. *Transactions of the Norfolk and Norwich Naturalists' Society* 14 (2): 116–130.

McLean, C (1954) *At Dawn and Dusk: Being my Record of Nearly Sixty Years of Wildfowling*. London: Batchworth Press.

McWilliam, N and Sekules, V (eds) (1986) *Life and Landscape: PH Emerson Art and Photography in East Anglia 1885–1900*. Norwich: Sainsbury Centre for Visual Arts.

Medhurst, A (2007) *A National Joke: Popular Comedy and English Cultural Identities*. London: Routledge.

Middleton, C (1978) *The Broadland Photographers*. Norwich: Albion.

Miller, AG (1937) *What to Do on the Norfolk Broads*. Wroxham: SS Miller.

Miller, D (1935) *Seen From a Windmill: A Norfolk Broads Revue*. London: Heath Cranton.

Mitchell, G (1948) *Holiday River*. London: Evans Brothers.

Mitchell, G (1978) *Wraiths and Changelings*. London: Michael Joseph.

Moeran, EJ (1924) *Six Folk Songs From Norfolk*. London: Augener.

Moeran, EJ (1927) *Songs Collected in Norfolk*. Mss, 63pp, Copy in Norfolk Heritage Centre, Millennium Library, Norwich.

Moeran, EJ (1948) Folk Singing of To-Day. *Journal of the English Folk Dance and Song Society* 5: 152–154.

Moeran EJ, Gilchrist, AG, Kidson, F, Vaughan Williams, R and Broadwood, L (1922) Songs Collected in Norfolk. *Journal of the Folk–Song Society* 7: 1–24.

Moeran, EJ, Karpeles, M, Freeman, AM, Gilchrist, AG and Howes, F (1931) Love Songs and Ballads. *Journal of the Folk–Song Society* 8: 257–269.

Moore, PD (1978) The Norfolk Broads Under Pressure. *Nature* 274 (5672): 644, 17 August.

Morning Post (1886) *The Goupil Gallery*. 10 June: 3.

Morris, D (1995) *Geography of the County of Norfolk*. Mulbarton: David Wright (facsimile reprint from 1870s, first published London: William Collins).

Mosby, JEG (1938a) *Norfolk*. London: Geographical Publications.

Mosby, JEG (1938b) Mapping the Flooded Area. *Transactions of the Norfolk and Norwich Naturalists' Society* 14 (4): 346–348.

Mosby, JEG (1939) The Horsey Flood, 1938: An Example of Storm Effect on a Low Coast. *Geographical Journal* 93: 413–418.

Mosby, JEG and Sainty, JE (1948) The Effect of a Lock. *Eastern Daily Press*, 31 August.

Mosley, N (1976) *Julian Grenfell: His Life and the Times of His Death 1888–1915*. London: Weidenfeld and Nicolson.

Moss, B (1977a) Conservation Problems in the Norfolk Broads and Rivers of East Anglia, England – Phytoplankton, Boats and the Causes of Turbidity. *Biological Conservation* 12: 95–114.

Moss, B (1977b) The State of the Norfolk Broads. *The Ecologist* 7: 324–326 (October).

Moss, B (1979) An Ecosystem out of Phase. *Geographical Magazine* 52: 47–50 (October).

Moss, B (1983) The Norfolk Broadland: Experiments in the Restoration of a Complex Wetland. *Biological Reviews* 58: 521–561.

Moss, B (2001) *The Broads: The People's Wetland*. London: HarperCollins.

Moss, B (2010) *Ecology of Freshwaters*. Chichester: Wiley-Blackwell.

Mottram, RH (ed) (1935) *A Scientific Survey of Norwich and District*. London: British Association for the Advancement of Science.

Mottram, RH (1951) *East Anglia*. London: Collins.

Mottram, RH (1952) *The Broads*. London: Robert Hale.

Murphy, RC (1963) Lord William Percy (Obituary). *The Auk* 80: 414–416.

Nahum, A (2010) Exhibiting Science: Changing Conceptions of Science Museum Display, in PJT Morris (ed) *Science for the Nation*. Basingstoke: Palgrave Macmillan, 176–193.

Nash, C (2000) Performativity in Practice: Some Recent Work in Cultural Geography. *Progress in Human Geography* 24: 653–664.

Nash, D (2010) *David Nash at Yorkshire Sculpture Park*. Wakefield: Yorkshire Sculpture Park.

National Farmers Union (2010) *Why Farming Matters to the Broads*. Newmarket: NFU East Anglia.

National Parks Committee (1947) *Report of the National Parks Committee*. London: HMSO, Cmd.7121.

Nature (1871) Norfolk and Norwich Naturalists' Society. *Nature* 5 (113): 175, 28 December.

Nature (1887) A Batch of Guide-Books to the Norfolk Broads. *Nature* 36 (933): 457–459, 15 September.

Nature (1916) Obituary: Frank Southgate. *Nature* 97 (2420): 64, 16 March.

Nature (1925) Bird Life on the Norfolk Broads. *Nature* 116 (2906): 42–43, 11 July.

Nature (1940) Miss EL Turner. *Nature* 146 (3700): 424, 28 September (authored By 'P B-S').

Nature (1950) Leverhulme Research Fellowships. *Nature* 166 (4212): 137–138, 22 July.

Nature Conservancy (1965) *Report on Broadland*. London: Nature Conservancy.

Nature Conservancy (1968) *Hoveton Great Broad Trail*. Norwich: Nature Conservancy.

Nature Conservancy (1970) *A Country Drive for Motorists in the Norwich District*. Norwich: Nature Conservancy.

Nature Conservancy Council (1977) *The Future of Broadland*. London: Nature Conservancy Council.

Navy Cuttings (1966) Player's No.6 on the Norfolk Broads. *Navy Cuttings* 6 (3): 6–7, December.

Naylor, S (2002) The Field, the Museum and the Lecture Hall: The Spaces of Natural History in Victorian Cornwall. *Transactions of the Institute of British Geographers* 27: 494–513.

Naylor, S (2005) Introduction: Historical Geographies of Science – Places, Contexts, Cartographies. *British Journal for the History of Science* 38: 1–12.

Naylor, S (2010) *Regionalizing Science: Placing Knowledges in Victorian England*. London: Pickering and Chatto.

Newhall, N (1975) *PH Emerson: The Fight for Photography as a Fine Art*. New York: Aperture.

News Chronicle (1949) *The Albion, 51-Year-Old Norfolk Wherry, Sails Again*. 14 October.

Nicholson, EM (1957) *Britain's Nature Reserves*. London: Country Life.

Nicholson, EM and Koch, L (1936) *Songs of Wild Birds*. London: Witherby.

Nicholson, EM and Koch, L (1937) *More Songs of Wild Birds*. London: Witherby.

Nicholson, WA (1906) A Preliminary Sketch of the Bionomical Botany of Sutton and the Ant District. *Transactions of the Norfolk and Norwich Naturalists' Society* 8 (2): 265–289.

Nicholson, WA (1908) Proposal for a Botanical Survey of the Broads District. *Transactions of the Norfolk and Norwich Naturalists' Society* 8 (4): 618–621.

Nicholson, WA (1914) *A Flora of Norfolk*. London: West, Newman and Co.

Nochlin, L (1988) *Women, Art and Power*. London: Harper and Row.

Norden, G (1997) *Landscapes under the Luggage Rack*. Northampton: GNRP.

Norfolk and Norwich Naturalists' Society (2008) *A Natural History of the Catfield Hall Estate*. Norwich: Norfolk and Norwich Naturalists' Society.

Norfolk Naturalists Trust (1976a) *Nature in Norfolk: A Heritage in Trust*. Norwich: Jarrold.

Norfolk Naturalists Trust (1976b) *Visit of Her Majesty the Queen and His Royal Highness Prince Philip, Duke of Edinburgh, to Attend the Opening of the Trust's Broadland Conservation Centre at Ranworth Broad: Programme of Proceedings*. Norwich: Norfolk Naturalists Trust.

Norfolk Wherry Trust (1952) *The Norfolk Wherry*. Norwich: Norfolk Wherry Trust.

Norfolk Windmills Trust and Broads Authority (1995) *A Project for the Millennium: Drainage Mills of the Broads*. Norwich: Norfolk Windmills Trust/Broads Authority.

Norgate, F (1918) The Lantern Man. *Transactions of the Norfolk and Norwich Naturalists' Society* 10 (4): 390–392.

Norris, JD (1963) A Campaign Against the Coypus in East Anglia. *New Scientist* 17: 625–626.

Norris, JD (1967) A Campaign Against Feral Coypus (*Mycastor Coypus Molina*) in Great Britain. *Journal of Applied Ecology* 4: 191–199.

Norwich Mercury (1835a) *Conversazione*. 31 January: 3.

Norwich Mercury (1835b) *Conversazione*. 7 March: 3.

Norwich Mercury (1835c) *Conversazione*. 28 March: 3.

Norwich Mercury (1836a) *Conversazione*. 3 December: 3.

Norwich Mercury (1836b) *Conversazione*. 12 December: 3.

Nyhart, L (2004) Science, Art, and Authenticity in Natural History Displays, in S De Chadarevian and N Hopwood (eds) *Models: The Third Dimension of Science*. Stanford: Stanford University Press, 307–335.

Oliver, FW and Salisbury, EJ (1913) The Topography and Vegetation of the National Trust Reserve Known as Blakeney Point, Norfolk. *Transactions of the Norfolk and Norwich Naturalists' Society* 9 (4): 485–542.

Olwig, K (1996) Recovering the Substantive Nature of Landscape. *Annals of the Association of American Geographers* 86: 630–650.

Olwig, K (2008) The Jutland Cipher: Unlocking the Meaning and Power of a Contested Landscape, in M Jones and K Olwig (eds) *Nordic Landscapes: Region and Belonging on the Northern Edge of Europe*. Minneapolis: University of Minnesota Press, 12–49.

O'Riordan, AM (1976) *A Broadland Bibliography*. Norwich: Nature Conservancy Council.

O'Riordan, T (1969) Planning to Improve Environmental Capacity: A Case Study in Broadland. *Town Planning Review* 40: 39–58.

O'Riordan, T (1971) Environmental Management. *Progress in Geography* 3: 173–231.

O'Riordan, T (1978) An Example of Environmental Education. *Journal of Geography in Higher Education* 2: 3–16.

O'Riordan, T (1979a) Signs of Disaster and a Policy for Survival. *Geographical Magazine* 52: 51–59 (October).

O'Riordan, T (1979b) Ecological Studies and Political Decisions. *Environment and Planning A* 11: 805–813.

O'Riordan, T (1979c) The Broads Authority. *Town and Country Planning* 18: 49–50 (May).

O'Riordan, T (1980a) A Case Study in the Politics of Land Drainage. *Disasters* 4: 393–410.

O'Riordan, T (1980b) *Lessons From the Yare Barrier Controversy*. Norwich: University of East Anglia.

O'Riordan, T (1986) Ploughing into the Halvergate Marshes, in P Lowe, G Cox, M MacEwen, T O'Riordan and M Winter (eds) *Countryside Conflicts*. Aldershot: Gower, 265–300.

O'Riordan, T (1993) An Insider's View of Managing the Broads, in S Glyptis (ed) *Leisure and the Environment: Essays in Honour of Professor JA Patmore*. London: Belhaven, 253–265.

Orwell, G (1961) The Art of Donald McGill, in G Orwell, *Collected Essays*. London: Secker and Warburg, 167–178 (first published 1941).

Paasi, A (1991) Deconstructing Regions: Notes on the Scales of Spatial Life. *Environment and Planning A* 23: 239–256.

Paasi, A (1996) *Territories, Boundaries and Consciousness: The Changing Geographies of the Finnish–Russian Border*. Chichester: John Wiley & Sons, Ltd.

Paasi, A (2003) Region and Place: Regional Identity in Question. *Progress in Human Geography* 27: 475–485.

Paasi, A (2008) Finnish Landscape as Social Practice, in M Jones and K Olwig (eds) *Nordic Landscapes: Region and Belonging on the Northern Edge of Europe*. Minneapolis: University of Minnesota Press, 511–539.

Paget, CJ and Paget, J (1834) *Sketch of the Natural History of Yarmouth and its Neighbourhood*. Yarmouth: F Skill.

Pallis, M (1911a) The River-Valleys of East Norfolk: Their Aquatic and Fen Formations, in AG Tansley (ed) *Types of British Vegetation*. Cambridge: Cambridge University Press, 214–245.

Pallis, M (1911b) On the Cause of the Salinity of the Broads and of the River Thurne. *Geographical Journal* 37: 284–301.

Pallis, M (1916) The Structure and History of Plav: The Floating Fen of the Delta of the Danube. *Journal of the Linnean Society: Botany* 43: 233–290.

Pallis, M (1939) *The General Aspects of the Vegetation of Europe*. London: Taylor and Francis.

Pallis, M (1952) *Tableaux in Greek History*. Glasgow: Robert Maclehose.

Pallis, M (1956) *The Impermeability of Peat and the Origin of the Norfolk Broads / A Note on Acorn Distributing Birds*. Glasgow: Robert Maclehose.

Pallis, M (1958) *An Attempt at a Statement Concerning a Vital Unit as Shown By the Reed in the Delta of the Danube*. Glasgow: Robert Maclehose.

Pallis, M (1961) *The Status of Fen and the Origin of the Norfolk Broads*. Glasgow: Robert Maclehose.

Pallis, M (1963) *The Species Unit, Unit III*. Glasgow: Robert Maclehose.

Palmer, R (2003) Neglected Pioneer: EJ Moeran (1894–1950). *Folk Music Journal* 8: 345–361.

Pargeter, V and Pargeter, L (1990) *Maud: A Norfolk Wherry*. Ingatestone: Vincent and Linda Pargeter.

Parsons-Norman, G (1912) *Broadland (Series 1)*. Norwich: Jarrold.

Patterson, AH (1887) *Seaside Scribblings for Visitors*. London: Jarrold.

Patterson, AH (1888) *Notes on Pet Monkeys and How to Manage Them*. London: L Upcott Gill.

Patterson, AH (1892) *Broadland Scribblings: A Leisure-Hour Book for the Holidays*. Norwich: P Soman.

Patterson, AH (1896) *A Protest by a Masculine Naturalist*. Croydon: Society for the Protection of Birds.

Patterson, AH (1904) *Notes of an East Coast Naturalist*. London: Methuen.

Patterson, AH (1905) *Nature in Eastern Norfolk*. London: Methuen.

Patterson, AH (1907) *Wild Life on a Norfolk Estuary*. London: Methuen.

Patterson, AH (1909) *Man and Nature on Tidal Waters*. London: Methuen.

Patterson, AH (1916) The January Flood of 1916 at Great Yarmouth. *Transactions of the Norfolk and Norwich Naturalists' Society* 10 (2): 162–167.

Patterson, AH (1920) *Through Broadland in a Breydon Punt*. Norwich: HJ Vince.

Patterson, AH (1923) *The Cruise of the 'Walrus' on the Broads*. Norwich: Jarrold.

Patterson, AH (1929) *Wild-Fowlers and Poachers*. London: Methuen.

Patterson, AH (1930a) *Through Broadland by Sail and Motor*. London: Blakes.

Patterson, AH (1930b) *A Norfolk Naturalist*. London: Methuen.

Patterson, AH (1930c) *In Norfolk Bird Haunts in A.D. 1755*. Holt: Rounce and Wortley.

Patterson, AH and Smith, AH (1903) *Charles H Harrison, Broadland Artist*. London: Jarrold.

Payne, I (2004) Parsley, Osbert. *Oxford Dictionary of National Biography*.

Pearson, M (2006) *'In Comes I': Performance, Memory and Landscape*. Exeter: University of Exeter Press.

Pearson, M and Shanks, M (2001) *Theatre/Archaeology*. London: Routledge.

Percy, W (1951) *Three Studies in Bird Character: Bitterns, Herons and Water Rails*. London: Country Life.

Perrin, PM (1956) The Commercialised Broads. *Eastern Daily Press*, 12 May.

Petch, CP and Swann, EL (1968) *Flora of Norfolk*. Norwich: Jarrold.

Pettitts of Reedham (1949) *In the Valley of the Yare*. Reedham: Pettitts.

Pevsner, N (1961) *Suffolk*. Harmondsworth: Penguin.

Pevsner, N (1962) *North-East Norfolk and Norwich*. Harmondsworth: Penguin.

Pevsner, N and Wilson, B (1997) *Norfolk I: Norwich and North-East*. London: Yale University Press.

Pite, R (2002) *Hardy's Geography: Wessex and the Regional Novel*. Basingstoke: Palgrave Macmillan.

Plutarch (1949) *Plutarch's Moralia III*. Cambridge, Mass.: Harvard University Press, trans. Frank Cole Babbitt.

Politiken (1956) *Er Broad-Soeme Skabt Af Gorm Den Gamles Bedstefar?* (Are the Broad-Lakes Made by Gorm the Old's Grandfather?). 17 July.

Pope, S (2008) *The Memorial Walks*. London: Film and Video Umbrella.

Port, MH (2004) Peto, Sir (Samuel) Morton. *Oxford Dictionary of National Biography*.

Post Office Magazine (1938) *Postman of the Flood*. 5: 204 (May).

Post Office Magazine (1939) *Midlands, East and Wales*. 6: 37 (January).

Preston, A (1913) Notes on the Great Norfolk Rainstorm of 25 and 26 August, 1912. *Transactions of the Norfolk and Norwich Naturalists' Society* 9 (4): 551–557.

Prytherch, D (2009) Elegy to an Iconographic Place: Reconstructing the Regionalism/Landscape Dialectic in L'Horta De Valencia. *Cultural Geographies* 16: 55–85.

Purdy, RJW (1908) The Occasional Luminosity of the White Owl (*Strix Flammea*). *Transactions of the Norfolk and Norwich Naturalists' Society* 8 (4): 547–551.

Purseglove, J (1988) *Taming the Flood: A History and Natural History of Rivers and Wetlands*. Oxford: Oxford University Press.

Radio Times (1957) *The Mystery of the Broads*. 8 February: 42.

Ramuz, LR (1956) A Broads Society? *Eastern Daily Press*, 21 March.

Randall, A and Seaton, R (1974) *George Formby*. London: WH Allen.

Ransome, A (1934) *Coot Club*. London: Jonathan Cape.

Ransome, A (1940) *The Big Six*. London: Jonathan Cape.

Ready, O (1910) *Life and Sport on the Norfolk Broads*. London: T Werner Laurie.

Reid, C (1913) *Submerged Forests*. Cambridge: Cambridge University Press.

Revill, G (2007) William Jessop and the River Trent: Mobility, Engineering and the Landscape of Eighteenth Century 'Improvement'. *Transactions of the Institute of British Geographers* 32: 201–216.

Rice, A (1938) Pity the Poor Fish! *Sight and Sound* 7: 165.

Riviere, BB (1943) Nature Reserve Investigation in Norfolk. *Transactions of the Norfolk and Norwich Naturalists' Society* 15 (5): 392–394.

Robberds, JW (1826) *Geological and Historical Observations on the Eastern Vallies of Norfolk*. Norwich: Bacon and Kinnebrook.

Robertson, F (1949) The Battle of Blackhorse Broad. *Picture Post*, 26 March: 14–17.

Robertson-Scott, JW (1911) *Sugar Beet: Some Facts and Some Illusions*. London: Horace Cox.

Robic, M-C (1994) National Identity in Vidal's *Tableau de la Geographie de la France*: From Political Geography to Human Geography, in D Hooson (ed) *Geography and National Identity*. Oxford: Blackwell, 58–70.

Robinson, J (1920) *Broadland Yachting*. Oulton Broad: Jack Robinson.

Robinson, J (1934) *Motor Cruising on the Broads*. Oulton Broad: Jack Robinson.

Rolt, LTC (1944) *Narrow Boat*. London: Eyre and Spottiswoode.

Rolt, LTC (1947) *High Horse Riderless*. London: Allen and Unwin.

Rooke, D (1948) *Let's be Broad-Minded! The Bunk-Side Book of Brighter Yachting*. London: Blakes.

Rooke, D (1964) *Let's be Broad-Based. The Bunk-Side Book of Brighter Motor Cruising*. London: Blakes.

Rudd, AJ (1943) Norfolk Fishes. *Transactions of the Norfolk and Norwich Naturalists' Society* 15 (5): 377–391.

Ruffell, J (1956) The Commercialised Broads. *Eastern Daily Press*, 12 May.

Ryan, J (2000) 'Hunting with the Camera': Photography, Wildlife and Colonialism in Africa, in C Philo and C Wilbert (eds) *Animal Spaces, Beastly Places*. London: Routledge, 203–221.

Rye, W (1885) *A History of Norfolk*. London: Elliot Stock.

Rye, W (1887) *A Month on the Norfolk Broads*. London: Simpkin, Marshall.

Rye, W (1893) *The Hickling Broad Case*. Norwich: Goose.

Rye, W (ed) (1901) *The Victoria History of the Counties of England: Norfolk (Volume 1)*. London: Constable.

Rye, W (1916) *An Autobiography of an Ancient Athlete and Antiquary*. Norwich: Roberts.

Sackett, C (2004) *Englshpublshng*. Exeter: Spacex.

Sackett, C (2006) *The True Line: The Landscape Diagrams of Geoffrey Hutchings*. Axminster: Colin Sackett.

Sackett, C (2008) *River Axe Crossings*. Axminster: Colin Sackett.

Sainty, JE (1938) Past History of Flooding and the Cause of the 1938 Flood. *Transactions of the Norfolk and Norwich Naturalists' Society* 14 (4): 334–345.

Sainty, JE (1948a) The Origin of the Broads. *Transactions of the Norfolk and Norwich Naturalists' Society* 16: 369–374.

Sainty, JE (1948b) The Future of the Broads. *Eastern Daily Press*, 18 September.

Sainty, JE (1950) The Real East Anglia. *Eastern Daily Press*, 26 January.

Sainty, JE (1952) Coast Protection. *Eastern Daily Press*, 24 January.

Sainty, JE (1953a) This "Sinking" Land. *Eastern Daily Press*, 20 April.

Sainty, JE (1953b) Norfolk's Waters. *Eastern Daily Press*, 13 August.

Sainty, JE (1954) The Melting Ice Cap. *Eastern Daily Press*, 25 August.

Sale, K (1984) Mother of All: An Introduction to Bioregionalism, in S Kumar (ed) *The Schumacher Lectures Volume II*. London: Blond and Briggs, 219–250.

Sale, K (1985) *Dwellers in the Land: The Bioregional Vision*. San Francisco: Sierra Club Books.

Sampson, C (1931) *Ghosts of the Broads*. London: Yachtsman Publishing Company.

Sampson, C (1973) *Ghosts of the Broads*. Norwich: Jarrold.

Samuel, R (1995) *Theatres of Memory*. London: Verso.

Sanders, J (2011) *The Cultural Geography of Early Modern Drama, 1620–1650*. Cambridge: Cambridge University Press.

Sandred, KI (1996) *The Place-Names of Norfolk. Part Two: The Hundreds of East and West Flegg, Happing and Tunstead*. Nottingham: English Place-Name Society.

Savory, A (1953) *Norfolk Fowler*. London: Geoffrey Bles.

Savory, A (1956) *Lazy Rivers*. London: Geoffrey Bles.

Savory, A (1960) *Thunder in the Air*. London: Geoffrey Bles.

Sawbridge, M (1949) Marsh Mills – and Wherries (Letter), *Eastern Daily Press*, 7 January.

Scharf, A (1986) PH Emerson: Naturalist and Iconoclast, in N McWilliam and V Sekules (eds) *Life and Landscape: PH Emerson Art and Photography in East Anglia 1885–1900*. Norwich: Sainsbury Centre for Visual Arts, 21–32.

Scott, P (1977) *The Restoration of Windmills and Windpumps in Norfolk*. Norwich: Norfolk Windmills Trust.

Seago, M (2000) Ornithology in Norfolk: History and Personalities, in M Taylor, P Allard, M Seago and D Dorling, *The Birds of Norfolk*. Mountfield: Pica Press, 22–43.

Sebald, WG (1998) *The Rings of Saturn* (trans. Michael Hulse, first published in German 1995). London: Harvill Press.

Self, G (1986) *The Music of EJ Moeran*. London: Toccata Press.

Sharp, D (1954) *Conflict of Wings*. London: Putnam.

Sheail, J (1976) *Nature in Trust: The History of Nature Conservation in Britain*. London: Blackie.

Sheail, J (1988) The Extermination of the Muskrat (*Ondatra Zibethicus*) in Inter-War Britain. *Archives of Natural History* 15: 155–170.

Sheail, J (2003) Government and the Management of an Alien Pest Species: A British Perspective. *Landscape Research* 28: 101–112.

Shoard, M (1980) *The Theft of the Countryside*. London: Temple Smith.

Simmonds, A (ed) (1978) *The Future of Broadland: 2 July 1977*. Norwich: University of East Anglia.

Skipper, K (2001) *Hev Yew Gotta Loight, Boy? The Life and Lyrics of Allan Smethurst 'The Singing Postman'*. Newbury: Countryside Books.

Skipsey, J (1976) *Selected Poems*, ed. B Bunting. Sunderland: Ceolfirth Press.

Smith, AC (1978) *Drainage Windmills of the Norfolk Marshes*. Stevenage: Stevenage Museum.

Smith, CT (1966) Dutch Peat Digging and the Origin of the Norfolk Broads. *Geographical Journal* 132: 69–72.

Smith, WJ and Turner, EL (1927) *The Fundamentals of Screen Negative Making*. London: Aldenham Press.

Smith, WJ, Turner, EL and Hallam, CD (1932) *Photo Engraving in Relief*. London: Pitman.

Snell, K (1998) *The Regional Novel in Britain and Ireland 1800–1990*. Cambridge: Cambridge University Press.

Snow, CP (1963) *Death Under Sail*. Harmondsworth: Penguin (first published 1932).

Snoxell, FH (1956) *Holidays in Boats*. London: Temple Press.

Soar, CD (1905) The Hydrachnids of the Norfolk Broads. *Transactions of the Norfolk and Norwich Naturalists' Society* 8 (1): 83–89.

South-Eastern Naturalist (1933) Congress Report. *South-Eastern Naturalist* 38: xlv.

Southwell, T (1871) On the Ornithological Archaeology of Norfolk. *Transactions of the Norfolk and Norwich Naturalists' Society* 1 (2): 14–23.

Southwell, T (1901) Aves (Birds), in W Rye (ed) *The Victoria History of the Counties of England: Norfolk (Volume 1)*. London: Constable, 220–245.

Spirn, AW (1998) *The Language of Landscape*. New Haven: Yale University Press.

Spooner, S (ed) (2012) *Sail and Storm: The Aylsham Navigation*. Aylsham: Aylsham Local History Society.

Stannard, J (1966) Haymaking on the Broads. *Eastern Daily Press*, 11 June.

Stark, J and Robberds, JW (1834) *Scenery of the Rivers of Norfolk Comprising the Yare, the Waveney, and the Bure*. Norwich: John Stacy.

Starr, L (1904) Fattening Swans for the Tables of Epicures. *New York Times* (Magazine Section), 21 August.

Steers, JA (ed) (1934) *Scolt Head Island*. Cambridge: Heffer/NNNS.

Steers, JA (1953a) The East Coast Floods. *Geographical Journal* 119: 280–298.

Steers, JA (1953b) *The Sea Coast*. London: Collins.

Steers, JA (1954) The Broads. *Geographical Review* 44: 147.

Steers, JA (1962) *The Sea Coast*. London: Collins.

Stephen, GA (1921) *Books on the Broads: A Chronological Bibliography.* Norwich: Norwich Mercury.

Stevenson, H (1866) *The Birds of Norfolk* (Volume One). London: Van Voorst.

Stevenson, H (1870) *The Birds of Norfolk* (Volume Two). London: Van Voorst.

Stevenson, H (1872) Address. *Transactions of the Norfolk and Norwich Naturalists' Society* 1 (3): 7–19.

Stevenson, H (1890) *The Birds of Norfolk* (Volume Three). London: Gurney and Jackson.

Stone, E (1988) *Ted Ellis: The People's Naturalist.* Norwich: Jarrold.

Storey, NR (2012) *Norfolk Floods: An Illustrated History of 1912, 1938 and 1953.* Wellington: Halsgrove.

Stringer, C (2006) *Homo Britannicus: The Incredible Story of Human Life in Britain.* London: Penguin.

Suffling, E (1893) *The Land of the Broads.* Stratford: Benjamin Perry (Second Edition, first published 1885).

Summers, D (1978) *The East Coast Floods.* Newton Abbot: David and Charles.

Swift, G (1983) *Waterland.* London: Heinemann.

Swinburne, AC (1927) *Swinburne's Collected Poetical Works: Volume II.* London: William Heinemann.

Tansley, AG (1911a) (ed) *Types of British Vegetation.* Cambridge: Cambridge University Press.

Tansley, AG (1911b) The International Phytogeographical Excursion in the British Isles. I The Inception, and II Details of the Excursion. *New Phytologist* 10: 271–291.

Taylor, J (1995) *A Dream of England: Landscape, Photography and the Tourist's Imagination.* Manchester: Manchester University Press.

Taylor, J (2006) *The Old Order and the New: PH Emerson and Photography 1885–1895.* London: Prestel.

Taylor, JE (1872) The Norfolk Broads and Meres Geologically Considered. *Transactions of the Norfolk and Norwich Naturalists' Society* 1 (3): 30–40.

Taylor, M, Allard, P, Seago, M and Dorling, D (2000) *The Birds of Norfolk.* Mountfield: Pica Press.

Thompson, L (1947) *Norwich Inns.* Ipswich: Harrison.

Ticehurst, NF (1921) Stevenson's 'Birds of Norfolk' – A Correction. *Transactions of the Norfolk and Norwich Naturalists' Society* 11 (2): 202–203.

Ticehurst, NF (1925) The Swan-Roll in the Norwich Castle Museum. *Transactions of the Norfolk and Norwich Naturalists' Society* 12 (1): 17–25.

Ticehurst, NF (1928) The Swan-Marks of East Norfolk. *Transactions of the Norfolk and Norwich Naturalists' Society* 12 (4): 424–460.

Ticehurst, NF (1929) The Swan-Marks of West Norfolk. *Transactions of the Norfolk and Norwich Naturalists' Society* 12 (5): 581–630.

Ticehurst, NF (1937) Some More Swan-Marks of East Norfolk. *Transactions of the Norfolk and Norwich Naturalists' Society* 14 (3): 229–246.

Ticehurst, NF (1957) *The Mute Swan in England.* London: Cleaver-Hume.

Times (1962) *The Coypu Campaign.* 3 October: 7.

Tolia-Kelly, DP (2010) *Landscape, Race and Memory.* Farnham: Ashgate.

Tomaney, J (2007) Keeping a Beat in the Dark: Narratives of Regional Identity in Basil Bunting's *Briggflatts. Environment and Planning D: Society and Space* 25: 355–375.

Tomaney, J (2010) Parish and Universe: Patrick Kavanagh's Poetics of the Local. *Environment and Planning D: Society and Space* 28: 311–325.

Tooley, B (1985) *John Knowlittle: The Life of the Yarmouth Naturalist Arthur Henry Patterson, ALS.* Sprowston: Wilson-Poole.

Tooley, B (2004) *Scribblings of a Yarmouth Naturalist.* Norwich: Beryl Tooley.

Transactions of the Norfolk and Norwich Naturalists' Society (1916) *CH Martin* (Obituary). 10 (2): 181.

Transactions of the Norfolk and Norwich Naturalists' Society (1922) *Wild Bird Protection in Norfolk in 1922*. 11 (3): 301–309.

Transactions of the Norfolk and Norwich Naturalists' Society (1928) *Sir Eustace Gurney 1876–1927* (Obituary). 12 (4): 519–521.

Transactions of the Norfolk and Norwich Naturalists' Society (1929) *Walter Rye 1843–1929*. 12: 722–726.

Transactions of the Norfolk and Norwich Naturalists' Society (1935a) *The Norfolk Room*. 14 (1): Plates.

Transactions of the Norfolk and Norwich Naturalists' Society (1935b) *Arthur H Patterson*. 14 (1): 111–113.

Transactions of the Norfolk and Norwich Naturalists' Society (1939a) *The Season, 1938–39*. 15 (1): 1–3.

Transactions of the Norfolk and Norwich Naturalists' Society (1939b) *The Junior Branch of the Norfolk and Norwich Naturalists' Society*. 15 (1): 4.

Trimmer, K (1866) *Flora of Norfolk*. London: Hamilton, Adams and Co.

Tully, C (2002) *The Broads: The Official National Park Guide*. Newton Abbot: Pevensey.

Turner, EL (1912) The Return of the Bittern. *Transactions of the Norfolk and Norwich Naturalists' Society* 9 (3): 433–436.

Turner, EL (1918) Notes on the Breeding of the Bittern in Norfolk. *Transactions of the Norfolk and Norwich Naturalists' Society* 10 (4): 319–334.

Turner, EL (1922) The Status of Birds in Broadland. *Transactions of the Norfolk and Norwich Naturalists' Society* 11 (3): 227–240.

Turner, EL (1924) *Broadland Birds*. London: Country Life.

Turner, EL (1928) *Bird Watching on Scolt Head Island*. London: Country Life.

Turner, EL (1929) *Stray Leaves From Nature's Notebook*. London: Country Life.

Turner, EL (1932a) *My Swans, The Wylly-Wyllys*. London: Arrowsmith.

Turner, EL (1932b) *Togo, My Squirrel*. London: Arrowsmith.

Turner, EL and Bahr, PH (1907) *Home Life of Some Marsh Birds*. London: Witherby.

Turner, EL and Gurney, R (1925) *A Book About Birds*. London: C Arthur Pearson.

Turner, P and Wood, R (1974) *PH Emerson: Photographer of Norfolk*. London: Gordon Fraser.

Valentin, H (1953) Present Vertical Movements of the British Isles. *Geographical Journal* 119: 299–305.

Vaughan Williams, R and Lloyd, AL (1959) *The Penguin Book of English Folk-Songs*. Harmondsworth: Penguin.

Vidal de la Blache, P (1928) *The Personality of France*. London: Christophers.

Vincent, J (1926) The Romance of the Bittern. *Country Life*, 28 August: 303–307.

Vincent, J (1929) Broadland. *Transactions of the Norfolk and Norwich Naturalists' Society* 12 (5): 684–687.

Vincent, J (1935) Coot Drive on Lord Desborough's Shoot. *Country Life* 23 February: 190–191.

Vincent, J (1980) *A Season of Birds: A Norfolk Diary 1911*. London: Weidenfeld and Nicolson.

Wailes, R (1954) *The English Windmill*. London: Routledge and Kegan Paul.

Wailes, R (1956) Norfolk Windmills: Part II, Drainage and Pumping Mills Including Those of Suffolk. *Newcomen Society Transactions* 30: 157–177.

Wailes, R (1957) *Berney Arms Mill*. London: HMSO.

Walker, S (1978) A Study of Boat Users of the Norfolk Broads, in MJ Moseley (ed) *Social Issues in Rural Norfolk*. Norwich: Centre of East Anglian Studies, UEA, 145–166.

Wall, CS (2006) *The Prose of Things: Transformations of Description in the Eighteenth Century*. Chicago: University of Chicago Press.

Wallace, D and Bagnall-Oakeley, RP (1951) *Norfolk*. London: Robert Hale.

Ward Lock (1924) *The Broads*. London: Ward, Lock.

Wardle, R (1988) *Arthur Ransome's East Anglia*. North Walsham: Poppyland Publishing.

Wardle, R (2013) *Arthur Ransome on the Broads*. Stroud: Amberley.

Warner, ST (1948) *The Corner That Held Them*. London: Chatto and Windus.

Watkins, C, Matless, D and Merchant, P (2003) Cultures of Nature: Botany in Herefordshire, England, 1945–1970, in JP Diry and N Walford (eds) *Innovations in Rural Areas*. Clermont Ferrand: Ceremac, 11–27.

Watkins, C, Matless, D and Merchant, P (2007) Science, Sport and the Otter, 1945–1978, in RW Hoyle (ed) *Our Hunting Fathers: Field Sports in England After 1850*. Lancaster: Carnegie, 165–186.

Watts, HD (1971) The Location of the Beet-Sugar Industry in England and Wales, 1912–36. *Transactions of the Institute of British Geographers* 53: 95–116.

Wayre, P (1965) *Wind in the Reeds*. London: Collins.

Wellbye, R (1921) *Road Touring in Eastern England*. London: EJ Larby.

Weston, D (2011) The Spatial Supplement: Landscape and Perspective in WG Sebald's *The Rings of Saturn*. *Cultural Geographies* 18: 171–186.

Whatmore, S (2002) *Hybrid Geographies*. London: Sage.

Wheeler, SG (1955a) Origins of the Broads. *Eastern Daily Press*, 19 April.

Wheeler, SG (1955b) Origins of the Broads. *Eastern Daily Press*, 23 April.

Whitehead, M (2003) From Moral Space to the Morality of Scale: The Case of the Sustainable Region. *Ethics, Place and Environment* 6: 235–257.

Whittle, T (1955) *Spades and Feathers*. London: Jonathan Cape.

Wild Life Conservation Special Committee (1947) *Conservation of Nature in England and Wales*. London: HMSO, Cmd.7122.

Wilkins-Jones, C (1986) One of the Hard Old Breed: A Life of Peter Henry Emerson, in N McWilliam and V Sekules (eds) *Life and Landscape: PH Emerson Art and Photography in East Anglia 1885–1900*. Norwich: Sainsbury Centre for Visual Arts, 2–6.

Williams, R (1975) *The Country and the City*. St Albans: Paladin (first published 1973).

Williams, R (1976) *Keywords*. London: Fontana.

Williams, R (1983) *Keywords* (Revised Edition). London: Fontana.

Williamson, T (1997) *The Norfolk Broads: A Landscape History*. Manchester: Manchester University Press.

Wilson, E (2000) *Bohemians*. London: IB Tauris.

Withers, CWJ (2010) *Geography and Science in Britain, 1831–1939: A Study of the British Association for the Advancement of Science*. Manchester: Manchester University Press.

Withers, CWJ and Finnegan, DA (2003) Natural History Societies, Fieldwork and Local Knowledge in Nineteenth-Century Scotland: Towards a Historical Geography of Civic Science. *Cultural Geographies* 10: 334–353.

Wonders, K (1993) *Habitat Dioramas: Illusions of Wilderness in Museums of Natural History*. Uppsala: Acta Universitatis Upsaliensis.

Wonders, K (2003) Habitat Dioramas and the Issue of Nativeness. *Landscape Research* 28: 89–100.

Woods, J (2002) *Herbert Woods: A Famous Broadland Pioneer*. Tauranga: Captains Locker Publications.

Woodward, H (1883) The Scenery of Norfolk. *Transactions of the Norfolk and Norwich Naturalists' Society* 3 (4): 439–466.

Wright, P (1995) *The Village That Died for England: The Strange Story of Tyneham*. London: Jonathan Cape.

Wylie, J (2002) An Essay on Ascending Glastonbury Tor. *Geoforum* 33: 441–454.

Wylie, J (2005) A Single Day's Walking: Narrating Self and Landscape on the South West Coast Path. *Transactions of the Institute of British Geographers* 30: 234–247.

Wylie, J (2007) *Landscape*. Oxford: Blackwell.

Wylie, J (2009) Landscape, Absence and the Geographies of Love. *Transactions of the Institute of British Geographers* 34: 275–289.

Wylie, J (2012) Dwelling and Displacement: Tim Robinson and the Questions of Landscape. *Cultural Geographies* 19: 365–383.

Yarmouth Independent (1886) *Lecture on Photography*. 27 March.

Yarmouth Independent (1938) *Post Office Secrets*. 12 November.

Yarmouth Mercury (1886) *Photography*. 27 March.

Yarmouth Mercury (1938a) *The Sea Conquers Flood Village*. 26 February.

Yarmouth Mercury (1938b) *Horsey Survives Second Sea Invasion*. 5 March.

Yarmouth Mercury (1938c) *Thousands Visit Sea Breach At Horsey*. 12 March.

Yarmouth Mercury (1938d) *Flood Relief*. 2 April.

Yarmouth Mercury (1938e) *Post Office Takes to the Screen*. 12 November.

Yeldham, C (2004) Osborn, Emily Mary. *Oxford Dictionary of National Biography*.

Index

In the Nature of Landscape: Cultural Geography on the Norfolk Broads, First Edition. David Matless.
© 2014 David Matless. Published 2014 by John Wiley & Sons, Ltd.